VOID

Library of
Davidson College

ATOMIC AND MOLECULAR COLLISIONS

ATOMIC AND MOLECULAR COLLISIONS

SIR HARRIE MASSEY, F.R.S.
University College, University of London

TAYLOR & FRANCIS LTD
LONDON
HALSTED PRESS
(a division of John Wiley & Sons Inc)
NEW YORK—TORONTO
1979

First published 1979 by Taylor & Francis Ltd., London and Halsted Press (a division of John Wiley & Sons Inc.), New York

© *1979 Sir H. Massey*
All rights reserved. No part of this publication may be reproduced, stored in a retrieval system, or transmitted, in any form or by any means, electronic, mechanical, photocopying, recording, or otherwise, without the prior permission of the copyright owner

Photosetting by Thomson Press (India) Ltd., New Delhi

Printed and bound in Great Britain by Taylor & Francis (Printers) Ltd., Basingstoke, Hampshire

Library of Congress Cataloging in Publication Data

Massey, Harrie Stewart Wilson, *Sir*
   Atomic and molecular collisions.
   "A Halsted Press book"
   1. Collisions (Nuclear physics) 2. Atoms
   3. Molecules
   I. Title
   QC794.6.C6M36   539.7'54   79-11716

ISBN 0-470-26742-9

preface

The preparation of a book on atomic collisions which can be read with profit by a first year undergraduate faces a number of difficulties. The mathematics used must be kept to a level in which, for example, partial differentiation does not occur, at least explicitly. On the other hand, it is necessary not only to introduce the concepts of wave mechanics but to make use of these concepts in interpreting experimental data. Equally well, the experimental study of atomic collisions involves a complex of techniques, the description of each one of which could form the basis of a considerable volume. Compromise has therefore been essential in presenting the subject. In the four introductory chapters, an account is given of classical mechanics, wave motion and wave mechanics as well as of the essential features of atomic and molecular structure. Certain experimental techniques have been described within the main text as appropriate. Others are described in principle in Appendices.

Only a limited selection of topics from such a large field as the physics of atomic collisions could be considered in a book of this nature. In making this selection, the aim has been to expose the scope of the subject rather than to discuss a few aspects in detail. The approach is physical rather than mathematical throughout.

H. S. W. Massey
London, November 1978

contents

Preface	v
Introduction	xv

Chapter 1 PARTICLES AND WAVES — 1
    1.1 Particles—Elastic Collisions — 2
    1.2 Centre of Mass and Laboratory Co-ordinates — 5
    1.3 The Specification of Collision Probabilities—the Concept of Collision Cross Section — 9
    1.4 The Impact Parameter in a Classical Collision — 10
    1.5 Angular Momentum — 12
    1.6 Waves — 13
    1.7 Interference and Diffraction — 15
    1.8 Scattering of Waves—Shadow Scattering — 17
    1.9 The Wave Equation — 18

Chapter 2 WAVE MECHANICS—WAVE–PARTICLE DUALISM—THE UNCERTAINTY PRINCIPLE— SCATTERING CROSS-SECTIONS — 20
    2.1 Wave–Particle Dualism — 20
    2.2 Material Waves — 23
    2.3 Allowed Values of the Energy and Other Dynamical Quantities — 24
    2.4 The Uncertainty Principle — 28
    2.5 General Remarks about Wave versus Classical Mechanics — 28
    2.6 Collision Cross Sections in Wave Mechanics — 29
    2.7 Quantum Theory of Scattering by a Centre of Force — 32
    2.8 The Differential Scattering Cross-Section in Wave Mechanics — 35
    2.9 The Variation of the Phase Shifts with the Angular Momentum Quantum Number — 36
    2.10 Generalization to Collisions Involving Atomic Systems — 37

Chapter 3 ATOMS — 39
    3.1 Atomic Structure—Some General Features — 39

	3.2	Electron Spin	41
	3.3	Combination of Angular Momenta	41
	3.4	The Hydrogen Atom	42
	3.5	Optical Radiation from Excited Hydrogen Atoms—Allowed and Forbidden Transitions	43
	3.6	Hyperfine Structure	45
	3.7	Continuous Spectra	45
	3.8	Spontaneous Emission Coefficients	46
	3.9	Absorption of Energy by Hydrogen Atoms	46
	3.10	Probability Charge Distributions in Different States of a Hydrogen Atom	47
	3.11	Electric Field of a Hydrogen Atom	49
	3.12	Deuterium	50
	3.13	The Structure of Complex Atoms—the Pauli Principle	51
	3.14	The Ground State of Helium	51
	3.15	The Ground State of Lithium	52
	3.16	The Ground States of Other Atoms	52
	3.17	Excited States of Atoms	54
	3.18	Excited States of Helium	55
	3.19	Doubly Excited States—Autoionization	55
	3.20	Negative Atomic Ions	57
	3.21	Doubly Excited States of Negative Ions—Autodetachment	58
Chapter	4	MOLECULES	60
	4.1	The Interaction between Atoms	60
	4.2	Interaction between Hydrogen Atoms	61
	4.3	Examples of Other Stable Diatomic Molecules	62
	4.4	Excited Electronic States of Molecules	62
	4.5	Interactions between Ions and Atoms	62
	4.6	Molecular Vibration and Rotation	64
	4.7	Behaviour of the Nuclei in a Transition between Electronic States—the Franck–Condon Principle	66
Chapter	5	THE SCATTERING OF SLOW ELECTRONS BY ATOMS—THE RAMSAUER–TOWNSEND EFFECT	68
	5.1	Diffusion of an Electron Swarm in a Gas under the Action of an Electric Field	69
	5.2	The Measurement of Drift Velocities	71
	5.3	The Measurement of Total Cross-Sections for Scattering of Electrons by Atoms using Beam Techniques—Ramsauer's Method	73
	5.4	Results for the Rare Gases	75

Contents

	5.5	Interpretation of Results for the Rare Gases in Terms of Wave Mechanics	75
	5.6	The Angular Distribution of the Scattered Electrons—Experimental Methods	80
	5.7	Results of Angular Distribution Measurements	83
	5.8	Interactions Effective in the Scattering of Slow Electrons by Atoms	86

Chapter 6 ELECTRON SCATTERING BY ATOMS AND MOLECULES—INELASTIC SCATTERING AND RESONANCE EFFECTS — 88

	6.1	Introduction	88
	6.2	Direct and Resonant Scattering	89
	6.3	Energy Analysers	91
	6.4	Experimental Methods for Studying Fine Structure in Electron Scattering	95
	6.5	Typical Measurements of Fine Structure (Resonance) Effects in Elastic Scattering	98
	6.6	Measurement of Inelastic Cross-Sections—Differential Cross-Sections	101
	6.7	Cross-Sections for Ionization	102
	6.8	Cross-Sections for Excitation of Bound States	104

Chapter 7 COLLISIONS INVOLVING ELECTRON CAPTURE—RECOMBINATION AND ATTACHMENT — 108

	7.1	Recombination and Attachment Coefficient	108
	7.2	Radiative Recombination and Attachment	109
	7.3	Dissociative Recombination and Attachment	111
	7.4	The Experimental Study of Dissociative Recombination	114
	7.5	Results of Recombination Measurements—the Rare Gases	120
	7.6	Result of Recombination Measurements—Molecular Gases	123
	7.7	The Experimental Measurements of Attachment Rates	124
	7.8	Some Results of Attachment Experiments—Dissociative Attachment in O_2	128
	7.9	Dissociative Attachment in CO	131
	7.10	Attachment to Polyatomic Molecules—Nitrous Oxide (N_2O)	132
	7.11	Attachment to Sulphur Hexafluoride (SF_6)	134
	7.12	Attachment to Other Polyatomic Molecules	135
	7.13	The Electron Attachment Detector	135

Contents

Chapter 8 COLLISIONS BETWEEN NEUTRAL ATOMIC AND MOLECULAR SYSTEMS—A GENERAL SURVEY — 139
- 8.1 Atom–Atom Collisions—Elastic Scattering — 139
- 8.2 Collisions Involving Molecules—Excitation and Transfer of Vibration and Rotation — 142
- 8.3 Dispersion and Absorption of High-Frequency Sound — 144
- 8.4 Persistence of Vibration in Shock Wave Experiments — 146
- 8.5 Some Results of Vibrational Relaxation Measurements — 146
- 8.6 Relaxation in H_2 and D_2—Rotational Persistence — 148
- 8.7 Inelastic Collisions Involving Electronic Excitation — 149
- 8.8 Excitation Transfer—Sensitized Fluorescence — 151
- 8.9 Ionization by Metastable Atom Impact — 152
- 8.10 Production of Ion Pairs in Neutral–Neutral Collisions — 153

Chapter 9 SEMI-CLASSICAL COLLISIONS BETWEEN ATOMS — 156
- 9.1 The Form of the Interaction Energy between Gas Atoms — 157
- 9.2 Classical Theory of Collisions between Gas Atoms—the Deflection Function — 158
- 9.3 Relation between Angle of Deflection and Angle of Scattering — 161
- 9.4 The Angles of Deflection and of Scattering for Interactions between Rare Gas Atoms — 163
- 9.5 The Optical Rainbow — 165
- 9.6 The Glory Singularity — 167
- 9.7 The Classical Differential Cross-Section — 167
- 9.8 Orbiting — 168
- 9.9 The Quantum Scattering Formula near the Classical Limit—The 'Semi-Geometrical' Wave Theory of the Optical Rainbow — 169
- 9.10 The Semi-Classical Wave Theory of the Scattering of Particles — 171
- 9.11 Glory Undulations — 174
- 9.12 Collisions between Similar Atoms—Symmetry Interference Effects — 175
- 9.13 Application to Determination of Atomic Interactions — 176

Chapter 10 THE EXPERIMENTAL STUDY OF ATOM–ATOM COLLISIONS AT THERMAL ENERGIES — 178
- 10.1 Experimental Methods—General Remarks — 178
- 10.2 The Production of Atomic Beams — 180
- 10.3 Velocity Selection — 183

10.4	Detection of Atomic Beams	185
10.5	Some Typical Experiments and Results—Scattering of Alkali Metal Atoms	186
10.6	Collisions between Helium Atoms—Symmetry Oscillations	193

Chapter 11 THERMAL COLLISIONS BETWEEN IONIZED AND NEUTRAL ATOMIC AND MOLECULAR SYSTEMS — 200

11.1	The Long-range Polarization Force	200
11.2	The Condition for Orbiting	201
11.3	The Mobility of Ions in Gases	202
11.4	Ionic Reactions	204
11.5	Orbiting and Ionic Reaction Rates	206
11.6	Measurement of Ionic Reaction Rates under Thermal or Near-thermal Conditions	207
11.7	Results of Measurements of Ionic Reaction Rates	210
11.8	Charge Transfer Reactions	211
11.9	Associative Detachment	211
11.10	Rearrangement Collisions	211
11.11	Cluster Formation	212

Chapter 12 THE COLLISIONS OF ENERGETIC IONS IN GASES—CHARGE TRANSFER — 213

12.1	Symmetrical Charge Transfer—Theoretical Discussion—H^+–H Collisions	213
12.2	The H^+–H Interactions	214
12.3	Elastic Scattering and Charge Transfer	215
12.4	Effect of the Identity of the Nuclei	216
12.5	Symmetrical Charge Transfer in General	217
12.6	Experimental Methods and Results—Ion Sources	218
12.7	Measurement of Total Cross-Sections for Charge Transfer	219
12.8	Results of Total Charge Transfer Cross-Section Measurements	221
12.9	Measurement of Charge Transfer Probabilities and of Differential Scattering Cross-Sections	224
12.10	Results for He^+–He Collisions and their Interpretation	226
12.11	Other Reactions of Energetic Ions in gases	227

Chapter 13 PHOTON COLLISIONS WITH ATOMS AND MOLECULES—PHOTOIONIZATION AND PHOTODETACHMENT — 230

13.1	Introduction	230

13.2	Line Absorption and Stimulated Emission	232
13.3	Stimulated Emission and Laser Operation	234
13.4	Excitation of Autoionizing and Autodetaching States	236
13.5	Relation between Absorption and Photoionization Cross-Sections	238
13.6	The Experimental Study of Ionization by Single Photons—Introduction	238
13.7	Principles Involved in the Measurement of Absorption and Photoionization Cross-Sections	239
13.8	Radiation Sources	240
13.9	Photoelectron Spectroscopy	242
13.10	Some Results of Photoionization Measurements	243
13.11	Photodetachment by Single Photons	244
13.12	Measurement of Photodetachment Cross-Sections—Typical Experimental Arrangements	247
13.13	Measurement of Photoelectron Spectra from Photodetachment	248
13.14	Photodetachment from H^-—Absorption in the Sun's Atmosphere	250
13.15	Photodetachment from O^-	252
13.16	Photodetachment from Se^- and Si^-	253
13.17	Window Resonances in Photodetachment from Cs^- and Rb^-	254
13.18	Photoelectron Spectrum of O_2^-	255
13.19	Multiphoton Processes	256

Chapter 14   ATOMIC COLLISIONS IN THE EARTH'S ATMOSPHERE, THE SOLAR CORONA AND INTERPLANETARY SPACE   259

14.1	The Earth's Ionosphere	260
14.2	Theory of the Ionosphere before 1950	263
14.3	Post-war Research and the Ionosphere	265
14.4	The Present Position about Ionospheric Photochemistry	269
14.5	The Lower Ionosphere	273
14.6	Ionization and Recombination in the Solar Corona	275
14.7	Molecule Formation in Interstellar Space	278
14.8	The Problem of the CH^+ and CH Production	281

Appendices

A1	The Kinetic Theory of Gases—Effusive Flow	285
A2	High Vacuum Technique	288
A3	Supersonic Flow through a Nozzle	290

Contents

A4	The Motion of Charged Particles in Magnetic Fields	292
A5	Electron and Ion Optics	294
A6	The Detection and Measurement of Small Fluxes of Electrons, Ions and Neutral Atoms	297
A7	Mass Analysers and Selectors	299

Index   303

introduction

Soon after the discovery of the electron by J. J. Thomson, in 1897, and the identification of anode rays as positively charged particles of atomic mass, experiments were initiated to observe the motion of electrons and ions in gases. These were largely concerned with the measurement of mobilities of electrons and ions drifting in a gas under the influence of a uniform electric field. However, attempts were also made at quite an early stage to measure recombination coefficients, which determine the rate at which positive ions recombine with electrons or negative ions to form neutral atoms or molecules. In all cases the electrons or ions did not posses a well-defined velocity but drifted as a swarm with a distribution of velocities determined by the electric field, the nature and density of the gas and, to a less important extent, the gas temperature. Despite this, these early experiments marked the beginning of the study of electronic and ionic collisions.

The first experiments in which the collisions of charged particles of definite velocity with atoms were observed were the classic ones on the scattering of alpha particles by thin foils carried out in 1911 by Geiger and Marsden to check the predictions of Rutherford's nuclear model of the atom. The kinetic energy of the alpha particles in these experiments was so high that they were not appreciably deflected by the outer structure of the atoms through which they passed. A few years later, Ramsauer and his collaborators devised a technique for studying the collisions with atoms of electrons of well-defined kinetic energy in the range from a few electron-volts[†] to some tens of electron-volts. Electrons of these low velocities are strongly influenced by the outer structure of atoms with which they collide. It would be expected that the slower the electrons the more readily they would be deviated from their initial path by an atom. However, Ramsauer in 1921 found that argon is nearly transparent to electrons with energy close too 0·7 eV whereas it is more absorbent for electrons of higher as well as lower energy. Similar evidence was also obtained at about the same time by Townsend and his collaborators at Oxford, who had developed the early swarm technique to give quantitative results. They found that the mean free path of electrons drifting in argon was a maximum

---

[†] An electron-volt, usually written as eV, is the kinetic energy acquired by an electron due to the application of a potential difference of 1 volt. It is equal to $1 \cdot 6 \times 10^{-19}$ J.

Introduction

at a certain value of the ratio of electric field to gas pressure. As this ratio determines the mean energy of the drifting electrons this meant that the rate at which the electrons collided with the gas was a minimum at a certain mean electron energy. While not so clear-cut as the Ramsauer experiments, these results pointed to the same surprising conclusion.

At the time, this was by no means the only aspect of atomic physics which seemed to violate the concepts of classical physics. The stability of the Rutherford nuclear atom, in the face of the predictions of classical electromagnetic theory that the constituent electrons revolving round the nucleus would rapidly radiate their energy away as electromagnetic waves, had already been associated by Bohr with the quantum of action first postulated by Planck in 1901 to overcome the difficulties of classical radiation theory. But Bohr's resulting model of the hydrogen atom was still highly empirical, and difficult to generalize to more complex atoms and particularly to molecules. Moreover, it was, at best, only able to deal with equilibrium situations. It offered no means of interpreting the Ramsauer–Townsend effect, for example.

Indeed, in 1924, the theory of electronic and ionic collision processes was in a rudimentary and generally ineffective stage, although, about that time, Thomas applied classical mechanics with remarkable virtuosity to obtain some results which are still used today.

The introduction of wave mechanics in 1926 changed this situation almost overnight. Theoretical methods became available for calculating the rates at which changes took place in atomic and molecular systems through collisions or the emission of radiation, while at the same time the arbitary, empirical character of Bohr's description of the equilibrium structure of atoms was replaced by a comprehensive theory applicable at least in principle to systems of any complexity.

In classical mechanics, the path of a particle of mass m moving under the action of prescribed forces from a definite initial position with a definite initial velocity is accurately predictable. This is no longer true according to wave mechanics. All that can be done is to determine the probability that the particle will be found at a particular position at a particular time. A particle of mass m moving with velocity v is associated with a wave motion of wavelength h/mv, where h is Planck's quantum of action. The probability of finding the particle at a particular point at time t is given by $|\psi|^2$ where ψ is the amplitude of the associated wave. ψ is not necessarily real, but may be determined from the basic equations of wave mechanics.

Just as the wave nature of light is only apparent when one is observing the passage of light through an aperture of dimensions comparable with, or smaller than, the wavelength of the light, classical mechanics remains a good description of phenomena involving systems with dimensions of the order a provided the associated 'particle' wavelength is much less than a. For atoms, a is of the order 10^{-10} m. The wavelength associated with an electron with energy E eV is

Introduction

$1.23 \times 10^{-9} E^{1/2}$ m. It follows that we would expect marked deviations from classical behaviour in the collisions of electrons of a few electron-volts energy with atoms. This suggested strongly that the Ramsauer–Townsend effect is a manifestation of the wave aspect of the electron, and naturally stimulated a great deal of experimental and theoretical investigations of the scattering of slow electrons by atoms. This work soon confirmed the validity of the suggestion and provided much further insight into the mechanism of electron–atom collisions. Ever since that time, this subject has maintained a leading position in atomic and molecular physics. Although it now has important applications, the initial drive was towards a verification and understanding of the new mechanics under conditions in which classical mechanics is quite inapplicable. For this reason, it deserves an early discussion in any book on atomic collisions and will form the material of Chapters 5, 6 and 7 of the present work.

The wavelengths associated with bodies of atomic mass, even when moving with kinetic energies as low as 0·1 eV, are much shorter than atomic dimensions, so that many aspects of collisions between such systems can be described by classical mechanics. This applies in particular to the mobilities of ions, as distinct from electrons, in gases as well as to the gas kinetic collisions which determine the viscosity and thermal conductivity of gases. With increasing precision and variety of measurements, some wave mechanical effects could be discerned even in these cases. However, most experimental work concerned with the study of collisions between systems both of atomic mass, as well as of electron–atom collisions, was carried out with some application in mind. These included the detailed interpretation of electric discharge and breakdown phenomena, the determination of the interactions between gas atoms and the dispersion and absorption of ultrasonic waves in molecular gases. The importance of a knowledge of the rates of a wide variety of atomic and electronic collision processes for the interpretation of the properties of the earth's ionosphere, of the solar atmosphere and many astrophysical situations was also being realized. So little information was available from experiment and the difficulties in extending the scope of experimental measurements were so great that it was necessary to rely on theoretical predictions which were, in turn, not reliable because they could not be subjected to observational check. For these reasons, the study of atomic collisions languished during the last few years before World War 2.

Immediately after the war, the situation was transformed. The great technical advances made during the war period enormoosly expanded the potential of experimental research on collision phenomena. At the same time the development of high-speed computers placed a new, extremely powerful tool in the hands of the theorists. All that was needed was the provision of the necessary resources and encouragement. This was soon forthcoming because some of the major large-scale technical enterprises undertaken with the new technology

Introduction

called for knowledge of atomic collision rates on a grand scale. The *in-situ* exploration of the Earth's upper atmosphere has provided an almost overwhelming mass of new data which depend very much on atomic collision rates for their interpretation. A similar situation is building up in relation to other planets of the solar system. Again, the scope of astronomical observation, not only in the visible, but at all wavelengths has grown greatly and again makes many demands on information from atomic collision physics. The extensive research programme in plasma physics directed towards the controlled release of energy from nuclear fusion also relies heavily on knowledge of atomic collision rates for the interpretation of the results of the elaborate experiments involved. One of the outstanding recent new developments has been that of the laser, the possibilities of which seem almost unbounded. Many important types of laser use gases as the working medium and operate through certain specific processes involving atomic and electronic collisions.

It is no wonder that the study of atomic (including electronic) collisions expanded explosively, so that at the present time there are thousands of research workers in the subject—at the biennial international conferences on this subject, the attendance is close to one thousand and these represent only a fraction of the total number concerned. Remarkable developments in technique have occurred and there are few of the many types of collision which are now beyond the reach of the experimenter. The interplay between experiment and theory which is the lifeblood of science has become very close, helped by the greatly increased power of computational techniques. As a result we now have not only a fascinating understanding in depth of the way things work at the atomic level, but also an amount of reliable data sufficient for useful detailed applications in many different branches of physics, including atmospheric physics and astrophysics as well as to the design of gaseous lasers and to controlled nuclear fusion.

CHAPTER 1
particles and waves

As we have pointed out in the Introduction, there was a period of roughly a quarter of a century from 1901 during which classical physics was found wanting to an increasing extent when applied to atomic phenomena. A particularly puzzling feature was the realization that the wave nature of light, which had seemed so firmly established, was not consistent with lately observed optical phenomena such as the photoelectric effect and, more obviously, the Compton effect. In the latter effect, X-rays of frequency v were observed to collide with electrons as if the rays were composed of streams of particle-like photons with energy $E = hv$ and momentum $p = hv/c = h/\lambda$, where h is Planck's constant[†] and c the velocity of light, $\lambda = c/v$, is the wavelength of the light. It was pointed out by L. de Broglie in 1925 that certain arbitrary features of Bohr's hypothesis about the allowed energies of electrons in atoms could be understood if electrons, classically thought of as particles, possessed a wave aspect, with the frequency and wavelength bearing the same relation to the energy and momentum as with photons and light waves. Thus electrons of energy E and momentum p would be associated with waves of frequency E/h and wavelength h/p. A year or so later, direct experimental evidence was obtained that a beam of electrons behaves in certain circumstances as if it were a train of waves with these properties. In the intervening period, however, Schrödinger discovered wave mechanics, from which the dualistic wave–particle character of matter and radiation followed directly. Wave mechanics proved completely successful when applied to atomic and molecular problems and must be used as a background for the interpretation of electronic and atomic impact phenomena in particular.

To appreciate sufficient of this background for the latter purpose and also to provide a means for specifying the rates at which different types of collision may occur, we must begin by discussing the properties of particles and of waves according to classical ideas. In doing so, we shall incidentally obtain a number of results which will be used in later sections. After this, we shall discuss how the two aspects may be associated in a single system provided one is prepared to sacrifice certain classical ideas which have been held very firmly hitherto. At the

---

[†] h is a quantity with the dimensions of action and is of magnitude $6 \cdot 62 \times 10^{-34}$ J s.

same time, we shall consider also the conditions under which a classical description is a good approximation.

1.1 Particles—Elastic Collisions

We think of a particle as a small bundle of matter possessing a mass which determines its reaction to an applied force. Associated with the motion of the particle are two principal mechanical quantities, the momentum and energy, both of which remain constant if the motion is unaffected by any outside agent. Whereas energy is a scalar quantity $\frac{1}{2}mv^2$, where m is the mass and v the velocity of the particle, the momentum mv is a vector with direction and sense the same as that of the velocity.

Consider now a collision between two particles as, for example, two billiard balls, one of which is initially at rest. The collision is said to be *elastic* if there is no change in the internal energy of either particle as a result of it. If such change does occur, as could for example arise if either or both of the particles were deformed by the collision, the collision would be *inelastic*. We shall suppose here that the collision is elastic, in which case the total kinetic energy and total momentum will be unchanged by it.

Figure 1.1 depicts the initial and final velocities in a collision in which particle A strikes another B which is initially at rest. Before the collision, A is moving with velocity of magnitude u in the direction shown so as to impinge on B. When the impact is over, A is moving with a velocity of magnitude v_1 in a direction making an angle Θ with that of u and with a velocity of magnitude v_2 in a direction making an angle χ with that of u. If M_1, M_2 are the masses of A and B respectively, the conservation of energy requires that

$$\tfrac{1}{2}M_1 u^2 = \tfrac{1}{2}M_1 v_1^2 + \tfrac{1}{2}M_2 v_2^2, \tag{1.1}$$

since, the collision being elastic, no energy is converted to internal energy of either particle.

Since momentum is a vector quantity, each one of the components of the total momentum must be conserved. First of all, the initial and final components of momentum of A perpendicular to the plane containing u and v_1 are zero, so that of B must also be zero. This means that u, v_1 and v_2 all lie in the same plane. Resolving next along the direction of u, we must have

$$M_1 u = M_1 v_1 \cos \Theta + M_2 v_2 \cos \chi \tag{1.2}$$

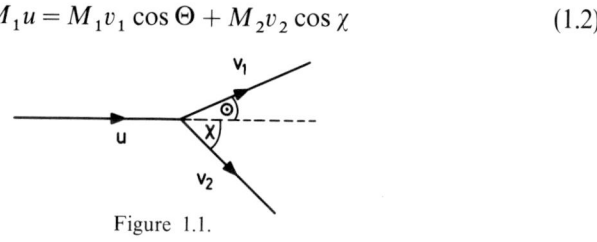

Figure 1.1.

and in the perpendicular direction in the plane of u, v_1 and v_2

$$0 = M_1 v_1 \sin \Theta - M_2 v_2 \sin \chi. \qquad (1.3)$$

If M_1, M_2 and u are known, these three equations are sufficient to determine any three of the quantities v_1, v_2, Θ and χ in terms of the fourth.

From (1.2) and (1.3) by squaring and adding, we have

$$M_1^2 u^2 = M_1^2 v_1^2 + M_2^2 v_2^2 + 2 M_1 M_2 v_1 v_2 \cos(\Theta + \chi). \qquad (1.4)$$

In the special case in which the particles are of equal mass, so that $M_1 = M_2 = M$ say, then

$$u^2 = v_1^2 + v_2^2 + 2 v_1 v_2 \cos(\Theta + \chi). \qquad (1.5)$$

As (1.2) reduces, when $M_1 = M_2$, to

$$u^2 = v_1^2 + v_2^2$$

it follows that $\cos(\Theta + \chi)$ must vanish so that

$$\Theta + \chi = 90° \quad \text{for } M_1 = M_2. \qquad (1.6)$$

This is a useful result in practice. It applies for example to two smooth[†] billiard balls as shown in figure 1.3 and to the collisions of alpha particles with helium atoms, as seen from the cloud chamber picture in figure 1.2. In this

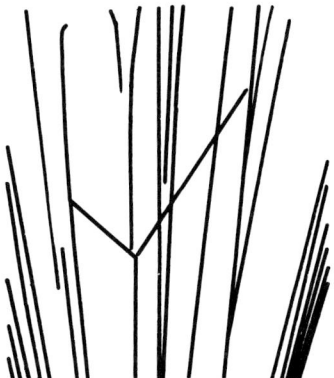

Figure 1.2. Reproduction of a cloud chamber photograph taken by Blackett and Champion showing tracks of alpha particles passing through helium gas. The forked track arises from an elastic collision between an alpha particle and the nucleus of a helium atom in the gas. After the collision the two helium nuclei, which are of the same mass, move off in directions at right angles. In the photograph this angle appears a little less than 90° because the plane of the photograph is not exactly coincident with the plane containing the tracks.

[†] We refer specifically to 'smooth' billiard balls so that no complications arise on impact from rotational energy. In fact, much of the art of the game of billiards depends on the use of rotational energy in the form of 'side', 'screw' or 'top' spin which is possible because real balls are not quite smooth.

case, the incident alpha particles are helium atoms which have lost two electrons, but this has a negligible effect on their mass, to the precision of the observations.

To determine all four quantities, more must be known about the details of the collision. For example, the striker in a game of billiards will project the ball so as to make a chosen degree of contact with the object ball. To deflect the cue ball through the so-called 'half-ball' angle, the direction of projection must be such that the path of the centre of the cue ball is a tangent to the object ball (see figure 1.3), and so on.

The half-ball angle may be readily calculated as follows. In any collision, the object ball is projected along the line of centres at the moment of impact, as anyone who plays snooker is at least implicitly aware. Referring to figure 1.3, we see that, under the conditions assumed, the angle of projection χ will be $30°$. It follows from (1.6) that the 'half ball' angle Θ is $90° - 30° = 60°$.

In contrast to collisions between billiard balls, which can be observed in detail throughout, including the geometry of the actual impact, on the atomic scale all that can be observed are the conditions after the collision at an effectively infinite distance from the impact. In that case, one of the four quantities Θ, χ, v_1 and v_2 is not immediately predictable, as we do not know the geometry of the impact. Atomic scattering experiments are then directed towards measur-

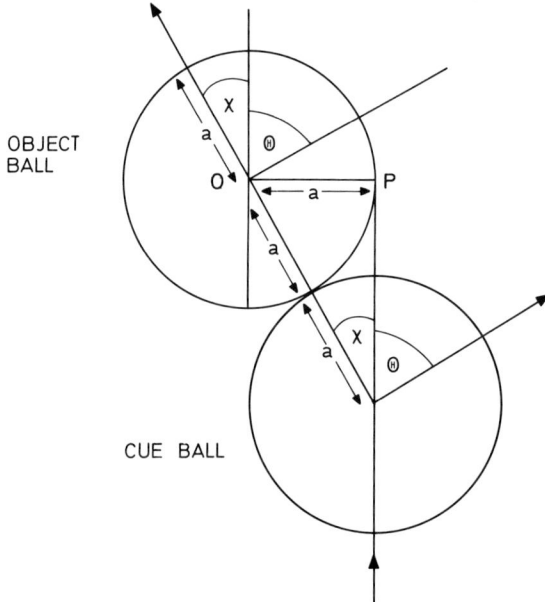

Figure 1.3. A 'half-ball' collision between two billiard balls. The object ball is initially at rest. It is struck by the cue ball the centre of which is moving in a direction which is tangent to the object ball. After the collision the cue ball moves off in a direction making an angle Θ with its initial direction of motion while the object ball is projected in a direction making an angle χ with this direction. Note that $\Theta + \chi = 90°$.

ing the *probability* that a collision will occur with a particular value of any chosen one, say Θ, of the four variables. From the observed probability as a function of Θ, we then hope to derive information theoretically about the details of the collision which cannot be directly observed.

1.2. Centre of Mass and Laboratory Co-ordinates

Collisions between atomic systems occur through the existence of a force between them which depends on their relative separation. Unlike billiard balls, which interact only when they are in contact, the force between atomic systems varies gradually with the distance between them. Because the force depends on the relative positions of the atoms and not on their separate locations in space, it is convenient to describe the mechanics of the collision in terms of the relative co-ordinates of the colliding systems rather than the ones we have used above, in which the struck particle is initially at rest relative to the laboratory.

Consider the motion in a straight line of two particles of mass M_1, M_2 respectively under the influence of an interaction force F which depends only on their distance apart. Thus if x_1, x_2 are the distances of the masses from some fixed point O in the line of motion, the force exerted by M_2 on M_1 is $F(x)$ and that by M_1 on M_2, $-F(x)$, where $x = x_1 - x_2$. We therefore have

$$M_1 \frac{d^2 x_1}{dt^2} = F(x), \tag{1.7}$$

$$M_2 \frac{d^2 x_2}{dt^2} = -F(x). \tag{1.8}$$

Adding these equations gives

$$\frac{d^2}{dt^2}\{M_1 x_1 + M_2 x_2\} = 0, \tag{1.9}$$

or, since $(M_1 x_1 + M_2 x_2)/(M_1 + M_2)$ is the co-ordinate $\bar{x}$ of the centre of mass, we see that the motion of the centre of mass is unaffected by the mutual interaction of M_1 and M_2.

On the other hand, multiplying (1.7) by M_2 and (1.8) by M_1 and subtracting gives

$$\frac{M_1 M_2}{M_1 + M_2} \frac{d^2 x}{dt^2} = F(x). \tag{1.10}$$

The relative co-ordinate x thus varies in exactly the same way as would that of a single particle of mass $M = M_1 M_2/(M_1 + M_2)$ subject to a force $F(x)$ depending only on the distance x from the fixed point O.

We note also that the total kinetic energy

$$\frac{1}{2}M_1\left(\frac{dx_1}{dt}\right)^2 + \frac{1}{2}M_2\left(\frac{dx_2}{dt}\right)^2 = \frac{1}{2}(M_1 + M_2)\left(\frac{d\bar{x}}{dt}\right)^2 + \frac{1}{2}M\left(\frac{dx}{dt}\right)^2 \quad (1.11)$$

so that once again the centre of mass and relative motions are completely separated.

There is no difficulty in generalizing these results to three dimensions. We need only replace $x_1, x_2, \bar{x}$ and x by the position vectors $\mathbf{r}_1, \mathbf{r}_2, \bar{\mathbf{r}}$ and $\mathbf{r}$. As the motion of the centre of mass will not change throughout a collision, we may concentrate attention on the relative motion. In three dimensions, this will be the same as that of a single particle of mass $M_1 M_2/(M_1 + M_2)$, known as the *reduced mass M*, moving under the influence of a force $F(\mathbf{r})$ which is a function of the relative position vector $\mathbf{r}$. The two-particle collision problem is then reduced to a single particle one.

Since the velocity of the centre of mass is unchanged by the collision, any change in total kinetic energy due to the collision arises from change in the magnitude of the relative velocity. If the collision is elastic, it follows that the magnitude of the relative velocity is unchanged, the only effect of the collision being to rotate the direction of this velocity through an angle θ. In an *inelastic* collision in which an amount ε of energy is transferred from kinetic energy to internal energy of either or both of the colliding particles, the magnitude of the relative velocity will be reduced from v to v', where

$$\tfrac{1}{2}Mv'^2 = \tfrac{1}{2}Mv^2 - \varepsilon. \quad (1.12)$$

In most of the cases with which we will be dealing, the force between the colliding systems will be a function only of the separation r between them. It is then convenient to specify the force in terms of a *potential energy function* $V(r)$, which is such that

$$F(r) = -\frac{dV(r)}{dr}. \quad (1.13)$$

$V(r)$ is then known as the *interaction energy* between the systems at separation r and is a central quantity in wave mechanics.

The relative co-ordinates $\mathbf{r}$ form the so-called centre of mass (CM) system of co-ordinates as essentially we are considering motion relative to the centre of mass. There is no difficulty in transforming results obtained in the CM system to those which will be observed in the laboratory system.

Thus let $\bar{\mathbf{v}}, \mathbf{v}$ be the (vector) velocities of the centre of mass and of the relative motion of the two particles, whose collision we have discussed above (see figure 1.1) before the impact. After the impact, $\bar{\mathbf{v}}$ will be unchanged but $\mathbf{v}$ will be changed to $\mathbf{v}'$ and we have

$$\bar{\mathbf{v}} = M_1 \mathbf{u}/(M_1 + M_2) = (M_1 \mathbf{v}_1 + M_2 \mathbf{v}_2)/(M_1 + M_2), \tag{1.14}$$

$$\mathbf{v} = \mathbf{u}, \tag{1.15}$$

$$\mathbf{v}' = \mathbf{v}_1 - \mathbf{v}_2. \tag{1.16}$$

Since the collision is elastic, $\mathbf{v}$ and $\mathbf{v}'$ will have the same magnitude but different directions.

Figure 1.4 gives the vector diagram relating these various velocities, it being noted that

$$\mathbf{v}_1 = \bar{\mathbf{v}} + M_2 \mathbf{v}'/(M_1 + M_2), \; \mathbf{v}_2 = \bar{\mathbf{v}} - \frac{M_1 \mathbf{v}'}{M_1 + M_2} = \frac{M_1}{M_1 + M_2}(\mathbf{v} - \mathbf{v}'). \tag{1.17}$$

θ, the angle between the directions of the initial and final relative velocities, is usually referred to as the angle of scattering in the CM system. It ranges from 0 to π.

In terms of θ the kinetic energy lost by the incident particle in the collision is given from (1.17) by

$$\tfrac{1}{2} M_2 v_2^2 = \frac{1}{2} \frac{M_1^2 M_2}{(M_1 + M_2)^2} |\mathbf{v} - \mathbf{v}'|^2,$$

$$= \frac{M_1^2 M_2}{(M_1 + M_2)^2} v^2 (1 - \cos \theta). \tag{1.18}$$

Since the initial kinetic energy is $\tfrac{1}{2} M_1 u^2 = \tfrac{1}{2} M_1 v^2$ from (1.15), the fractional energy loss $M_2 v_2^2 / M_1 u^2$ is given by

$$\frac{2 M_1 M_2}{(M_1 + M_2)^2} (1 - \cos \theta). \tag{1.19}$$

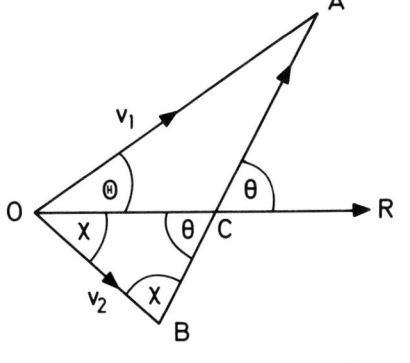

Figure 1.4.

Also, from the geometry of figure 1.4,
$$[\sin(\theta - \Theta)]/M_1 = [\sin \Theta]/M_2,$$
so that
$$\tan \Theta = M_2 \sin \theta/(M_1 + M_2 \cos \theta), \quad (1.20)$$
giving the relation between the angle of scattering in the laboratory and centre of mass systems.

When $M_1 \ll M_2$, as for example in the collision of electrons with atoms at rest, the reduced mass $\simeq M_1$ and the fractional energy loss is nearly equal to
$$\frac{2M_1}{M_2}(1 - \cos \theta), \quad (1.21)$$
which for many purposes is negligible. The relation (1.20) gives
$$\tan \Theta \simeq \tan \theta, \quad (1.22)$$
so that the angle of scattering is effectively the same in both centre of mass and laboratory systems. For many purposes, we may discuss the scattering of electrons by atoms as if the latter were simply centres of force of infinite mass.

When $M_1 = M_2$, $M = \tfrac{1}{2} M_1$ and the relation (1.20) gives
$$\tan \Theta = \frac{\sin \theta}{1 + \cos \theta},$$
$$= \tan \tfrac{1}{2} \theta,$$
so that
$$\Theta = \tfrac{1}{2} \theta, \quad (1.23)$$
and thus ranges between 0 and $\pi/2$. The fractional energy loss is
$$\tfrac{1}{2}(1 - \cos \theta) = \sin^2 \tfrac{1}{2} \theta,$$
$$= \sin^2 \Theta. \quad (1.24)$$

In the other limit, $M_1 \gg M_2$, (1.20) gives
$$\tan \Theta \simeq \frac{M_2}{M_1} \sin \theta, \quad (1.25)$$
so that the maximum value of Θ is only M_2/M_1.

The angle χ of projection of the struck particle may also be obtained in terms of θ from figure 1.4. Thus, because $OC = BC$, $\angle OBC = \chi$ and hence
$$2\chi + \theta = \pi, \quad (1.26)$$
showing that χ ranges from 0 to $\pi/2$ irrespective of the ratio M_1/M_2.

These relations are of considerable importance in practice and will frequently be referred to in later chapters.

1.3. The Specification of Collision Probabilities—the Concept of Collision Cross-Section

We now discuss convenient ways in which the rates at which collisions occur between assemblies of particles may be specified. Suppose, for simplicity, that the target particles are all rigid impenetrable spheres (rather like billiard balls) at rest. Let a be the radius of each sphere and n the number of the spheres per unit volume. The incident particles we suppose are of negligible extension and moving with a uniform velocity v into the assembly of spheres. In passing a small distance δx into the assembly, a particle will make a collision if it passes within a distance a of the centre of a sphere; in other words if a sphere is to be found within the cylindrical volume $\pi a^2 \delta x$ (figure 1.5). Thus the chance of collision will be $n\pi a^2 \delta x$.

We may call πa^2 the cross-section Q exposed by a target sphere towards collision with an incident particle. In this case, Q has a very simple and definite geometrical significance and is a constant independent of the velocity of the particles. We may readily generalize to collisions of any kind between a stream of particles and an assembly of target systems. Thus we write for the probability of a collision taking place in passing a small distance δx through the assembly

$$nQ\delta x \qquad (1.27)$$

where Q, which is called the *effective cross-section* for collisions of the type involved, will in general be a function of the relative velocity of the colliding systems.

The number of collisions in a time δt will be

$$nQ\frac{dx}{dt}\delta t = nQu\delta t \qquad (1.28)$$

where u is the incident velocity of the particles in the beam. nQu is usually referred to as the *rate coefficient* for the particular collisions and denoted by k.

When the targets are impenetrable spheres, there is no doubt that Q is finite. If the interaction energy between the colliding systems falls off gradually, to vanish only at an infinite separation, as in the case in electronic and atomic collisions, this no longer applies. No matter how widely separated the colliding systems are at their closest approach, they will experience some mutual interaction causing deviation in their relative motion. It would seem that the concept

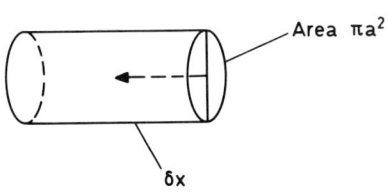

Figure 1.5.

of effective collision cross-section is not very useful under these circumstances, but wave-particle dualism comes to the rescue. As will be discussed on p. 31, in wave mechanics the effective collision cross-section is a definite finite function of the relative impact velocity provided the interaction energy $V(r)$ of (1.13) falls off faster than $1/r^2$ at large r.

Even in classical mechanics, the difficulty only arises when we are considering all possible types of collisions which can occur. We can, on the other hand, specify the chance that, in traversing a distance δx within the target assembly, a collision will occur in which the direction of relative motion is rotated through an angle between θ and $\theta + \delta\theta$ as

$$nJ(\theta)\delta\theta\delta x, \qquad (1.29)$$

where $J(\theta)$ is referred to as the *differential collision cross-section per unit angle* in the CM system. Except in the limit $\theta \to 0$ $J(\theta)$ will be finite in classical as well as well as wave mechanics. The total cross-section is related to $J(\theta)$ by

$$Q = \int_0^\pi J(\theta)\,d\theta. \qquad (1.30)$$

In classical theory $J(\theta)$ tends to infinity as $\theta \to 0$ at least as fast as $1/\theta$ so Q does not exist. As we shall see, according to wave mechanics, provided that the interaction energy falls off faster than r^{-2}, $J(\theta)$ behaves as $\theta \to 0$ so that the integral (1.29) is finite.

For many purposes it is convenient to work not in terms of $J(\theta)$ but $I(\theta)$, where

$$J(\theta) = 2\pi I(\theta)\sin\theta. \qquad (1.31)$$

$I(\theta)$ is then referred to as the *differential scattering cross-section per unit solid angle* in the CM system.

1.4. The Impact Parameter in a Classical Collision

We have already seen how the details of the collision between two billiard balls depend on the distance at which the centre of the cue ball would pass that of the object ball if it were undeflected. The corresponding quantity is of importance in the classical description of elastic collisions in general. As we have explained, we can always reduce the problem to that of the collision of a particle of finite mass with an infinitely massive centre of force. The *impact*

Figure 1.6.

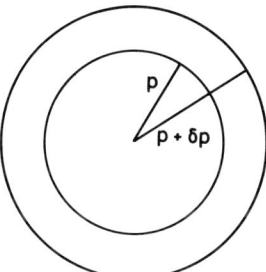

Figure 1.7.

parameter p in such a collision is the closest distance the particle would approach the centre if it were undeflected by the interaction (see figure 1.6).

Because the value of p determines the path of the particle through the field of the centre we expect a definite relation between p and the angle of scattering θ, i.e. $p = p(\theta)$. Then to produce scattering between θ and $\theta + \delta\theta$ the impact parameter must lie between p and $p + \delta p$, where $\delta p = (\mathrm{d}p/\mathrm{d}\theta)\delta\theta$. This means that the particle must be incident on an annulus of radius between p and $p + \delta p$ about the scattering centre. The corresponding target area is $2\pi p|\delta p| = 2\pi p|\mathrm{d}p/\mathrm{d}\theta|\delta\theta$ (see figure 1.7). The differential cross-section per unit angle is therefore given by

$$J(\theta)\delta\theta = 2\pi I(\theta)\sin\theta\delta\theta = 2\pi p\left|\frac{\mathrm{d}p}{\mathrm{d}\theta}\right|. \tag{1.32}$$

In fact, it is not necessarily true that a unique value of p exists for a given θ (see chapter 9 p. 165). If there are several different values p_1, p_2 etc. which yield the same value of θ, then

$$J(\theta) = 2\pi\left\{p_1\left|\frac{\mathrm{d}p}{\mathrm{d}\theta}\right|_{p=p_1} + p_2\left|\frac{\mathrm{d}p}{\mathrm{d}\theta}\right|_{p=p_2} + \ldots\right\}. \tag{1.33}$$

Calculation of the differential cross-section in classical theory depends on the evaluation of θ as a function of p. This may be done by the standard methods of orbit theory described in Chapter 9.

The special case of the impact of two similar impenetrable spheres of radius a may be worked out very quickly geometrically. In the 'half-ball' impact shown in figure 1.3, the impact parameter is equal to the radius a of either ball. In the general case when the impact parameter is p, the length of OP in figure 1.3. will be changed to p but will remain perpendicular to the initial direction of motion of the cue ball. OC will remain equal to $2a$ so that, in the same notation as on p. 4,

$$\sin\chi = p/2a.$$

and hence, using (1.26),
$$2a \cos \tfrac{1}{2}\theta = p,$$
$$-a \sin \tfrac{1}{2}\theta = dp/d\theta,$$
so, referring to (1.32),
$$I(\theta) = a^2. \qquad (1.34)$$

The differential cross-section per unit solid angle is therefore a constant, independent both of scattering angle and impact velocity. The total collision cross-section is

$$2\pi \int_0^\pi I(\theta) \sin \theta \, d\theta = 4\pi a^2, \qquad (1.35)$$

as it should be.

1.5. Angular Momentum

The relation of the classical to the wave mechanical theory of scattering of particles by a centre of force is most readily studied not in terms of the impact parameter but the *angular momentum* of the particle about the centre of force.

The angular momentum of a particle of mass m, moving at a point P with velocity v, about a centre O is the vector of magnitude $mv.\text{OP}.\sin \psi$, where ψ is the angle between the direction of v and OP (see figure 1.8), and of direction perpendicular to the plane of v and OP—it is, in other words, the moment about O of the momentum of the particle. Its magnitude can be written as mv times ON, where ON is the perpendicular distance from the centre to the line of motion of the particle. In particular, for a particle pursuing an orbit with impact parameter p and initial momentum of magnitude mv, the angular momentum about the centre when the particle is a great distance from the centre is of magnitude

$$L = mvp \qquad (1.36)$$

and direction perpendicular to the plane of motion.

The special importance of the angular momentum is that it remains constant throughout the motion of the particle in the field of the centre. Thus it is a

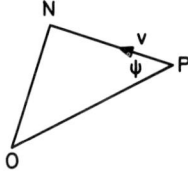

Figure 1.8.

well-known result of classical mechanics that the time rate of change of the moment of momentum of a particle about a fixed point is equal to the moment of the force acting on the particle, about the same point. But in motion in the field of a fixed centre with interaction energy $V(r)$, the force is directed along the radius vector from the centre to the particle and so has no moment about the centre. The moment of momentum of the particle about the centre therefore remains constant.

Taken together with the conservation of energy, the conservation of angular momentum for a particle moving under the action of a fixed centre provides a rapid way of obtaining the form of the orbit for fixed initial conditions, as well as the $p - \theta$ or $J - \theta$ relation from which the differential cross-section (1.31) may be derived (see Chapter 9, p. 158). It is worth noting that, because of (1.36), the differential cross-section (1.31) becomes

$$\frac{2\pi}{m^2 v^2} L \left| \frac{dL}{d\theta} \right| \delta\theta. \tag{1.37}$$

1.6. Waves

Energy transmission by a stream of particles involves a flow of matter, a mass transport. On the other hand, through wave motion, energy may be transmitted without any accompanying flow of matter. All that is involved is the passage of a disturbance in the medium through which the wave propagates.

Thus we have ocean waves in which the disturbance is one of oscillation of the sea water at each point along the path of the wave, sound waves in which the disturbance is a deformation of the medium along the direction of the wave, and electromagnetic waves in which the disturbance consists of oscillations of the electric and magnetic fields.

The simplest and most basic wave motion is a simple harmonic wave. All more complicated wave motions may be built up from combinations of such waves. Suppose a simple harmonic wave is transmitting energy along the direction of the x-axis. At any point on the axis, the disturbance will consist of a simple harmonic oscillation, of frequency v say, of the quantity concerned — an elastic displacement of the medium along the direction of propagation of a sound wave, electric and magnetic fields, perpendicular to each other and to the direction of propagation for electromagnetic waves, and so on. We may therefore write the disturbance at time t and displacement x as

$$A \cos(2\pi v t + \varepsilon), \tag{1.38}$$

where the amplitude A and frequency v are the same for all x but the phase ε varies with x. For simple harmonic waves we have

$$\varepsilon = 2\pi \frac{x}{\lambda} + \varepsilon_0, \tag{1.39}$$

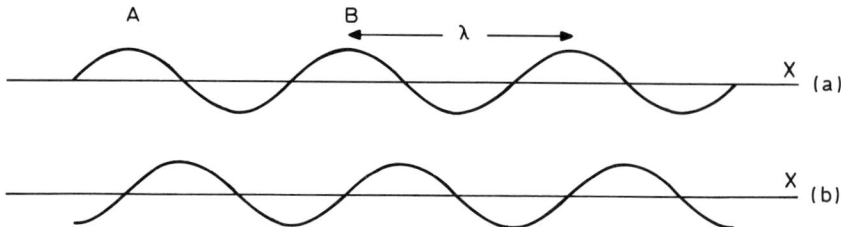

Figure 1.9. (a) The variation of the disturbance in a simple harmonic wave of wavelength λ with distance x at some instant of time.
(b) The variation of the disturbance with distance at an instant one quarter of a period later than for (a).

where λ and ε_0 are constant. Thus the disturbance at any time t at the point x is

$$A \cos\left\{2\pi\left(vt + \frac{x}{\lambda}\right) + \varepsilon_0\right\}. \tag{1.40}$$

At a particular instant a plot of the disturbance as a function of x will be a cosine curve (see figure 1.9). The distance between successive crests or successive troughs or, in fact, between successive points at which the phase of the motion is the same, is λ which is known as the *wavelength*.

As time changes, the phase of the disturbance travels in the direction of the wave. The velocity at which the phase travels is known as the *phase velocity* of the wave. Suppose that at one instant of time t, successive crests are at A and B as in figure 1.9. One whole period T, $= 1/v$, later the wave will appear to be of the same form but in fact the crest that was at A has moved to B i.e. it has travelled a distance λ in a time $1/v$. The phase velocity is therefore given by

$$v = v\lambda. \tag{1.41}$$

It is important to note that a simple harmonic wave is determined once its amplitude A, phase ε, frequency v and wavelength λ or phase velocity v are given. To avoid the presence of the 2π in (1.40), it is often written

$$A \cos(\omega t + kx + \varepsilon_0) \tag{1.42}$$

where ω is called the *angular frequency* and k the *wave number*.

In general the rate at which energy is transmitted by the wave is proportional to the square of the amplitude and, for simple harmonic waves, it propagates with the phase velocity.

As referred to earlier, we may build up a wave disturbance of any shape by suitably combining simple harmonic waves of different amplitude, wavelength, phase and phase velocity. Of particular interest in this connection is a wave in which the profile at any interest is a single crest or *wave group* as shown in figure 1.10. To build up such a group, simple harmonic waves with a large range of properties must be combined. In fact, measuring the width Δx of the

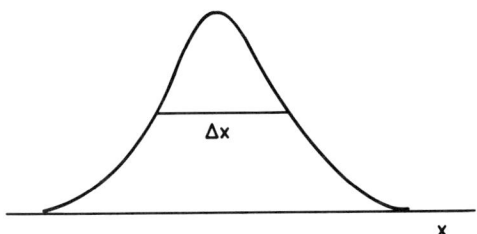

Figure 1.10. Typical shape of a wave group of width Δx.

group as shown in figure 1.10, we have

$$\Delta x \Delta k \simeq 1 \qquad (1.43)$$

where Δk is the range of values of the wave number involved. As time changes, the width of a wave group will, in general, increase while preserving nevertheless a definite crest (see figure 1.10). The rate at which the crest travels is known as the *group velocity* and is the velocity at which energy is transmitted by the group. If the medium through which the wave is propagating is dispersive, so the phase velocity depends on the wavelength, the group velocity will be very different from the phase velocity.

So far we have considered only waves in one dimension, whereas in practice we must deal with waves in three dimensions. Neighbouring points at which the phase of the disturbance is the same at any instant then trace out a surface known as the *wave front*. The direction of propagation of the wave at any point is normal to the wave front at that point.

A wave in which the wave fronts are infinite parallel planes normal to the direction of propagation is known as a *plane wave*. In a simple harmonic plane wave, the wavelength is the shortest distance between two wave fronts at which, at any time, the phase of the disturbance is the same. Similarly, all the other features of linear simple harmonic waves may readily be extended to apply to simple harmonic plane waves.

A spherical wave is one in which the wave fronts are concentric spheres. The wave is said to be outgoing or ingoing respectively according as the sense of propagation is outwards away from, or inwards towards, the centre. An example of a spherical wave analogous to the wave group described above (see figure 1.10) is the shock wave spreading outwards from an explosion in the deep sea or in the atmosphere. There is again no difficulty in defining the wavelength, frequency etc. for a simple harmonic spherical wave, but because the area of the wave front increases as the distance r from the centre increases the amplitude of an outgoing wave falls off inversely as $1/r$.

1.7. Interference and Diffraction

We turn now to consider that feature of wave motion which was of particular importance in distinguishing between waves and particles in the days before wave mechanics. This is the phenomenon of *interference*. Suppose that some point in space is subject to disturbance by two wave motions. The net disturbance may be greater or less than that due to either motion separately. Thus, if the waves are both simple harmonic of the same amplitude, frequency and wavelength propagating in the x-direction, the separate disturbances can be written

$$A \cos\left\{2\pi\left(vt + \frac{x}{\lambda}\right) + \varepsilon_1\right\} \text{ and } A \cos\left\{2\pi\left(vt + \frac{x}{\lambda}\right) + \varepsilon_2\right\},$$

where ε_1 and ε_2 are the phases. The sum of the two gives a net disturbance

$$2A \cos\left\{2\pi\left(vt + \frac{x}{\lambda}\right) + \bar{\varepsilon}\right\} \cos\{\tfrac{1}{2}(\varepsilon_1 - \varepsilon_2)\}, \qquad (1.44)$$

where $\bar{\varepsilon} = \tfrac{1}{2}(\varepsilon_1 + \varepsilon_2)$. (1.44) will vanish if ε_1 and ε_2 differ by 180°, i.e. if the waves are completely out of phase with each other. On the other hand, if $\varepsilon_1 = \varepsilon_2$ the net disturbance will have twice the amplitude of either separate disturbance.

The fact that the 'addition', as it were, of two or more wave motions can lead to smaller as well as larger net disturbances, depending on the phase relations, is a special characteristic of wave motion—no such interference effects can arise when combining the energy flows at a particular point due to the passage past it of separate streams of particles.

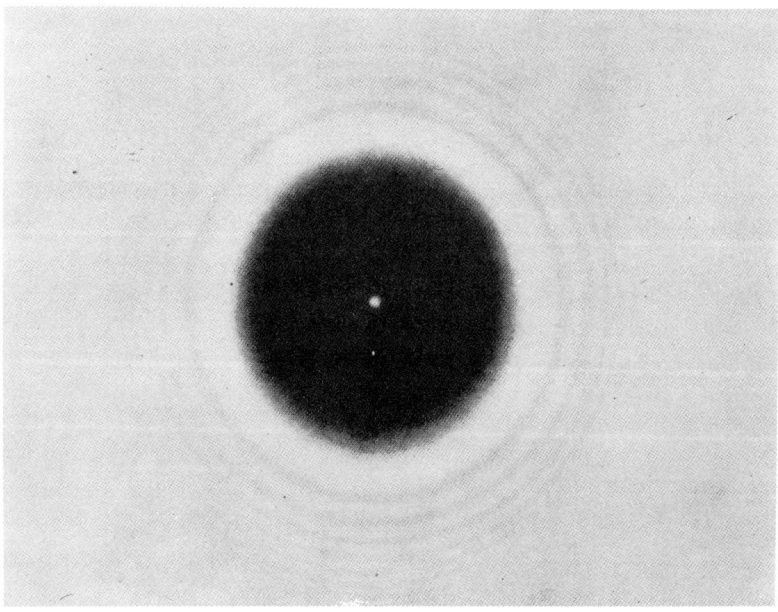

Figure 1.11. Photograph of a light shadow cast by a circular disc showing the fine structure due to diffraction.

Particles and Waves

The maxima and minima of intensity arising from interference between waves of different phase but the same frequency and wavelength, form a pattern whose scale is determined by the wavelength because it is over this distance that the phase goes through a complete cycle.

Interference effects were observed with light as long ago as 1807 and are made great use of in technical optics. A particular manifestation of interference which is of major importance for the wave mechanical theory of scattering is the phenomenon known as *diffraction* which again was observed with light over 150 years ago.

Consider a plane light wave impinging on an obstacle. Secondary waves will be scattered off the obstacle and will interfere with the primary rays to produce a more or less regular pattern of disturbance maxima and minima. This is known as a *diffraction pattern* and because of it the shadow cast by the obstacle is never completely sharp. However, the sharpness increases as the ratio λ/a of the wavelength λ, which determines the scale of the pattern, to the dimension a of the obstacle, decreases. For visible light λ lies between 4×10^{-7} and 7×10^{-7} m so blurring of shadows by diffraction is not observable except with comparatively sophisticated equipment (see, however, figure 1.11). On the other hand, the ability of sound waves, of much greater wavelength, to bend round obstacles is well known.

1.8. *Scattering of Waves—Shadow Scattering*

The existence of diffraction effects leads to features in the scattering of waves by obstacles which do not occur in particle collisions. Consider, for example, the effect of a spherical object of radius a, the surface of which completely absorbs any waves incident upon it. From a classical viewpoint, every particle colliding with a completely sticky sphere would be absorbed. None would be scattered elastically. With an incident plane wave, however, the situation is not so clear cut.

The interaction between the wave and the sphere will be such as to cancel out, as completely as possible, the wave in the shadow region behind the sphere. This can only occur through interference between the incident wave and secondary waves scattered from the sphere. But, as we have noted, destructive interference of this kind can never lead to a sharp boundary. There will be scattered waves which blur the edge. Such waves will be confined within a cone of semi-angle of order of magnitude λ/a about the incident direction, λ being the wavelength. We thus see that the completely absorbing sphere leads to elastic scattering as well as to absorption.

When the situation is analysed in detail, it is found that, when $\lambda/a \ll 1$, the scattering, while confined to very small angles about the incident direction, is such that the integrated effect contributes a target area πa^2 for elastic scattering equal to that for absorption. This elastic scattering, which is essentially a diffraction effect, is usually referred to as *shadow scattering*.

Not only will this scattering arise in collisions with a completely absorbing sphere, but it must also occur in collisions with a rigid, impenetrable sphere which scatters elastically but does not absorb. When $\lambda/a \ll 1$, the total target area, which will now be that for elastic collisions alone, will not be πa^2 but $2\pi a^2$. However, once again the shadow scattering, which doubles the target area, is confined to angles of order λ/a.

It is noteworthy that we have referred to cases in which the wavelength is small compared with the radius of the obstacle. Under these conditions, for light waves, we would expect geometrical optics to be a good approximation and it is so for scattering through angles much greater than λ/a. When λ/a is approximately equal to or much greater than 1, geometrical optics is not valid even for scattering through large angles.

These considerations are very important for wave mechanics as we shall see below.

1.9. The Wave Equation

The disturbance (1.40), generated by passage of a linear simple harmonic wave, can be written in the form

$$B\psi(x)\exp(2\pi i v t) + B^*\psi^*(x)\exp(-2\pi i v t), \tag{1.45}$$

where

$$B = \tfrac{1}{2}A\exp(i\varepsilon_0),\ \psi = \exp(2\pi i x/\lambda).$$

B^*, ψ^* are the corresponding conjugate complex quantities. $\psi(x)$ and $\psi^*(x)$ satisfy the equation

$$\frac{d^2\psi}{dx^2} + \frac{4\pi^2}{\lambda^2}\psi = 0, \tag{1.46}$$

which is often referred to as the *time-independent wave equation*.

Those readers familiar with the notation of partial differentiation will appreciate that the disturbance as a whole satisfies the time-dependent partial differential wave equation

$$\frac{\partial^2 \Psi}{\partial x^2} - \frac{1}{v^2}\frac{\partial^2 \Psi}{\partial t^2} = 0, \tag{1.47}$$

where $v = \nu\lambda$ is the phase velocity of the wave.

For a light wave in free space, the velocity of propagation is $c = 3 \times 10^8\ \mathrm{m\ s^{-1}}$, but in a medium of refractive index μ, the phase velocity is c/μ. For waves of frequency ν, the wavelength is therefore changed to $\lambda' = c/\mu$ and the equation (1.46) becomes

$$\frac{d^2\psi}{dx^2} + \frac{4\pi^2\mu^2}{\lambda^2}\psi = 0. \tag{1.48}$$

Particles and Waves

In most optical situations, the refractive index is a constant in a given medium, but even in optics there are conditions in which the refractive index varies with position. Under these conditions the equation (1.48) still applies, but now μ is a function of x, so

$$\frac{d^2\psi}{dx^2} + \frac{4\pi^2}{\lambda^2}\{\mu(x)\}^2\psi = 0. \tag{1.49}$$

These considerations may be generalized to three dimensions, but even the time-independent equation then becomes a partial differential equation

$$\frac{\partial^2\psi}{\partial x^2} + \frac{\partial^2\psi}{\partial y^2} + \frac{\partial^2\psi}{\partial z^2} + \frac{4\pi^2}{\lambda^2}\{\mu(x,y,z)\}^2\psi = 0. \tag{1.50}$$

While the case of variable refractive index is of relatively small importance in optics, it is central in wave mechanics, as we shall see. It is of some interest in this connection to note the conditions which now correspond to geometrical optics. We can largely neglect the wave nature of the propagation if we can regard $\lambda/\mu(x)$ as a 'local' wavelength $\lambda_l(x)$ say, and then apply the usual condition $\lambda_l(x) \ll a$ where a is the dimension defining the structure of the region through which the wave is propagating. The concept of a local wavelength will be valid provided the change of $\lambda_l(x)$ due to a change of x comparable with λ_l is small compared with λ_l, i.e.

$$\frac{d\lambda_l(x)}{dx} \ll 1. \tag{1.51}$$

These considerations may be extended without difficulty to waves in three dimensions. They are very relevant for applications of wave mechanics in atomic physics, as we shall see in the next chapter.

CHAPTER 2
wave mechanics— wave–particle dualism— the uncertainty principle— scattering cross-sections

2.1. Wave–Particle Dualism

For a long period, it was thought that an observed flow of energy resulted either from the passage of a stream of particles—such as the cathode rays (electrons) and anode rays (positive ions) from an electric discharge, and the alpha and beta rays from a radioactive substance—or from a wave motion, as with light, X-rays and gamma rays. The two possibilities were regarded as mutually exclusive and definite experimental tests were available to identify one or the other.

The first suspicions that the situation was not quite so clear-cut came when Planck in 1901 suggested that, in order to interpret the observed results on the frequency distribution of radiation from a black body at a given temperature, it must be supposed that radiation of frequency v can only be emitted in discrete energy bundles or quanta hv, where h is a small constant (of magnitude $6 \cdot 6 \times 10^{-34}$ J s)—a corpuscular aspect of what had been well established as a wave motion! The interpretation of the observed features of the photoelectric effect by Einstein in 1905 depended even more on the concept of quanta, or particles, of light. Finally the discovery of the Compton effect, in 1924, appeared to establish the corpuscular character of light as firmly as it was for electrons. In short, collisions were observed between light quanta and electrons which satisfied all the conditions appropriate to impacts between particles.

Thus, the energy of a light particle or quantum as hv and its momentum as of magnitude hv/c in the direction of propagation, the equations (1.1), (1.2)

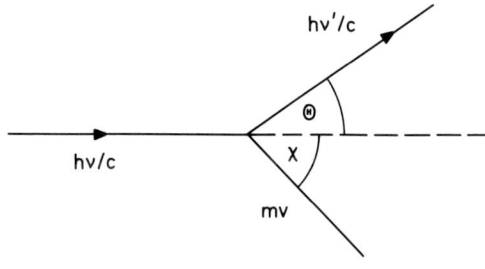

Figure 2.1.

Wave Mechanics and Scattering Cross-Sections

and (1.3) of Chapter 1 were found to be valid. Corresponding to figure 1.1, we have now figure 2.1, the collision resulting in the projection of an electron, initially at rest, with velocity v in the direction shown, and the scattering of the quantum, with reduced frequency v', into the direction shown. Corresponding to (1.1), (1.2) and (1.3) we have

$$hv = hv' + \tfrac{1}{2}mv^2, \qquad (2.1)$$

$$\frac{hv}{c} = \frac{hv'}{c} \cos \Theta + mv \cos \chi, \qquad (2.2)$$

$$0 = \frac{hv'}{c} \sin \Theta - mv \sin \chi. \qquad (2.3)$$

The validity of these relations was established by observation of v', Θ, v and χ. What more convincing evidence could there be that light consists of a stream of particles, called light quanta or photons? And yet, all the observed results about interference and diffraction of light, showing it to be a wave motion, still stand.

As mentioned earlier, the possibility that electrons might have a wave aspect was raised by L. de Broglie in 1925. Assuming the same reciprocal relations between the mechanical quantities E and p and the wave quantities v and λ as for photons, he found it possible to interpret some of the otherwise purely empirical features of Bohr's non-classical atom model. A year later the discovery of wave mechanics by Schrödinger confirmed the existence of the wave aspect, not only of electrons but of all material particles, and made it possible to understand how wave–particle dualism is possible, how in fact we should not think of waves alone or of particles alone but of 'wavicles' possessing both aspects. A little later still, in 1927, the postulated wave aspect of electrons was confirmed by the observation by Davison and Germer of the diffraction of electrons by a nickel crystal.

How does one reconcile the combination of both wave and particle aspects is one system? When does one aspect predominate and when the other? To answer these questions, it is necessary to give up one of the principal features of classical mechanics, its determinism. Thus classical mechanics provides the means for calculating where a particle will be and what will be its velocity, at some subsequent time t, when its position and velocity at some initial time t_0 are known, as well as the force acting on the particle throughout the period t_0 to t. With wave mechanics, however, the situation is less definite. All that can be done, given the same initial conditions, is to determine the *probability* that, at the time t, the particle will have any particular location and velocity. This probability may be only close to unity within a small range of locations and velocities about the classical values, in which case classical mechanics is a good approximation but, under other conditions, the probability distribution may be very different.

Crudely speaking, we can say that the wave aspect determines the probability that the 'wavicle' will be at a particular position, moving with a particular velocity, while the particle aspect essentially determines what will occur if it does happen to possess these values.

To make this clearer, we may consider a 'thought' experiment, one that can be conceived in principle but not readily carried out in practice for technical reasons. Suppose that a beam of light, of definite wavelength, passes normally through a pinhole in a screen to impinge subsequently on a second screen coated with material which glows when the light falls upon it (figure 2.2). If the pinhole has a diameter comparable with the wavelength of the light, the glow produced on the fluorescent screen will map out the diffraction pattern of the light. It will consist of a bright centre of approximately the same diameter as the pinhole, and a series of fainter concentric rings rather as in figure 1.11. But suppose we gradually reduce the intensity of the light until there is a considerable interval between the passage of successive quanta through the pinhole. Will the full pattern persist but merely grow fainter and fainter as the light intensity is reduced? The answer is quite definitely No! Each quantum will produce a single splash of light when it hits the screen but where it hits cannot be predicted. It will be most probable where the intensity in the diffraction pattern is highest. Hence, when the radiation is weak enough, the full diffraction pattern will degenerate to single flashes where each single quantum hits the screen. It is not possible to predict where any particular flash will occur. However, if a record is kept of the number of flashes which occur within different small areas of the fluorescent screen then it will be found, when a sufficiently large number have been recorded, that the distribution is exactly the same as that given by the intensity distribution in the diffraction pattern.

In this experiment, we see clearly how the wave aspect determines the probability of a light quantum reaching a particular part of the screen, while the particle aspect determines what happens when a quantum arrives there.

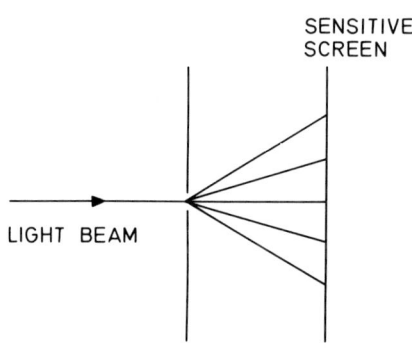

Figure 2.2.

Wave Mechanics and Scattering Cross-Sections

An exactly similar experiment can be conceived in which the light beam is replaced by a beam of electrons or of any other particles.

2.2 Material Waves

For electrons moving in free space with energy E and momentum of magnitude p, the frequency and wavelength of the associated waves are E/h and h/p respectively. By analogy with light waves, the wave function ψ, corresponding to (1.48), for electrons moving in the x-direction with energy E, will satisfy the time-independent wave equation

$$\frac{d^2\psi}{dx^2} + \frac{4\pi^2}{\lambda^2}\psi = 0, \tag{2.4}$$

or, since $\lambda = h/p$ and $p^2 = 2E/m$, where m is the electron mass,

$$\frac{d^2\psi}{dx^2} + \frac{8\pi^2 mE}{h^2}\psi = 0. \tag{2.5}$$

So far so good, but how do we deal with the situation when the electron is not moving freely, but is under the action of a force derived from a potential $V(x)$, i.e. the force is $-\frac{dV}{dx}$? In such a case, we replace E in (2.5) by $E - V(x)$, the kinetic energy of the electron at x, to give

$$\frac{d^2\psi}{dx^2} + \frac{8\pi^2 m}{h^2}\{E - V(x)\}\psi = 0. \tag{2.6}$$

This corresponds in optics to propagation in a medium whose refractive index varies with position. This is an unusual situation in optics, but is the rule rather than the exception with matter waves.

As discussed on p. 19, we may use the concept of local wavelength under the same conditions as stated. Thus the local wavelength λ_1 is given by

$$\frac{4\pi^2}{\lambda_1^2} = \frac{8\pi^2 m}{h^2}\{E - V(x)\}, \tag{2.7}$$

so

$$\lambda_1 = h/\{2m(E - V)\}^{1/2}. \tag{2.8}$$

The concept will be useful when $d\lambda_1/dx \ll 1$. If, in addition $\lambda_1/a \ll 1$, where a is the dimension defining the scale of the structure through which the wave is passing, the behaviour will be nearly classical—geometrical mechanics will be a good approximation to wave mechanics.

Even when the second condition $\lambda_1/a \ll 1$ is not well satisfied but $d\lambda_1/dx \ll 1$, the local wavelength concept may be useful. Thus suppose it is required to calculate the phase change η produced in a train of matter waves by passage through a region extending from x_1 to infinity, due to the action of a potential $V(x)$. Since, in proceeding a distance equal to one wavelength, the phase changes by 2π, the phase change required will be 2π times the change in the number of wavelengths between x_1 and infinity, due to $V(x)$. With the local wavelength approximation, this is given by

$$\eta = 2\pi \int_{x_1}^{\infty} \left\{ \frac{1}{\lambda_1} - \frac{1}{\lambda} \right\} dx. \tag{2.9}$$

We shall use this approximation applied to spherical waves in the discussion of Chapter 8, p. 171.

Having obtained the wave function ψ by solution of the equation (2.6), the probability of finding the particle between x and $x + \delta x$ is given by

$$|\psi|^2 \delta x. \tag{2.10}$$

We have used the modulus $|\psi|$ here because ψ is not necessarily real, but, as long as we are dealing with a stationary situation, as we shall be, this need not concern us further. $|\psi|^2$ is usually called the *probability density*.

There is no difficulty in generalizing to three dimensions. Those familiar with partial derivatives will realize that (2.6) generalizes to

$$\frac{\partial^2 \psi}{\partial x^2} + \frac{\partial^2 \psi}{\partial y^2} + \frac{\partial^2 \psi}{\partial z^2} + \frac{8\pi^2 m}{h^2} \{E - V(x, y, z)\} \psi = 0, \tag{2.11}$$

which is the famous time-independent Schrödinger wave equation. Having obtained the solution of this equation, the probability of finding the particle in the element of volume $\delta x\, \delta y\, \delta z$ around the point x, y, z will be

$$|\psi|^2 \delta x\, \delta y\, \delta z. \tag{2.12}$$

2.3. Allowed Values of the Energy and Other Dynamical Quantities

The probability of finding a particle in the volume element $\delta x\, \delta y\, \delta z$ must be finite or zero and have a definite unique value. This requires that $|\psi|^2$ should be finite throughout space and be everywhere single-valued. At the same time it must satisfy certain prescribed boundary conditions. In general, these requirements cannot be satisfied for all values of the total energy E which appears in (2.11). Those values of E for which they can be satisfied are known as *eigenvalues* and the corresponding solutions ψ as *eigenfunctions*. A physical system in a stationary state—one in which the probability distribution is independent of the time so that (2.11) applies—can only possess one of the allowed values of E. This selection of allowed states is the way quantization appears naturally in wave mechanics.

A simple illustrative case is that of a harmonic oscillator in one dimension, of natural frequency v and mass m. The force acting will be $-kx$ where x is the displacement from the equilibrium position and the elastic constant k is related to v by

$$(k/m)^{1/2} = 2\pi v. \qquad (2.13)$$

The time-independent wave equation is

$$\frac{d^2\psi}{dx^2} + \frac{8\pi^2 m}{h^2}\{E - \tfrac{1}{2}kx^2\}\psi = 0 \qquad (2.14)$$

and ψ must vanish as $x \to \pm\infty$. For ψ to be finite and single-valued everywhere, it is found that

$$E = (n + \tfrac{1}{2})hv, \quad n = 0, 1, 2, \ldots \qquad (2.15)$$

giving a set of allowed energies which are equally spaced at intervals hv apart.

For our purposes the three-dimensional case of motion in the field of a centre of force is most important. For this problem it is convenient to work in terms of the spherical polar coordinate system shown in figure 2.3. The position of the particle is then denoted by (r, θ, ϕ) where

$$x = r\sin\theta\cos\phi,\ y = r\sin\theta\sin\phi,\ z = r\cos\theta,$$
$$0 \leq r \leq \infty,\ 0 \leq \theta \leq \pi,\ 0 \leq \phi \leq 2\pi,$$

and

$$\delta x \delta y \delta z = r^2 \sin\theta\, \delta r \delta\theta \delta\phi.$$

For a system possessing symmetry about an axis, as in the case of a stream of particles incident on a scattering centre, we may choose this axis as the z-axis. The wave function ψ will then depend only on r and θ. It can be shown that it may be written in the form

$$\psi = F(r)G(\theta). \qquad (2.16)$$

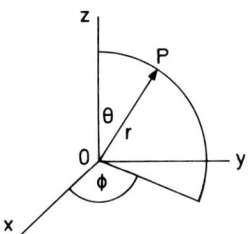

Figure 2.3.

The functions $F(r)$ and $G(\theta)$ then satisfy the ordinary differential equations

$$\frac{d^2 F}{dr^2} + \frac{2}{r}\frac{dF}{dr} + \left\{\frac{8\pi^2 m}{h^2}(E-V) - \frac{J^2}{r^2}\right\} F = 0, \tag{2.17a}$$

$$\frac{1}{\sin\theta}\frac{d}{d\theta}\left(\sin\theta \frac{dG}{d\theta}\right) + J^2 G = 0. \tag{2.17b}$$

Consideration of the second equation shows that G does not satisfy the required conditions of finiteness etc., for $0 \leq \theta \leq \pi$ unless

$$J^2 = l(l+1), \quad l = 0, 1, 2, \ldots \tag{2.18}$$

This means that the first equation may be written

$$\frac{d^2 F}{dr^2} + \frac{2}{r}\frac{dF}{dr} + \frac{8\pi^2 m}{h^2}(E-V')F = 0, \tag{2.19}$$

where

$$V' = V + \frac{l(l+1)}{r^2}\frac{h^2}{8\pi^2 m}. \tag{2.20}$$

Now, in classical theory, a particle of mass m possessing angular momentum L about the centre will be acted on by a centrifugal force mv^2/r, where $v = L/mr$, i.e. by an outward force L^2/mr^3. This may be written $-dV_c/dr$, where

$$V_c = L^2/2mr^2. \tag{2.21}$$

The extra term in (2.20) can therefore be interpreted as arising from the centrifugal force, provided the angular momentum is given by

$$\{l(l+1)\}^{1/2} h/2\pi, \quad l = 0, 1, 2, \ldots \tag{2.22}$$

In other words the angular momentum is quantized.

Corresponding to each integral value of l there will be a solution G_l of (2.18) which is such that, for each value of r, the probability that the radius vector from the centre to the particle lies in a direction between θ and $\theta + \delta\theta$, is given by

$$|G_l|^2 \sin\theta\, \delta\theta. \tag{2.23}$$

The probability that the radius vector lies between 0 and π must be unity so that

$$\int_0^\pi |G_l|^2 \sin\theta\, d\theta = 1. \tag{2.24}$$

The functions G_l are in fact multiples of well-known functions $P_l(\cos\theta)$ which are called *zonal harmonics*. Thus

$$G_l = \left(\frac{2l+1}{2}\right)^{1/2} P_l(\cos\theta). \tag{2.25}$$

Wave Mechanics and Scattering Cross-Sections

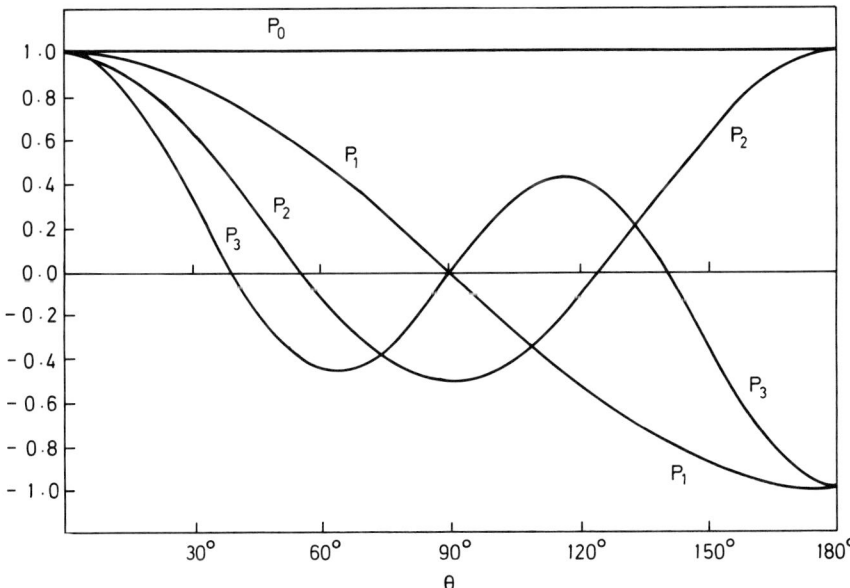

Figure 2.4. The zonal harmonics $P_n(\cos\theta)$ with $n = 0, 1, 2$ and 3, illustrated as functions of θ between 0 and π.

The first few harmonics are given by

$$P_0 = 1, P_1 = \cos\theta, P_2 = \tfrac{1}{2}(3\cos^2\theta - 1), P_3 = \tfrac{1}{2}(5\cos^3\theta - 3\cos\theta), \quad (2.26)$$

and their forms as functions of θ are illustrated in figure 2.4. They satisfy the condition

$$\int_0^\pi P_l(\cos\theta) P_m(\cos\theta) \sin\theta \, d\theta = 0, \qquad l \neq m$$
$$= (2/2l + 1) \quad l = m. \qquad (2.27)$$

The quantization of angular momentum has important consequences for the determination of differential and total collision cross-sections (see p. 32).

For each value of l, there will in general exist a series of allowed values of the energy determined from the solution of the equation (2.17a) with $J^2 = l(l+1)$. In all cases of practical interest, no matter what the form of $V(r)$, all positive values of the total energy are allowed, forming the so called *continuous spectrum* of energy values. When the total energy is negative, the particle is bound or, in classical terms, pursues a closed orbit round the centre—energy must be given to it to remove it to infinity. There may exist a continuous spectrum of such energies, a discrete spectrum in which only certain values are allowed, or no allowed values at all, depending on the form and magnitude of $V(r)$.

For the important case in which $V = -C/r$, corresponding to motion of an electron in the field of a positive charge, the allowed negative values of E are

$$-2\pi^2 mC^2/n^2h^2, \quad n = 1, 2, \ldots \geq l \tag{2.28}$$

The application of this result to the analysis of atomic structure is discussed on p. 42.

2.4. The Uncertainty Principle

To avoid unnecessary complication, we return again to the linear motion of a particle along the x-axis with energy E and momentum p. If the probability of finding the particle near the point x is close to unity the wave function ψ must have a shape similar to that of the wave group shown in figure 1.10. To build up such a wave group requires a large number of simple harmonic waves covering a wide frequency or wave number range (see p. 15). Thus, if we regard Δx, as shown in figure 1.10, as measuring the accuracy with which the particle may be located, the range of wave numbers Δk required is given, according to (1.43) by

$$\Delta x \, \Delta k \simeq 1.$$

Since $k = 2\pi/\lambda = 2\pi p/h$, we have

$$\Delta x \, \Delta p \simeq h/2\pi. \tag{2.29}$$

This means that, in order to determine the location of the particle within an uncertainty Δx, an uncertainty Δp in the momentum of magnitude $h/2\pi\Delta x$ must be accepted, and vice versa. In other words to measure the position of a particle with high accuracy, we must sacrifice precise measurement of its momentum and vice versa.

This is an example of the *uncertainty principle*, which is a consequence of wave mechanics and wave–particle dualism. It applies not only to the measurement of position and momentum but also to other pairs of complementary quantities. The most important of these is that between energy and time measurements

$$\Delta E \, \Delta t \simeq h/2\pi. \tag{2.30}$$

This implies that an energy fluctuation ΔE in a system can only persist for a time of order $h/2\pi\Delta E$. Equally well, the energy of a system of finite lifetime τ must be uncertain by an amount of the order $h/2\pi\tau$. The importance of this relation in certain collision phenomena is discussed in Chapter 6, p. 90.

2.5. General Remarks about Wave versus Classical Mechanics

Wave–particle dualism can be thought of as blurring, by diffraction, the classical picture of mechanical events to an extent which depends on the ratio of the wavelength to the scale of the system concerned. When this ratio is large,

Wave Mechanics and Scattering Cross-Sections

the blurring will almost totally obscure the classical background, but as the ratio decreases, the picture sharpens up until, when the wavelength is very small, wave effects are hardly apparent.

Even at very small wavelengths, the wave mechanical description may include rapid oscillations about the classical which are only discerned with detecting equipment of high resolution. Otherwise, they are averaged out in observations so that only a classical picture appears. The oscillations are due to interference effects which occur because two or more wave motions differing in phase are contributing to the phenomenon under observation. If ψ_1 and ψ_2 are the wave functions corresponding to each, then the net wave function will be $\psi_1 + \psi_2$. Hence the probability density (the probability of finding the system in unit volume) will be

$$|\psi|^2 = |\psi_1 + \psi_2|^2. \tag{2.31}$$

If we write

$$\psi_1 = \phi_1 \exp(i\eta_1), \psi_2 = \phi_2 \exp(i\eta_2), \tag{2.32}$$

where ϕ_1, ϕ_2, η_1 and η_2 are real, then

$$|\psi|^2 = \phi_1^2 + \phi_2^2 + 2\cos(\eta_1 - \eta_2)\phi_1\phi_2. \tag{2.33}$$

When the wavelength is small, ϕ_1^2 and ϕ_2^2 will be closely equal to the classical values, but because of the phase difference $\eta_1 - \eta_2$, the probability density will oscillate about the classical value $\phi_1^2 + \phi_2^2$. With observing equipment which cannot resolve the oscillations, the measurements made will agree with classical theory, but equipment with much higher resolving power will detect them.

The consequence of wave mechanics, that the combined effect of a number of different motions must be calculated by *first* combining the separate wave functions, including the phase differences, and only then taking the square of the modulus of the combined function to obtain the probability density, is very important to bear in mind throughout. Many instances will arise in applications to collision theory in which the interference effects which are produced in this way yield information about the colliding systems which is not obtainable in other ways (see Chapter 9).

2.6. Collision Cross-Sections in Wave Mechanics

Consider first the scattering of particles of mass m and velocity v by a rigid, immovable spherical obstacle of radius a. The considerations of p. 18 on the scattering of light from a perfectly reflecting sphere may then be applied. Thus if the wavelength $\lambda(=h/mv)$ is $\ll a$, the scattering will be classical except through angles $< \lambda/a$. The total cross-section will be $2\pi a^2$ of which πa^2 comes from classical scattering and πa^2 from the shadow scattering. Figure 2.5 illustrates the differential cross-section per unit solid angle for the case $\lambda/a = \pi/10$. As

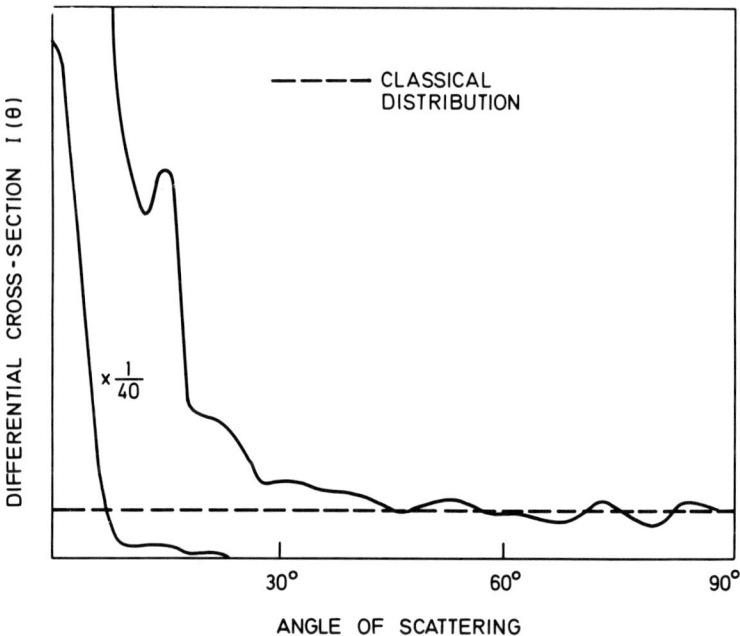

Figure 2.5. Differential cross-section per unit solid angle for scattering of particles, of wavelength $\pi a/10$, from a rigid, impenetrable and immovable sphere of radius a.
——, according to wave mechanics;
- - - -, according to classical theory.

shown on p. 12, the classical differential cross-section per unit solid angle is $\frac{1}{4}a^2$ independent of scattering angle. For angles $\theta > 30°$ the scattering oscillates fairly closely about this value while for $\theta < 30°$ it increases rapidly as θ falls due to the shadow effect.

It may seem paradoxical that the total cross-section for collision between two billiard balls of radius a should not be $\pi(2a)^2$ but $2\pi(2a)^2$. However, for ordinary billiard balls moving with normal speeds, the wavelength of relative motion is around 10^{-24} m. Since a is about 3×10^{-2} m, departure from classical scattering only sets in at scattering angles $< 3 \times 10^{-23}$ radians, many orders of magnitude too small to be significant in a game of billiards!

For λ comparable with, or greater than a, the differential cross-section departs from the classical over most if not all of the angular range while the total cross-section increases slowly as λ increases to a limit of $4\pi a^2$ when $\lambda \to \infty$ ($v \to 0$).

Again, for collisions with a completely absorbing sphere of radius a, when $\lambda \ll a$, the cross-section for collisions leading to absorption is πa^2 as given classically but, at the same time, the cross-section for elastic scattering, due to shadow effects, is also πa^2.

Interesting and important as these examples are, they are special in that the range of interaction between the incident particle and the target is finite, so

Wave Mechanics and Scattering Cross-Sections

there is no doubt of the existence of a finite total cross-section. On p. 9 we discussed, in classical theory, the situation which arises when the interaction $V(r)$ tends only gradually to zero at infinite separation r. Under these circumstances the total cross-section is unbounded and the results obtained in experiments designed to measure it would depend on the resolving power of the apparatus, i.e. the smallest angle of deflection which could be detected. In wave mechanics, the uncertainty principle makes it impossible to observe with precision deviations through sufficiently small angles. This blurs out the differential cross-section at these angles, so that, provided $V(r)$ tends to zero as $r \to \infty$ faster than $1/r^2$, the total cross-section is finite and determined only by $V(r)$, the energy and momentum of the relative motion and the initial relative velocity v.

Thus, consider a collision in which the classical impact parameter is P and the initial relative velocity is v along the x-direction (see figure 2.6). For the collision to be describable in classical terms, the uncertainty ΔP in P must be small compared with P. Also, if the angle of scattering is to be well defined, the uncertainty Δv_y in the component of velocity perpendicular to Ox must be small compared with θ. Thus

$$\frac{\Delta v_y}{v} \ll \theta, \quad \frac{\Delta P}{P} \ll 1. \tag{2.34}$$

But from the uncertainty principle applied to motion along Oy

$$m \Delta P \Delta v_y \simeq h/2\pi,$$

mv_y being the component of momentum in this direction. It follows that, for the classical description to be valid,

$$P\theta \gg h/mv, \tag{2.35}$$

or

$$\theta \gg \lambda/P,$$

where λ is the wavelength of relative motion. This is just the result we would

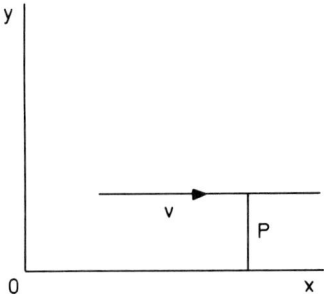

Figure 2.6.

have expected from consideration of the scale of the structure in the diffraction pattern. Now in classical orbit theory (see Chapter 9, p. 159), for $V(r)$ tending to zero faster than r^{-2} at infinity, $P\theta \to 0$ as $\theta \to 0$. It follows that, for sufficiently small angles of scattering, the collision is no longer describable by classical mechanics, no matter how small the wavelength may be. It is because of this that the total cross-section calculated from (1.30) remains finite in quantum theory.

The case in which $V(r) = C/r$, where C is a constant, is a very special one. It is the only one in which the classical formula for the differential cross-section

$$I(\theta) = (C/2mv^2)^2 \operatorname{cosec}^4 \tfrac{1}{2}\theta, \tag{2.36}$$

agrees at all angles of scattering with the wave mechanical one. It was derived by Rutherford as a basis for the experiments on the scattering of alpha particles (of charge $2e$ where $-e$ is the electron charge) by thin foils of material of atomic number Z, designed to test his nuclear theory of the atom. In this case $C = 2Ze^2$. It is fortunate that this formula gives the correct results, for otherwise confirmation of the nuclear theory of the atom could not have come so early—the experiments on alpha particle scattering were carried out in 1911, 15 years before the introduction of wave mechanics!

2.7. Quantum Theory of Scattering by a Centre of Force

Whereas according to classical theory the particle will be deviated from its initial direction of motion unless the force due to the centre vanishes everywhere along its path, in quantum theory we assign a probability $\rho(p)$ that a particle, incident with impact parameter between p and $p + dp$, will suffer an observable deviation. The total collision area will then be

$$Q = 2\pi \int_0^\infty \rho(p) p \, dp. \tag{2.37}$$

Since $mvp = L$, the angular momentum of the particle about the centre of force, we may rewrite (2.37) as

$$Q = (2\pi/m^2v^2) \int_0^\infty L\sigma(L) \, dL, \tag{2.38}$$

where σ is now the chance that a particle with angular momentum between L and $L + dL$ will suffer an observable deviation.

We must further take account of the fact that, in quantum theory, as explained on p. 26, L can only have the discrete values

$$L = \{l(l+1)\}^{1/2} h/2\pi, \tag{2.39}$$

where $l = 0, 1, 2, \ldots$. To allow for this, we first transform the integral by the

Wave Mechanics and Scattering Cross-Sections

substitution (2.39) assuming l to be a continuous variable. This gives

$$Q = (h^2/4\pi m^2 v^2) \int_0^\infty (2l+1)\sigma(L)\,dl.$$

If l is confined to positive integral values or zero, we need only replace the integral by a sum to give

$$Q = (h^2/4\pi m^2 v^2) \sum_{l=0}^\infty (2l+1)\gamma(l), \tag{2.40}$$

where $\gamma(l)$ may be taken as the probability for an observable deviation when the angular momentum quantum number is l.

It is now necessary to consider $\gamma(l)$. In terms of the wave aspect of matter, since the frequency and wavelength are fixed, the scattering will be specified in terms of the amplitude and phase of the scattered waves (at infinity) relative to the incident. As there is no absorption, the amplitude will be unchanged, so that the probability of a collision, for a fixed angular momentum, will depend on the phase change η_l and vanish with it.

To examine this further, we consider the special case in which $V(r)$ is given by

$$V(r) = D, \quad r < a,$$
$$= 0, \quad r > a, \tag{2.41}$$

as in figure 2.7. For head-on collisions, $l = 0$ and $\gamma(0)$ is certainly unity on classical theory. According to quantum theory, the motion of the particles will be represented by a train of waves of wavelength $\lambda (= h/mv)$ outside the obstacle $(r > a)$ and of wavelength $\lambda' = h/(m^2v^2 - 2Dm)^{1/2}$ inside. These two trains must join smoothly at $r = a$ (see figure 2.8). For this to occur and at the same time keep the amplitude finite at $r = 0$, a phase change must be introduced in the train for $r > a$ relative to that which would exist in the absence of the obstacle. This phase change is in principle observable at infinity and alone indicates the presence of the obstacle. An important point now arises, because it is not possible experimentally to count the difference in the number of waves between the obstacle and the point of observation due to the obstacle, and so a phase change which is an integral multiple of 2π will not be detectable. This means that there will be no observable effect of the obstacle when this is so.

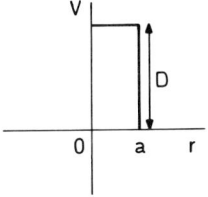

Figure 2.7.

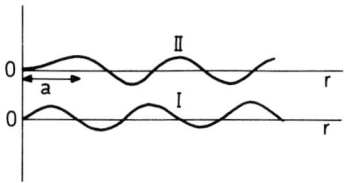

Figure 2.8. The phase shift produced in a train of spherical waves by a potential barrier.
I, Wave train in the absence of a potential barrier.
II, Wave train with the potential barrier present. The effect of the barrier is to reduce the magnitude of the disturbance in the range 0 to a over which it extends. As a result the first maximum occurs at a higher value of r than in I, so producing phase change.

Hence $\gamma(0) = 0$ according to quantum theory when the obstacle either introduces or eliminates a whole number of complete waves.

On this basis, we expect that $\gamma(0)$ will be a periodic function of η_0 with period 2π, vanishing when $\eta_0 = 2s\pi$, with $s = 0, 1, 2 \ldots$. Furthermore it can never be negative. Similar arguments apply to $\gamma(l)$. The simplest function satisfying these conditions is

$$\gamma(l) = A \sin^2 \eta_l, \tag{2.42}$$

where η_l is the phase shift produced in the waves associated with angular momentum $\{l(l+1)\}^{1/2}h/s\pi$. This would lead to

$$Q = (Ah^2/4\pi m^2 v^2) \sum_{l=0} (2l+1) \sin^2 \eta_l. \tag{2.43}$$

To fix the value of A, we consider the application of the analysis to scattering of particles by a rigid spherical obstacle of radius a which is such that $a \gg h/mv$ the wavelength of the particle's motion, a case which we have already discussed (see p. 30).

Under the nearly classical conditions, L/mv will be nearly equal to the classical impact parameter p and, classically, no scattering will occur for $p > a$. We would therefore expect that the phase shifts will be very small for $l > l_0$, where

$$l_0 h/2\pi = L/mv > a. \tag{2.44}$$

Since $2\pi amv/h \gg 1$, l_0 will be large and we may approximately reconvert (2.43) to the integral

$$Q = (Ah^2/2\pi m^2 v^2) \int_0^{l_0} l \sin^2 \eta_l \, dl. \tag{2.45}$$

For $l < l_0$ we expect η_l to be large because the impenetrable obstacle will eliminate a number $a/\lambda \gg 1$, of the incident waves. We may therefore replace $\sin^2 \eta_l$ in the integrand by its mean value $\frac{1}{2}$ to give

$$Q = \tfrac{1}{2} A\pi a^2.$$

Wave Mechanics and Scattering Cross-Sections

As described on p. 30, when allowance is made for shadow scattering $Q = 2\pi a^2$. It follows that $A = 4$ and

$$Q = (4\pi/k^2)\sum_l (2l + 1) \sin^2 \eta_l, \qquad (2.46)$$

where $k_, = 2\pi mv/h = 2\pi/\lambda$, is the wave number of the motion.
A detailed mathematical analysis shows that (2.46) is indeed correct.

2.8. The Differential Scattering Cross-Section in Wave Mechanics

According to the discussion on p. 26, for particles possessing a quantized angular momentum $\{l(l+1)\}^{1/2} h/2\pi$ about a centre of force, the probability of finding the particle at points where the polar angle θ lies between θ and $\theta + \delta\theta$ is

$$(l + \tfrac{1}{2})\{P_l(\cos\theta)\}^2 \sin\theta\, \delta\theta, \qquad (2.47)$$

where the $P_l(\cos\theta)$ are the functions discussed on p. 26 and illustrated in figure 2.4. We might therefore expect that the differential cross-section $I(\theta)$ defined in (2.5) will be given by

$$I(\theta) = \frac{2\pi}{k^2} \sum_l (2l + 1)^2 \sin^2 \eta_c \{P_l(\cos\theta)\}^2. \qquad (2.48)$$

However, this is not correct, because it makes no allowance for interference between the waves scattered with different angular momenta. We should instead write

$$I(\theta) = |f(\theta)|^2 \qquad (2.49)$$

where

$$f(\theta) = \sum_l (2l + 1) \exp(i\eta_l) \sin\eta_c P_c(\cos\theta). \qquad (2.50)$$

In other words, we associate with each angular momentum a scattered amplitude with the correct phase factor $\exp(i\eta_l)$ included. The total scattered amplitude (2.50) is then obtained by summing these separate amplitudes and thence the total scattered intensity by taking the square of the modulus.
This gives

$$I(\theta) = \frac{2\pi}{k^2} \sum_l (2l + 1)^2 \sin^2 \eta_l \{P_l(\cos\theta)\}^2 + \text{interference terms}. \qquad (2.51)$$

When $I(\theta) \sin\theta$ is integrated between 0 and π to obtain the total cross-section, the interference terms make no contribution because (see (2.27))

$$\int_0^\pi P_l(\cos\theta) P_m(\cos\theta) \sin\theta\, d\theta = 0, \; l \neq m.$$

2.9. The Variation of the Phase Shifts with the Angular Momentum Quantum Number

In classical theory, the deflection of a particle with impact parameter p will be small if the potential energy exerted by the centre when the particle is at a distance p from it is small compared with the kinetic energy of the particle, i.e.

$$V(p) \ll \tfrac{1}{2}mv^2. \tag{2.52}$$

Noting that classically $p = L/mv$, where L is the angular momentum, and that in wave mechanics $L = \{l(l+1)\}^{1/2}h/2\pi$, we may translate (2.52) into quantum theory terms by stating that $\sin^2 \eta_l$ will be small for such values of l that

$$V(L/mv) \ll \tfrac{1}{2}mv^2, \tag{2.53}$$

where $L = \{l(l+1)\}^{1/2}h/2\pi$. This may be expressed as

$$V(r_0) \ll \tfrac{1}{2}mv^2, \tag{2.54a}$$

where

$$kr_0 = \{l(l+1)\}^{1/2}. \tag{2.54b}$$

To examine the consequences of this, suppose $V(r)$ is of the form C/r^s. Then (2.54) requires that, if η_l is small,

$$C/r_0^s \ll \tfrac{1}{2}mv^2, \tag{2.55}$$

so that

$$r_0 \gg (2C/mv^2)^{1/s}. \tag{2.56}$$

and hence from (2.54b)

$$l \gg (2C)^{1/s}(2\pi/h)m^{1-1/s}v^{1-2/s} = l_0, \text{ say}. \tag{2.57}$$

When $s > 2$, $l_0 \to 0$ when $v \to 0$, so that all phase shifts except perhaps η_0 must tend to zero as the velocity tends to zero. Again, the higher the velocity, the larger the number of important phases in the series (2.46) and (2.50). The same conclusions apply to cases in which $V(r)$ falls of exponentially with distance at large distances r.

Since l_0 will be the only important phase shift at low velocities of impact, it follows that, under these conditions, $I(\theta)$ will behave as $\{P_0(\cos \theta)\}^2$; that is to say, it will be constant, independent of angle.

The formulae (2.46) and (2.50) for the total and differential cross-sections are of even wider generality. No matter how complicated the colliding systems may be, if the only collisions which can occur between them are elastic, the cross-sections can always be written in the forms (2.46) and (2.50) with the phase shifts all real numbers. Even when inelastic collisions are also possible the formulae may still be used to give the sum of the elastic and inelastic cross-sections provided the phase shifts are complex numbers with negative imaginary

Wave Mechanics and Scattering Cross-Sections

parts, but this result, though useful in certain contexts will not be used henceforward. We shall however discuss in some detail in Chapter 5, p. 75, the interpretation of the experimental data on the elastic scattering of electrons by rare gas atoms in terms of (2.46) and (2.50) and the expected behaviour of the phase shifts as functions of electron velocity.

2.10. Generalization to Collisions Involving Atomic Systems

The colliding systems we have been discussing have been essentially structureless particles which suffered no deformation or other change of internal energy during the collision. In practice, we must deal in atomic physics with collisions in which one or both of the colliding systems is composite so that, during the impact, we need to take account of the fact that the motion of more than two particles is concerned. Thus, even in the relatively simple case of collisions between electrons and hydrogen atoms, we need to take into account the motion of two electrons and a relatively massive proton. Approximate methods of treating these cases, based on the simple two-body problems we have discussed, must be developed, but it is not our intention at this stage to discuss these methods. Some account will be given in Chapter 5 and 6. Here we wish to show how the concept of effective cross-section may be generalized to cover these cases.

The presence of structure in the colliding systems introduces many new possible consequences of a collision. Thus energy may be transferred from that of relative motion to internal motion in one or both of the systems. In such cases, the collisions are said to be *inelastic*. As a result of the transfer of energy to internal motion, one or more of the constituents of either or both systems may acquire enough energy to escape. Thus, in collisions of electrons with atoms, sufficient energy may be given to the atom so that an electron is ejected from it leaving a positively charged ion. To specify the rates at which such processes occur, we may first of all retain the concept of the *total collision cross-section* Q_t which includes all possibilities. If P_i is the probability that, in collision, a particular process such as ionization, may occur, then we may call

$$Q_i = P_i Q_t, \tag{2.58}$$

the effective cross-section for that process. Thus if a system A is traversing a distance δx through a gas containing N systems B per unit volume, the chance that the system A will experience a collision with a system B in which the particular process occurs will be

$$NQ_i \delta x. \tag{2.59}$$

Similarly the chance that such a collision will occur in a time δt is

$$NQ_i v \delta t, \tag{2.60}$$

where v is the velocity of the system A. $NQ_i v$ is often referred to as the *collision frequency* for the particular process.

It is also readily possible to generalize the concept of differential cross-section per unit angle or unit solid angle. Thus if $p_i(\theta) \sin \theta \, d\theta$ is the probability that, in a collision between systems A and B in which a particular process i occurs, the direction of relative motion is turned through an angle between θ and $\theta + \delta\theta$ then

$$I_i(\theta) = p_i(\theta) Q_i,$$

is the differential cross-section per unit solid angle for scattering between θ and $\theta + \delta\theta$ in which the particular process occurs.

CHAPTER 3
atoms

In preceding chapters, we have distinguished between elastic and inelastic collisions. Whereas in the former case, the impact does not change the internal energy of either colliding system, in an inelastic collision such changes do occur. Obviously, in order to interpret experimental data on these collisions, it is necessary to understand in some detail the internal structure of the colliding systems. In fact, even for elastic collisions, the nature of the internal structure is important. For our purposes, we must therefore describe the salient features of atomic and molecular structure which are relevant to the discussion of the collisions with which we are concerned.

3.1. *Atomic Structure—Some General Features*

Atoms are all composed of a central nucleus which possesses a positive electric charge Ze, where e is the electron charge and Z is a positive integer known as the *atomic number*. Held bound in orbits round the nucleus are Z electrons whose total negative charge balances the positive charge on the nucleus. The probability distribution of the electrons is determined from wave mechanics. The nucleus has a radius of order 10^{-14} m, while the closest electron orbit has a radius of $(0.53/Z) \times 10^{-10}$ m.

The nucleus is composed of protons, each with charge $+e$ and mass M, 1837 times that of an electron, and neutrons, possessing no charge but with mass only a little greater than that of the proton. If N is the number of nuclear neutrons, the total mass of the nucleus is to a good approximation equal to $(N+Z)M$. $N+Z$ is usually referred to as the mass number A. Thus for an atom of this mass number and an atomic number Z, we have:

number of electrons $= Z$, number of protons $= Z$, number of neutrons $= A - Z$.

The chemical properties of an atom are determined only by the atomic number. It is therefore possible to have atoms of different mass number which nevertheless possess the same chemical properties. They differ only in the number of nuclear neutrons and are referred to as *isotopes* of each other.

Atoms with each value of Z are distinguished by names, abbreviated symbolically to one or two letters. Separate isotopes are distinguished by adding the

Period	1	2	3	4	5	6	7	8			0
1	Hydrogen H 1										Helium He 2
2	Lithium Li 3	Beryllium Be 4	Boron B 5	Carbon C 6	Nitrogen N 7	Oxygen O 8	Fluorine F 9				Neon Ne 10
3	Sodium Na 11	Magnesium Mg 12	Aluminium Al 13	Silicon Si 14	Phosphorus P 15	Sulphur S 16	Chlorine Cl 17				Argon Ar 18
4	Potassium K 19	Calcium Ca 20	Scandium Sc 21	Titanium Ti 22	Vanadium V 23	Chromium Cr 24	Manganese Mn 25	Iron Fe 26	Cobalt Co 27	Nickel Ni 28	
	Copper Cu 29	Zinc Zn 30	Gallium Ga 31	Germanium Ge 32	Arsenic As 33	Selenium Se 34	Bromine Br 35				Krypton Kr 36
5	Rubidium Rb 37	Strontium Sr 38	Yttrium Y 39	Zirconium Zr 40	Niobium Nb 41	Molybdenum Mo 42	Technetium Tc 43	Ruthenium Ru 44	Rhodium Rh 45	Palladium Pd 46	
	Silver Ag 47	Cadmium Cd 48	Indium In 49	Tin Sn 50	Antimony Sb 51	Tellurium Te 52	Iodine I 53				Xenon Xe 54
6	Caesium Cs 55	Barium Ba 56	Rare Earths 57–71	Hafnium Hf 72	Tantalum Ta 73	Tungsten W 74	Rhenium Re 75	Osmium Os 76	Iridium Ir 77	Platinum Pt 78	
	Gold Au 79	Mercury Hg 80	Thallium Tl 81	Lead Pb 82	Bismuth Bi 83	Polonium Po 84	Astatine At 85				Radon Rn 86
7	Francium Fr 87	Radium Ra 88	Actinides[†] 89–103								

Figure 3.1. Periodic table of the elements.

The chemical symbol and atomic number are given under the name of each element.
[†] Includes thorium (Th 90) and uranium (U 92) as well as other elements.
Elements 43, 85, 87 and 93–103 are not known in nature but have been produced artificially.

Note that after the third period the columns are divided into left-hand and right-hand halves. Whether an element is placed on the left- or right-hand side of a column is indicated by the position of its chemical symbol and mass number. The arrangement is such that elements in the left-hand halves of one column show similarity of chemical and physical properties as do elements in the right-hand halves. There is usually little resemblance between elements placed in different halves of the same column.

mass number as a superscript. Thus atoms with $Z = 2$ are known as helium atoms and symbolized as He. The two isotopes with $A = 4$ and $A = 3$ respectively are then distinguished as ^{4}He and ^{3}He.

Figure 3.1 shows a classification of atoms according to their chemical properties, in the so-called Periodic Table.

The simplest atom is that of hydrogen with $Z = 1$, $A = 1$. Two other isotopes are known one, deuterium, with $Z = 1$, $A = 2$ which is stable and the other, tritium, with $Z = 1$, $A = 3$ which is radioactive. We shall begin our description of atomic structure with the hydrogen atom, which is not only important for its own sake but also is the basis for theoretical interpretation of the properties of all other atoms and even molecules.

In addition to neutral atoms, we may also have *positive ions*, which are atoms from which one or more electrons have been removed. When isolated such ions are stable. The positive ions of helium are denoted by He^+, He^{2+} respectively according as one or two electrons is missing, with similar notation for other cases.

Negative ions also exist in which an additional electron is attached to an atom. Not all atoms form stable negative ions. Thus H^- is stable but He^- is not.

3.2. Electron Spin

Before proceeding to describe briefly the structure and properties of specific atoms we must draw attention to the existence of electron spin—an electron behaves not as a point charge but as a rotating system with angular momentum closely prescribed by the requirements of wave mechanics. The magnitude of this angular momentum is $\{s(s+1)\}^{1/2} h/2\pi$ where s can only have the single value $\frac{1}{2}$. It may be shown that the existence of electron spin arises naturally when account is taken of the need to satisfy special relativity as well as wave mechanics.

The interaction between the electron spin and its translational motion is weak but by no means insignificant. It nevertheless happens, because of certain symmetry requirements which must be met by the wave function describing a system of similar particles, that the existence of electron spin exerts a much larger indirect influence on the allowed energies of the system then would be expected.

3.3. Combination of Angular Momenta

In considering the total energies of composite systems such as atoms, it is often necessary to use the rules established in wave mechanics for the combination of angular momenta. Suppose, for example, we have two sub-systems which can possess allowed angular momenta $\{l_1(l_1+1)\}^{1/2}h/2\pi$ and $\{l_2(l_2+1)\}^{1/2}h/2\pi$ respectively. If the two sub-systems are combined to form a composite system, what then are the allowed angular momenta for this system?

The answer is that these will be given by

$$\{L(L+1)\}^{1/2}h/2\pi,$$

where L has one of the values

$$l_1+l_2, l_1+l_2-1, l_1+l_2-2, \ldots |l_1-l_2|. \tag{3.1}$$

This is a general result of much value in discussing atomic structure.

3.4. The Hydrogen Atom

This atom is a two-body system involving a proton of mass M and an electron of mass m where $M \simeq 1837m$. If we neglect electron spin, the relative motion is the same as that of a particle of mass $m' = (mM/m+M) \simeq m$, moving under the influence of an electrical attraction $V(r) = -e^2/r$ where r is the relative separation.

The wave equation (2.11) may then be set up and solved to determine the allowed energy values E for the system (see p. 24 of Chapter 2). Taking as zero the energy of the system when the electron is at rest relative to the proton and at an infinite distance from it, it follows that, for $E > 0$ the electron is not bound, but pursues, in classical terms, an open orbit relative to the proton. If $E < 0$, however, the electron is bound—a positive energy must be added to extract the electron to affinity.

As already summarized on p. 28 of the preceding chapter, all positive values of the energy are allowed, but the only allowed negative values are given by

$$E_n = \frac{-2\pi^2 m' e^4}{n^2 h^2}, n = 1, 2, \ldots \tag{3.2}$$

When electron spin is included, these allowed energies are modified to a relatively small extent. The energy is found to depend not only on the number n but also on the angular momentum quantum number l of the relative motion of electron and proton. The possible values of l for a given n are

$$l = 0, 1, \ldots n-1.$$

States with $l = 0, 1, 2, 3 \ldots$ are usually referred to as s, p, d, f... states respectively, a notation which arose because of a connection with different types of spectrum lines observed in the hydrogen spectrum—s for sharp, p for principal, d for diffuse and f for fundamental. We shall use this notation frequently henceforward.

Since the electron spin quantum number is $\frac{1}{2}$, the total angular momentum

Atoms

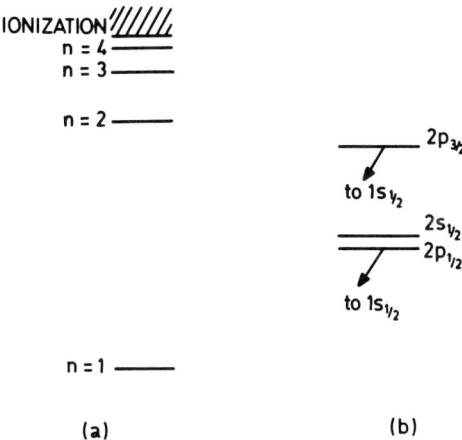

Figure 3.2. (a) Energy levels of the hydrogen atom. On the scale used, the dependence of the energy on the quantum numbers l and j is too small to be seen.
(b) The energy levels for which $n = 2$ are shown on a greatly expanded scale so that the dependence on l and j is apparent. Optically allowed transitions are indicated. As the hydrogen atom contains only one electron, the spin quantum number s is always equal to $\frac{1}{2}$.

(orbital and spin) quantum number j, according to (3.1), for given l, is given by

$$j = l \pm \tfrac{1}{2}. \tag{3.3}$$

It turns out that, for the hydrogen atom, the total energy for a given value of n depends mainly on j and only very slightly on l. Figure 3.2 shows the so-called energy level diagram for the hydrogen atom with associated quantum numbers, n, j and l. We employ the usual notation for the states of different orbital and total angular momenta, the quantum number for the latter being indicated as a subscript.

3.5. Optical Radiation from Excited Hydrogen Atoms—Allowed and Forbidden Transitions

A hydrogen atom in the lowest or ground energy state with $n = 1, l = 0, j = \tfrac{1}{2}$, if left alone, is absolutely stable. In any other state, referred to as an excited state, the atom possesses an energy greater than in the ground state. It may get rid of this energy by making a transition to a lower state, emitting a quantum of radiation in the process. Thus if E_n is the energy of the initial excited state, E_m of the final state, the frequency v_{nm} of the emitted quantum will be given by

$$E_n - E_m = h v_{nm}. \tag{3.4}$$

It follows that the light emitted by an assembly of hydrogen atoms, which have been raised to excited states by some external energy source, will consist of a series of spectrum lines whose frequencies satisfy the condition (3.4) with E_n, E_m the energies of excited states of the atom.

The chance that an atom in a state n will make a transition to a lower state m in a time δt depends on the difference between the orbital angular momentum quantum numbers l_n, l_m of the two states. If $l_n - l_m = \pm 1$ the chance is greatest and will be of the order 10^7 to 10^8 times δt. In other words, the mean time which will elapse before the transition occurs will be of order 10^{-7} to 10^{-8} s. Under these circumstances the transition is said to be *optically allowed*.

If $l_n - l_m = \pm 2$, the mean time before transition occurs will be of the order 10^{-3} to 10^{-4} s. It will be even longer for other values of $l_n - l_m$. In all such cases, the transition is *optically disallowed*.

A state from which no transition is optically allowed is referred to as a *metastable* state. An atom, if left alone in such a state, will have a mean lifetime of order 10^{-4} s or longer. The $2s_{1/2}$ state of hydrogen (see figure 3.2) is effectively metastable. The transition to the $1s_{1/2}$ state is optically disallowed and, while that to the $2p_{1/2}$ is allowed in principle, the very small energy separation between them reduces the probability far below 10^4 s^{-1}. In fact, the $2s_{1/2}$ state has a natural lifetime of 0·12 s.

However, because of the small energy separation from the $2p_{1/2}$ state, which may make an optically allowed transition to the ground state, the presence of an electron field as low as 100 V m^{-1} reduces the lifetime of the metastable state to the order 10^{-7} s by exciting transitions from the $2s_{1/2}$ to the $2p_{1/2}$ state.

When account is taken of the selection rules, $l_n - l_m = \pm 1$, the allowed spectrum of transitions in hydrogen atoms is somewhat simplified. The well known series of lines known as the Balmer, Lyman and Paschen series correspond to the series of transitions as follows.

	Final states	Initial states	
Lyman	$n = 1, l = 0$	$n', l' = 1$	$n' = 2, 3, 4, \ldots$
Balmer	$n = 2, l = 0, 1$	$n', l' = 0, 1, 2$	$n' = 3, 4, 5, \ldots$
Paschen	$n = 3, l = 0, 1, 2$	$n', l' = 0, 1, 2, 3$	$n' = 4, 5, 6, \ldots$

Thus the frequences of the lines of the Lyman series are given by

$$h\nu_L = \frac{2\pi^2 me^4}{h^2}\left(1 - \frac{1}{n'^2}\right) \tag{3.5}$$

if the electron spin is ignored. When this is included, each line is split into a number of neighbouring lines. For the Lyman series, the splitting will be into two lines, because the $l' = 1$ initial state is double with $j' = l' \pm \tfrac{1}{2} = \tfrac{1}{2}, \tfrac{3}{2}$.

This splitting, which occurs in all the spectral series, is referred to as the *fine structure* of the spectrum lines.

In each case, the series converges to a limit when $n' \to \infty$. In the Lyman series, the longest wavelength ($Ly\alpha$) is 121·6 nm and that at the limit 91·2 nm, both in the ultraviolet. The longest wavelength in the Balmer series is in the visible at 656·3 nm and in the Paschen series in the infrared at 1875·1 nm.

3.6. Hyperfine Structure

There is one further dynamical feature which has not been allowed for and that is the internal spin angular momentum of the proton. As for the electron, it has the unique allowed magnitude $\{i(i+1)\}^{1/2} h/2\pi$ with $i = \frac{1}{2}$. Its interaction with the relative transitional motion of electron and proton and with the electron spin is very small but leads to a further splitting of the hydrogen spectrum, considerably smaller than the fine structure. It is accordingly known as the *hyperfine structure*.

We refer to this here because of the astrophysical importance of the hyperfine splitting of the ground $1s_{1/2}$ state. When the proton spin with spin quantum number $\frac{1}{2}$ is allowed for, two states arise with different total angular momentum (orbital + electron spin + proton spin) associated with the originally single state. One will have a total angular momentum quantum number 1 and the other 0. The first of these has a slightly higher allowed energy than the second, which is the true ground state of hydrogen. The wavelength associated with a transition between them is 21 cm, in the radio region of the spectrum. Even though this transition is highly disallowed optically, the mean lifetime of the upper state being 10^7 years, radiation resulting from it is observed from hydrogen atoms in our galaxy and from extragalactic regions. This is because in interstellar space hydrogen atoms exist for very long periods, comparable with or greater than the mean lifetime of the upper states, before colliding with any other system. Because of this, atoms in the upper state have little chance of losing their excess energy other than by radiation.

The importance of the observability of the 21 cm radiation is that it makes it possible to study the distribution of atomic hydrogen, the most abundant element, in the galaxy and beyond. In particular, it has been shown from such studies that our galaxy has a spiral structure like many external galaxies.

3.7. Continuous Spectra

We have, in discussing the hydrogen spectrum, made no allowance so far for the fact that the electron may possess any positive energy E_c, say. Transitions from such states to a final allowed state of negative energy will yield, not a series of isolated spectrum lines, but a continuous range of frequencies, forming what is known as a *continuous spectrum*.

Thus, associated with the Lyman series, there will be a Lyman continuum with frequencies given by

$$\nu_{Lc} = \nu_{L\infty} + \nu_c, \qquad (3.6)$$

where $\nu_{L\infty}$ is the frequency of the Lyman limit and ν_c ranges from 0 to ∞.

Similarly we have the Balmer and Paschen continua and so on.

Figure 3.3 illustrates diagrammatically the Lyman series and continuum.

3.8. Spontaneous Emission Coefficients

The chance per second that an atom in a state n will make a transition to state m with emission of radiation is known as the spontaneous emission coefficient A_{nm}. The mean lifetime of the state towards emission of this radiation will then be $1/A_{nm}$. An assembly of a large number N_n of atoms in the excited state n will radiate $N_n A_{nm}$ quanta of frequency $(E_n - E_m)/h$, per second.

3.9. Absorption of Energy by Hydrogen Atoms

A hydrogen atom in its ground state, of energy E_1, can only absorb an amount of energy δE if $E_1 + \delta E$ is an allowed energy for the atom. Remembering that $E_1 < 0$, we see that if $\delta E < |E_1|$, so that $E_1 + \delta E$ is < 0, δE can have only one of the discrete set of values

$$\delta E = E_n - E_1, n = 2, 3, \ldots \qquad (3.7)$$

where E_n is the energy of an allowed negative energy state. In particular, the least energy which can be absorbed is

$$E_2 - E_1,$$

where E_2 is the energy of the lowest excited state. This energy is known as the *excitation energy*. Ignoring hyperfine structure, it has the value 10·15 eV for hydrogen.

If $\delta E > |E_1|$, $E_1 + \delta E > 0$ and any value is allowed. In this case the electron is unbound in the final state and the atom is then said to be ionized. The minimum energy to produce ionization, the *ionization energy*, is simply $|E_1|$. For hydrogen it is 13·54 eV.

These considerations apply no matter how the energy is communicated,

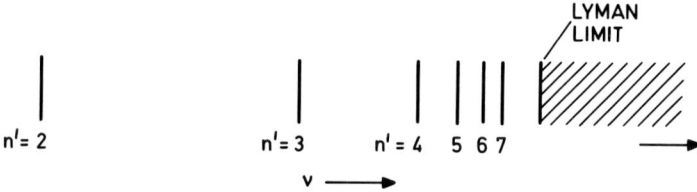

Figure 3.3. The Lyman series and Lyman continuum in the spectrum of atomic hydrogen.

by absorption of radiation or in collision with some other systems. Thus in a collision of an electron with an atom in which the relative translational energy is E, the impact will be elastic if

$$E < E_2 - E_1. \tag{3.8}$$

It may be inelastic if (3.8) is not satisfied. However, when

$$E < |E_1|,$$

the energy lost must have one of the discrete values

$$\delta E = E_n - E_1, n = 2, 3, \ldots \tag{3.9}$$

If

$$E > |E_1|$$

the energy loss may have not only one of the discrete values (3.9), but also a continuous set of values from $|E_1|$ to E.

The energy loss spectrum for electrons scattered under these different conditions will therefore have the general form shown in figure 3.4. It was from the experimental study of such spectra, initiated by Franck and Hertz, that a great deal of supporting evidence for quantized atomic energy levels was obtained in the 1920s.

Absorption of radiation will be discussed further in Chapter 13, p. 230.

3.10. Probability Charge Distributions in Different States of a Hydrogen Atom

Corresponding to each allowed state n of an atom, there is a wave function ψ_n which will in general be a function of the spherical polar co-ordinates r, θ and ϕ (see p. 25). $|\psi_n|^2 \delta V$ is then the probability of finding the electron in the elementary volume δV at (r, θ, ϕ). By integrating over θ and ϕ, we may obtain the probability $P_n(r)\delta r$ of finding the electron at a distance between r and $r + \delta r$ from the proton, when the atom is in the state concerned.

In general, the lower the state the more concentrated is this distribution.

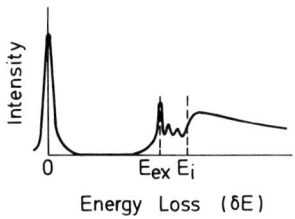

Figure 3.4. Typical form of the energy loss spectrum for electrons with initial kinetic energy greater than the ionization energy, scattered by hydrogen atoms. The diagram shows the probability of an energy loss δE as a function of δE. E_{ex}, E_i are the excitation and ionization energies of an atom.

For the ground $1s_{1/2}$ state we have, ignoring the small difference between m' and m,

$$P_{1s}(r) = (4/a_0^3)r^2 \exp(-2r/a_0), \qquad (3.10)$$

where

$$a_0 = h^2/4\pi^2 me^2$$
$$= 0{\cdot}53 \times 10^{-10}\,\text{m}.$$

This probability is a maximum where $r = a_0$, which is often known as the Bohr radius of the hydrogen atom.

For higher s states, the maximum moves out to larger values of r. Thus for 2s

$$P_{2s}(r) = (r^2/8a_0^3)(2 - r/a_0)^2 \exp(-r/a_0), \qquad (3.11)$$

with a maximum at $r = 5{\cdot}24 a_0$. For 3s it moves out further to $r = 13 a_0$ and so on. The distribution corresponding to the lowest p state is given by

$$P_{2p}(r) = (r^4/24a_0^5) \exp(-r/a_0) \qquad (3.12)$$

with a maximum at $r = 4a_0$. For 3p it is at $r = 10{\cdot}5 a_0$ and so on.

Again for the lowest d state

$$P_{3d}(r) = (8r^6/15 \times 81^2 a_0^7) \exp(-2r/3a_0), \qquad (3.13)$$

with a maximum at $r = 9a_0$, moving to $21{\cdot}2 a_0$ for 4d.

Figure 3.5 shows these distributions as functions of r. It is important to notice that, not only do the maxima move out as n increases but also that, for a given n, the chance of finding the electron very close to the proton falls off as l increases.

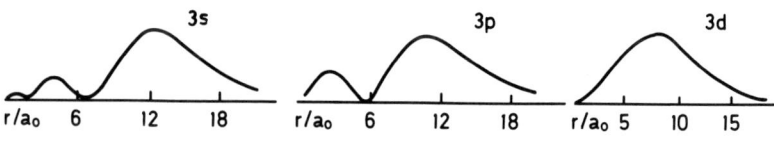

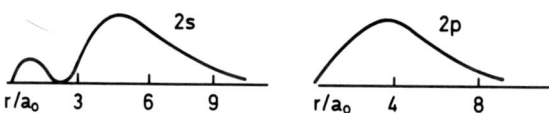

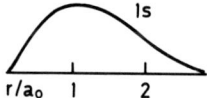

Figure 3.5. The radial probability densities $P(r)$ for different states of a hydrogen atom.

Atoms

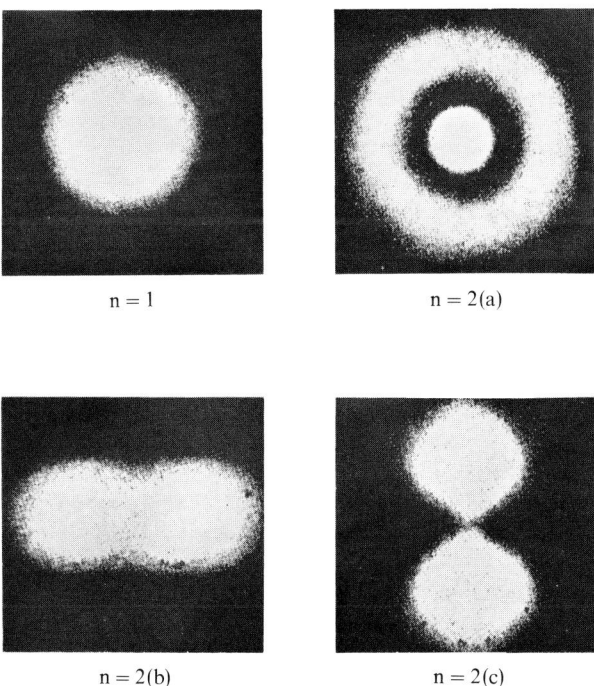

Figure 3.6. Probability distributions for electrons in the lowest stationary states of a hydrogen atom. The probability of finding the electron at any point is proportional to the brightness of the picture there.
The state with $n = 1$ is the ground 1s state which is spherically symmetrical ($l = 0$). For $n = 2$ there are 4 states. 2(a) is the 2s state which is again spherically symmetrical. 2(b) and 2(c) are 2p states, the third of which is of the same shape as (b) and (c) but with long axis perpendicular to the plane of the paper.
The proton is at the centre of the pattern in all cases. The magnification is about 10^8 for $n = 2$ and 2×10^8 for $n = 1$.

The angular momentum determines the form of the dependence of the probability distribution as a function of θ and ϕ (see Chapter 2, p. 26). For s states, the distributions are spherically symmetric (i.e. do not depend on θ and ϕ). On the other hand, for p states there are three possible distributions, corresponding to the same energy. These can be taken as varying as $\cos^2 \theta$, $\sin^2 \theta \cos^2 \phi$ and $\sin^2 \theta \sin^2 \phi$ respectively. In general, there are $2l + 1$ possible distributions associated with angular momentum quantum number l.

Figure 3.6 illustrates some typical distributions in such a way as to bring out the angular dependence.

3.11. *Electric Field of a Hydrogen Atom*

Although an atom is electrically neutral in that it possesses no net electric charge, it does however possess an external electric field because the proton and electron are not located at the same point. The electrostatic potential of this field is given by

$$v(r) = e\left(\frac{1}{r} + \frac{1}{a_0}\right)\exp(-2r/a_0), \qquad (3.14)$$

and so is negligible at distances very much larger than the Bohr radius a_0 from the proton. It follows that the potential energy of interaction of a hydrogen atom with an electron, when the latter is at a distance r from the proton, would be

$$V(r) = -ev(r).$$

This gives rise to an attractive force on a negatively charged particle because the atomic electron can only imperfectly screen off the charge of the proton.

This interaction is said to be that of the electron with the static (i.e., undisturbed) field of the hydrogen atom. In fact, the atomic field is disturbed by the presence of the additional electron. Thus the atomic electron will be repelled so that the atom is polarized—the atomic electron will tend to be located on the opposite side of the proton to the second electron. When the second electron is at a large distance r from the atom, the polarization energy may be determined as follows.

The electric dipole moment induced in an atom by an electric field of strength F is along the field and of magnitude proportional to F. It may therefore be written αF, where α is a constant for a given atom known as the *atomic polarizability*. The electric field at the atom due to an electron at a large distance r will be of magnitude e/r^2 so that the induced electric dipole moment will be of magnitude $D = \alpha e/r^2$ and in a sense to attract the electron. This dipole will exert a force of magnitude $2De/r^3$ on the electron. Writing this as $-dV_p/dr$, where V_p is the interaction energy due to the polarization, we have

$$\begin{aligned}V_p &= -2e\int_r^\infty \frac{D\,\mathrm{d}r}{r^3} \\ &= -2\alpha e^2 \int_r^\infty \frac{\mathrm{d}r}{r^5} \\ &= -\tfrac{1}{2}\alpha e^2/r^4.\end{aligned}$$

This result will be used in Chapter 5, p. 86 and Chapter 11, p. 201.

Other effects also arise due to the disturbance of the atom by an incident electron, as discussed further in Chapter 5, p. 86.

3.12. Deuterium

The binding energy of the electron in a deuterium atom is very nearly the same as in hydrogen. It differs only because of the difference in reduced mass of the electron–deuteron and electron–proton systems. If m is the electron mass, the respective reduced masses are $(1 - 5\cdot4 + 10^{-4})$ and $(1 - 2\cdot7 \times 10^{-4})m$. Substitution of these values for the reduced mass m' in (3.4) shows that the difference in binding energy is $2\cdot7 \times 10^{-4}$ of that of the hydrogen atom, i.e. $3\cdot6 \times 10^{-3}$ eV. For many purposes, this small energy difference can be ignored and in most respects deuterium atoms behave chemically in the same way as hydrogen atoms.

3.13. The Structure of Complex Atoms—The Pauli Principle

The hydrogen atom and its isotopes, deuterium and tritium, are the only atoms that contain only a single electron. For all other atoms, the problem of calculating the structural properties is complicated by the fact that allowance must be made for the interaction between the atomic electrons as well as that of each with the nucleus. It is still possible, however, to obtain at least a qualitative idea about the structure of these atoms by using our knowledge of that of hydrogen. Before discussing this, we must draw attention to a most important principle which has a profound effect on the allowed energy values for a system of more than one electron.

To the approximation in which we can assign allowed states and energies to the electrons separately, the principle states that no two electrons can occupy the same quantum state. In applying this, allowance must be made for electron spin and for the fact that there are $2j + 1$ states associated with the total angular momentum quantum number j. Let us now see how the principle applies to the helium and lithium atoms with two and three electrons respectively.

To avoid confusion between states of the atom as a whole and states of orbital motion for single electrons, we shall refer to the latter states henceforth as *orbitals*.

3.14. The Ground State of Helium

For the helium atom, since there are two spin states, both electrons may occupy the same orbital. The ground state can then be described as one in which each electron occupies a $1s_{1/2}$ orbital. Since the nuclear charge is $2e$ compared with $1e$ for hydrogen, the magnitude of the energy of each electron would be four times as large as that of the electron in the ground state of hydrogen. Furthermore the probability electron density would be a maximum at $a_0/2$ instead of a_0. The helium atom would therefore be more compact than that of hydrogen and much more firmly bound. A more accurate treatment, allowing for the repulsion between the helium electrons, shows that each electron has a binding energy about 2·9 times that in hydrogen and the probability density is a maximum at $a_0/1\cdot7$.

The minimum energy required to remove an electron from a helium atom leaving a singly ionized atom (He$^+$) is given by

total energy of He$^+$ − total energy of He.

If E_H is the ionization energy of H, the total energy of He$^+$ = $-4E_H$ and of He = $-2 \times 2 \cdot 9 E_H = -5 \cdot 8 E_H$. The ionization energy of He is therefore $1 \cdot 8 E_H = 24 \cdot 3$ eV. An accurate calculation gives 24·581 eV, in agreement with observation.

3.15. *The Ground State of Lithium*

Let us look now at the ground state of lithium. In this case, the Pauli principle prevents the third electron from occupying the same quantum $1s_{1/2}$ state as the first two. Instead it will occupy the lowest vacant state which is the $2s_{1/2}$[†]. In this state it will spend most of its time at a considerably greater distance from the nucleus than the two electrons in the $1s_{1/2}$ state (see p. 48). The latter will therefore screen off almost $2e$ of the total nuclear charge, leaving only a net positive charge $+e$ to hold in the outer electron. It will thus be in a state very like the $2s_{1/2}$ state of hydrogen with a maximum probability separation of $5 \cdot 24 a_0$ from the nucleus. The lithium atom will therefore be much less compact than that of helium.

Not only is the lithium atom much larger than that of hydrogen, but it also has a much lower ionization energy. Since the magnitude of the negative energy of a $2s_{1/2}$ electron in H is only $\frac{1}{4}$ of that of the ground $1s_{1/2}$ electron, the ionization energy of lithium is comparable with $\frac{1}{4} E_H = 3 \cdot 38$ eV. In fact it is 5·39 eV.

3.16. *The Ground States of Other Atoms*

It is not our intention here to follow these arguments through to describe the salient features of the different atoms in their ground states. We will, however, draw attention to the periodicity of properties which leads to the heavier rare gas atoms being similar in many ways to helium, and the alkali metal atoms to lithium (see figure 3.1).

According to the Pauli principle, we can build up the structure of an atom containing Z electrons by feeding in the electrons one by one. To form the ground state of the atom as a whole, each successive electron occupies the lowest energy level accessible to it.

Consider now the case of neon, an atom with 10 electrons. The first two occupy the 1s orbitals, but as described above the next must go into the 2s orbital. A second further electron can also occupy this orbital, but the next must go to a 2p orbital. There are three of these orbitals which lie quite close to the 2s orbital and correspond to probability distributions which differ little

---

[†] In all atoms other than hydrogen, the 2s orbital lies below the 2p.

in range from it. The remaining electrons can completely occupy these 2p orbitals. Neon is thus the heaviest atom in which no electron occupies a state with total quantum number greater than 2.

The next atom, sodium, contains an additional electron which must go to a 3s orbital, in which it spends much of its time well outside the inner 10 electrons and is relatively weakly bound.

Neon and sodium thus correspond respectively to helium and lithium, account being taken of the fact that there are eight available states with $n = 2$ as contrasted with two for $n = 1$.

It is in this kind of way that the electronic structure of the ground states of atoms may be traced out, although allowance must be made for the fact that interelectronic interactions may change the energy order of the states of higher n, 3d falling above 4s and 4p in some atoms, etc.

The assignment of electrons to the individual orbitals of different n and l determines the electron *configuration* of the atom. Thus for helium it would be written $(1s)^2$, for lithium $(1s)^2\,2s$, neon $(1s)^2\,(2s)^2\,(2p)^6$, sodium $(1s)^2\,(2s)^2\,(2p)^6\,3s$ and so on.

It is usual to refer to the totality of states available to electrons through occupation of orbitals of specified n and l as a *shell*. Thus we have the 1s shell, the 2s shell etc. A shell in which all the vacancies are filled is said to be *closed*. Thus helium and neon have configurations in which all the shells are closed but in lithium and sodium the outer shell is incompletely filled.

Atoms with similar outer configuration have similar properties. For the heavier rare gases, which are relatively compact atoms with high ionization energies similar to helium and neon, the configurations are:
argon $(1s)^2\,(2s)^2\,(2p)^6\,(3s)^2\,(3p)^6$,
krypton $(1s)^2\,(2s)^2\,(2p)^6\,(3s)^2\,(3p)^6\,(3d)^{10}\,(4s)^2\,(4p)^6$,
xenon $(1s)^2\,(2s)^2\,(2p)^6\,(3s)^2\,(3p)^6\,(3d)^{10}\,(4s)^2\,(4p)^6\,(4d)^{10}\,(5s)^2\,(5p)^6$,
again with completely closed shells. The heavier alkali metal atoms, which are large and possess low ionization energies, similar to lithium and sodium, are potassium, rubidium and caesium, containing respectively a 4s, 5s and 6s

H	He	Li	C	O	N	F
13.54	24.58	5.39	11.26_5	13.61_5	14.54_5	17.42
	Ne	Na				Cl
	21.56	5.14				12.96
	Ar	K				Br
	15.76	4.34				11.84
	Kr	Rb				I
	14.00	4.18				10.44
	Xe	Cs				
	12.13	3.89				

Table 3.1. Ionization energies (in electron-volts) of certain atoms.

Atomic and Molecular Collisions

electron additional to the corresponding rare gas atoms argon, krypton and xenon.

Table 3.1 gives the ionization energies of a few selected atoms.

3.17. Excited States of Atoms

An excited electron configuration is one in which one or more electrons does not occupy the lowest accessible orbital. Thus $(1s)^2 (2s)^2 (2p)^5$ 3s is an excited configuration of neon.

While the energy of any state of a complex atom depends strongly on the configuration, it also depends to a lesser extent on other quantum numbers. For many atoms, these may be taken as the quantum numbers of the total orbital angular momentum L, the total spin S and the total angular momentum J.

For a given configuration, there will, in general, be a number of possible values for these quantum numbers calculable from the rules for combining angular momenta (see (3.1)), with allowance for the Pauli principle.

For a configuration in which all the shells are closed, only one state can arise, that with $L = 0, S = 0, J = 0$. It follows that in determining the states arising from a particular configuration, we can ignore all shells which are fully occupied.

As an example, consider a configuration in which all shells are fully occupied apart from an outer p shell which contains two electrons. According to the angular momentum combination rules, the possible values of L are 0, 1, 2 and of S, 0 and 1. The limitations imposed by the Pauli principle reduce these six possibilities to three, namely

$$L = 0, S = 0; L = 1, S = 1 \text{ and } L = 2, S = 0.$$

In shorthand notation, these states are denoted as $^1S, ^3P, ^1D$, respectively, the foreindex being the value of $2S + 1$.

So far we have ignored the quantum number J which, in fact, has only a slight effect on the energy. It determines the fine structure of the energy level system. For a given L and S, there will be $2S + 1$ substates with different J if $L \geq S$. $2S + 1$ is therefore called the *multiplicity* of the states with given L and S. Because these states are not really single but include the $2S + 1$ substates they are usually referred to as *terms* of given L and S i.e. from a p^2 configuration there arise $^1S, ^3P$ and ^1D terms.

For terms arising from a given configuration in which there is only one shell not fully occupied the following rules apply:

(a) Terms with largest S lie deepest and of these the lowest is the one with the greatest L.

(b) If the partly occupied shell is the outermost one and is less than half filled, the energy for given L and S increases with J and the multiplet is said to be *normal*. On the other hand, if the shell is more than half filled, the energy decreases with J and the multiplet is said to be *inverted*.

Thus of the terms arising from a p^2 configuration, the ^3P will be the lowest. Also, since J ranges from $L+S$ to $L-S$, for the ^3P terms J will have the values 2, 1 or 0 giving, in the usual notation, $^3P_2$, $^3P_1$ and $^3P_0$ states. According to the above rule, the $^3P_0$ state will be lowest and so will be the lowest state arising from the p^2 configuration.

We have taken no account of nuclear spin, which will impose a hyperfine structure on the level system, but for our purposes this may be ignored.

Just as for atomic hydrogen, energy changes within complex atoms must involve transitions between allowed energy levels. As for hydrogen atoms, some of these transitions, satisfying certain selection rules, are optically allowed, with spontaneous emission coefficients of order 10^4 to 10^8 s^{-1} (see p. 46).

For most atoms, metastable states exist from which no optically allowed transitions can occur to any lower state. Except for heavy atoms, transitions which involve a change of the total electron spin quantum number S, a change of multiplicity, are optically forbidden. This is a consequence of the weak interaction between electron spin and orbital motion.

3.18. Excited States of Helium

For the helium atom, containing two electrons only, the lowest excited configuration is the 1s 2s, in which one electron is excited to a 2s orbital. From this configuration there arise only ^1S and ^3S terms, the latter lying lower. These terms, distinguished as 2^1S and 2^3S respectively, are both metastable and their existence often has an important influence on reactions in gaseous helium.

Next in order come the (1s2p) 2^1P and 2^3P terms, which are not metastable — the 2^1P may make optically allowed transitions to the 2^1S and 1^1S (ground) states and the 2^3P to the 2^3S.

In this way, a series of excited states are generated associated with configurations in which one electron remains in the lowest, ground, orbital, while the other is excited. As we increase the total quantum number of the excited orbital, the energy approaches that of an ionized helium atom and a free electron. This is illustrated in figure 3.8.

3.19. Doubly Excited States — Autoionization

Excited states of helium with energy in excess of the ionization energy lie in the continuous spectrum, just as for hydrogen atoms. One new factor appears, however, because of the presence of more than one electron. So far we have considered excited configurations in which only one electron is in an excited

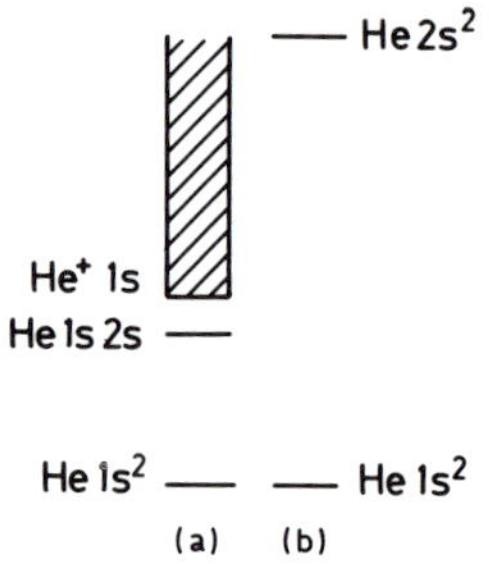

Atomic and Molecular Collisions

Figure 3.7. Singly and doubly excited states of a helium atom.
(a) The singly excited states. The shaded region represents the continuum of states extending upwards from the first ionization limit which occurs at the energy of the ground 1s state of the ionized helium atom (He$^+$).
(b) The lowest doubly excited state. He 2s^2 has an excitation energy above that of the first ionization limit.

orbital. But what about, for example, (2s)^2 in which *both* electrons are in excited orbitals? If $E(2s)$ is the energy of a 2s orbital and $E(1s)$ of a 1s,

$$2|E(2s)| < |E(1s)|.$$

It follows that the energy of the doubly excited configuration lies above the ionization limit for singly excited configurations (see figure 3.7). Since (2s)^2 is the lowest doubly excited configuration, this will apply to all such configurations.

What this means is that, if an atom is formed in a doubly excited configuration, it will have a finite probability per second of making a transition to the single excited continuum configuration of the same total energy, i.e.

$$(2s)^2 \to (1s)(Es)$$

where Es denotes an orbital in which the electron is moving freely with kinetic energy E. Such a transition leads to ionization of the atom and the process is known as *autoionization*.

Because states in doubly-excited configurations are unstable towards autoionization they have a finite mean lifetime τ. According to the uncertainty principle, this implies that the energy of the state is uncertain by an amount ΔE where

$$\Delta E \simeq h/2\pi\tau.$$

In typical cases, τ is of order 10^{-14} s so that $\Delta E \simeq 0.05$ eV as shown in detail on p. 90. This is about 10^6 times larger than the uncertainty due to the possibility of an allowed radiative transition.

The existence of autoionizing states for atoms with more than one electron

Atoms

leads to interesting and important effects in atomic collision phenomena. In Chapter 13 we shall describe effects of this kind which occur in the absorption of light by rare gas atoms.

3.20. Negative Atomic Ions

We have already referred to the ionization of an atom involving the removal of one or more electrons to leave a positively charged core. If, however, we use the term *ion* to apply to an atom which possesses an electric charge, we must allow also for the fact that many atoms can attach an additional electron in a stable state, so forming a system with a net negative charge. We call such a system a *negative atomic ion* as distinguished from the *positive atomic ion* which is left when one or more electrons are removed.

Whereas all positive ions, apart from that of hydrogen which is just a bare proton, possess an infinite set of allowed energy states, negative atomic ions may possess no bound (i.e., negative energy) states at all. Moreover, most which do so possess only one bound configuration.

This major difference arises from the fact that, whereas an electron in a positive ion of charge ze, when at a large distance r from the nucleus, sees an unscreened positive charge $(z+1)e$, in a negative ion of net charge $-e$ the nucleus is almost completely screened from it by the inner atomic electrons. The interaction which binds the electron therefore falls off for large r as $(z+1)e^2/r$ for the positive ion but much more rapidly for the negative ion (see p. 50). In the latter case the number of bound states for the electron will be finite and may even be zero if the interaction is weak enough (see p. 27).

The energy required to detach an electron from a negative ion in its ground state is known as the *detachment energy* of the ion. For an atom it is also equal to the energy given out when an electron of zero kinetic energy is captured by the atom from the negative ion—the *electron affinity* of the atom.

The atom with the largest electron affinity is chlorine, for which it is 3·61 eV. Comparable values are found for the other halogen atoms F(3·40 eV), Br(3·36 eV), and I(3·06 eV).

The negative ion H^- of hydrogen is stable with an electron affinity of 0·754 eV, but helium does not form a stable negative ion. This applies also to the other rare gas atoms.

Of importance for atmospheric physics is the fact that the ion O^- is stable, the electron affinity of O being 1·462 eV. On the other hand N^- is not stable.

The alkali metal atoms all form stable negative ions, the electron affinities for Li, Na, K, Rb and Cs being respectively, in eV, 0·62, 0·55, 0·50, 0·49 and 0·47.

Other examples of stable negative ions with the associated electron affinities in eV, are S (2·077) Se (2·0206), P (0·74), C (1·27), Si (1·38), Cu (1·226), Ag (1·303) and Au (2·3086).

Among the stable negative ions, there are few which have bound excited

Atomic and Molecular Collisions

states if one ignores the fine structure of the ground term. Of the cases we have referred to above, C^- and Si^- are the only ones in which the energy of an excited term associated with the ground configuration is less than that of the ground state of the neutral atom.

3.21. Doubly excited States of Negative Ions—Autodetachment

Just as with neutral atoms, doubly excited configurations of negative ions are unstable towards a transition in which one of the electrons drops to a ground orbital and the other leaves with positive kinetic energy. This is a process analogous to autoionization (see p. 56) and is called *autodetachment*.

For example, attachment of an electron to a helium atom in a 2^3S excited state can occur with release of about 0·08 eV energy. That is to say, the energy of the negative ion with configuration

$$1s(2s)^2$$

is about 0·08 eV less than that of the atom in the $1s2s\ 2^3S$ state. Nevertheless, because the $1s\ 2s\ 2^3S$ state lies 19·72 eV above the ground $(1s)^2 1^1S$ state of the neutral atom, the energy of the ion in the $1s(2s)^2$ state will be 19·64 eV higher than that of the neutral atom. This surplus energy is released in autodetachment (see figure 3.8).

As described on p. 56 for neutral atoms, the energy of a doubly excited state is uncertain by an amount of order $h/2\pi\tau$ where τ is the lifetime towards autodetachment which is, in general, of the same order, 10^{-14} s, as for autoionization. Despite their instability towards autodetachment, the doubly excited states of negative ions, just as for neutral atoms, play an important part in collision phenomena. We shall discuss the effects they produce in electron collisions with atoms in Chapter 6 and in the absorption of light by negative ions in Chapter 13 p. 254.

Ions in certain doubly excited states possess a lifetime towards autodetachment which is orders of magnitude larger than the usual lifetimes. An example is the $(1s)(2s)(2p)\ ^4P$ term of the He^- ion. The $^4P_{5/2}$ state has a mean lifetime

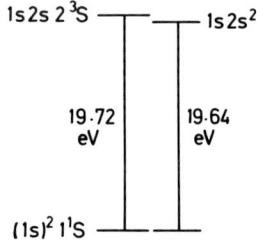

Figure 3.8. Illustrating the relation of the energy of the $1s\ 2s^2$ level of He^- to that of the ground $(1s)^2\ 1S$ state and the $1s\ 2s\ 2^3S$ excited state of He.

as long as 5×10^{-4} s while the $^4P_{3/2}$ and $^4P_{1/2}$ are only a little more ephemeral, their lifetimes being around 10^{-5} s. Because of these long lifetimes, He$^-$ ions in these states with kinetic energies of a few hundred eV or more can pass through normal experimental equipment without suffering autodetachment.

CHAPTER 4
molecules

4.1. *The Interaction between Atoms*

As explained on p. 50, although an atom is electrically neutral, it does produce an external electrical field. The electrons and the positively charged nucleus of a second atom, at a finite distance from the first, will be acted on by this field. Because the electrons and the nucleus are not coincident, the net force on the second atom will not vanish. The fact that atoms may combine to form stable molecules shows that, under many circumstances, the force must be attractive.

Consider two atoms A and B a fixed distance R apart, this being the separation between the nuclei. In principle, we may calculate the allowed energies for the total system, which we may refer to as the molecule AB, by the usual methods of wave mechanics. This may be done at each separation R. We are concerned with the way the lowest energy varies with R.

As $R \to \infty$ it will be equal to $E_1(A) + E_1(B)$, where $E_1(A)$ and $E_1(B)$ are the energies of the ground states of the isolated atoms A and B. For large but finite R, the energy falls below the limit by an amount which varies as R^{-6}. This is known as the *Van der Waals energy* and results from a phasing of the motions of the electrons in the two atoms so that the repulsion between them is minimized. At smaller distances, the total energy will have either the form (a) or (b) of figure 4.1.

If it is of the form (a) then, apart from the weak long-range Van der Waals attraction, the total energy of the molecule is always greater than that of the separated atoms. This means that the interaction is repulsive except at very large separations. No stable molecule AB will be formed. Thus the mean kinetic energy of molecules in a gas at a temperature of 300 K is 0·03 eV, so any molecules with binding energy much less than this are broken apart in collisions. The maximum binding produced by the Van der Waals attraction is usually considerably less than 0·03 eV (between argon atoms it is about 0·012 eV, and much smaller between helium atoms, 0·0009 eV).

With form (b), however, the total energy has a deep minimum at a nuclear separation R_0. Whereas the maximum attraction energy due to the Van der Waals force is only a fraction of an eV, the maximum for form (b) will be of the order of a few eV. Under these circumstances, a molecule AB will have a stable ground state with a binding energy close to E_0 as shown in figure 4.1(b). However,

Molecules

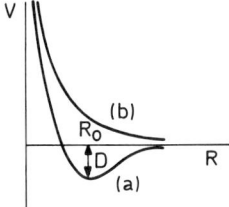

Figure 4.1. Typical form of the variation with nuclear separation R of the interaction energy V between two atoms.
(a) Interaction with a deep attractive minimum.
(b) Repulsive interaction.

because the electrons are so much lighter than the nuclei, they can rapidly readjust their total energy as the nuclear separation varies. The motion of the nuclei may therefore be worked out just as if they interacted according to the curves derived as above. For this reason, the curves are usually referred to as the *potential energy curves* for the different molecular states. Because of the uncertainty principle, we can never have the nuclei as rest in the equilibrium position, for this would imply exact knowledge of both the relative position and momentum of the nuclei—the lowest vibrational state lies above the minimum in curve of figure 4.1(b) by an amount $\frac{1}{2}h\nu$ where ν is the classical vibration frequency. The effective binding energy is therefore $E_0 - \frac{1}{2}h\nu$ (see figure 4.1(b). This is known as the *dissociation energy* of the molecule. The energy $\frac{1}{2}h\nu$ is often referred to as the *zero point energy* of the system.

Before discussing further the vibration and rotation of the nuclei, we must say a little about the conditions under which the interaction between atoms in their ground states is of the repulsive form (a) or attractive form (b), in figure 4.1.

Atoms with completely closed shells only interact repulsively with other atoms, so that the rare gas atoms do not form chemically stable diatomic molecules. Helium, neon and probably argon do not even participate in stable polyatomic molecules, although krypton and xenon do form a limited number of complex compounds. Atoms with incomplete shells interact, in general, in more than one way. Of these hydrogen is a simple and instructive example.

4.2. Interaction between Hydrogen Atoms

If the spin states of the elctrons in the two atoms are the same, then they interact repulsively, but if the spin states are different they attract. Figure 4.2 shows the two interactions. It will be seen that the attractive interaction has a depth of 4·74 eV at the minimum, which occurs at a nuclear separation 0·076 nm. The zero point energy is 0·26 eV, so the binding energy of the hydrogen molecule (H_2) is 4·48 eV.

The existence of the repulsive interaction is important because under certain conditions transitions may take place between the attractive and repulsive curves. Such a transition produced in a stable molecule would lead to break up or dissociation of the molecule.

The chance that, in an assembly of normal hydrogen atoms, any pair will interact attractively is 1/4, this being the chance that the electrons will be in different spin states.

4.3. Examples of Other Stable Diatomic Molecules

Other well known examples of stable diatomic molecules with the same nuclei, so-called *homonuclear molecules*, are those of oxygen (O_2), nitrogen (N_2) and the halogen molecules F_2, Cl_2, Br_2 and I_2 with respective dissociation energies of 5·09, 9·76, 1·58, 2·475, 1·97, and 1·54 eV. The neutral diatomic molecule with the highest binding energy is carbon monoxide (CO) for which it is 11·11 eV. That for nitric oxide (NO) is much lower (6·49 eV).

In most of these cases, the constituent neutral atoms in their ground states may interact in more than two ways, so that there may be more than one attractive as well as repulsive type of interaction.

4.4. Excited Electronic States of Molecules

Although the repulsive interaction cannot lead to stable molecule formation, the repulsive curve is said to correspond to a repulsive molecular state. In that sense, the ground states of diatomic molecules formed by rare gas atoms are repulsive. For molecules such as hydrogen, the lowest repulsive state will be an excited molecular state. It is, however, distinguished by the fact that at very large nuclear separations its energy becomes equal to the total energy of the two isolated atoms in their ground states. A great number of other excited molecular states will arise from the interaction between atoms one or both of which is in an excited state. Figure 4.2 (curves III and IV) illustrates some of these interactions for H_2.

It is clear that for pairs of even comparatively simple atoms, a great number of molecular electronic states will arise. The relation of the corresponding interaction curves to each other often has a profound effect on the outcome of collisions between atomic systems.

4.5. Interactions between ions and atoms

Similar considerations apply to the interactions between ions and atoms. In such cases, there is an additional long-range attractive force due to the polarization of the atom by the net electric charge on the ion. This gives rise to an energy $-\frac{1}{2}\alpha e^2/R^4$, where α is the polarizability of the atom (see Chapter 3). This attraction is small compared with the usual chemical attractions, but is larger than the Van der Waals interaction.

Molecules

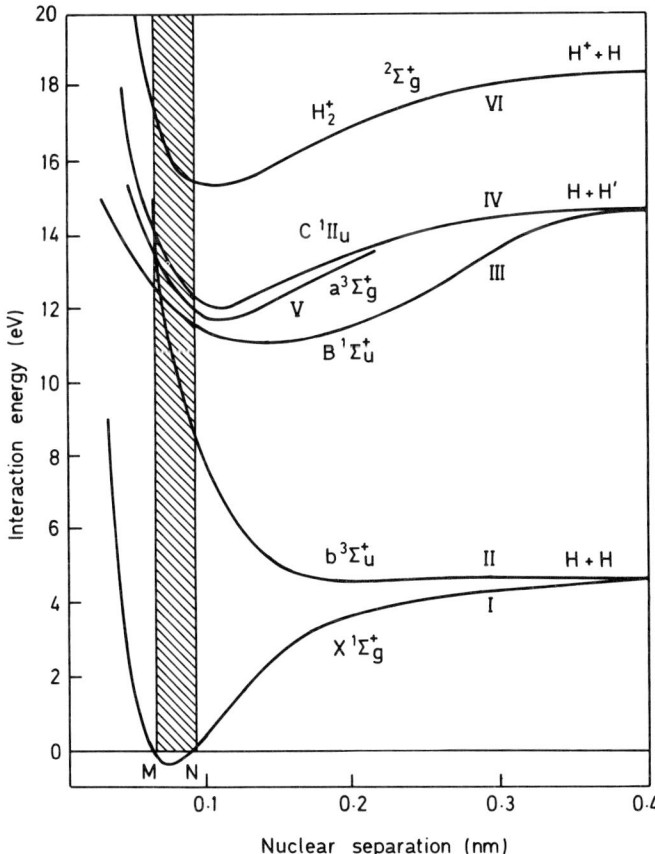

Figure 4.2. Interaction energy between two hydrogen atoms in different states.
Curves I and II are the two interactions which arise between hydrogen atoms in their ground states. The attractive curve I arises when the total electron spin quantum number is zero (case of antiparallel spins in the two atoms), while the repulsive curve II arises when it is unity (case of parallel spins in the two atoms).
Curves III, IV and V are examples of interactions between one atom (H) in the ground state and the other (H) in an excited state.
Curve VI is the potential energy curve for the lowest state of the ionized molecule H_2^+ (see figure 12.1).

For the simplest case of the interaction of a hydrogen atom in its ground state with a proton, two interactions, one repulsive and one attractive, arise as shown in figure 12.1. The attractive interaction leads to a stable molecular ion H_2^+ with a dissociation energy of 2·69 eV. There is an equal chance that a proton and a hydrogen atom will interact in one or other way. The two molecular electronic states differ in their symmetry with respect to interchange of the two protons—the electronic wave function for the attractive state remains unchanged in this process, whereas that for the repulsive is changed in sign. The existence of the two states has a decisive effect on collisions between H and H^+ in which an electron is transferred from one proton to the other, a so-called charge transfer collision (see Chapter 12, p. 215).

The attraction which leads to formation of a stable molecular ion H_2^+ is essentially a consequence of the fact that the electric field exerted by either proton on the electron is of the same form and strength—the energy of the system is unchanged when the electron is transferred from one proton to the other. To a very good approximation, the equality of the electric fields remains true if one of the protons is replaced by a deuteron (see Chapter 12, p. 214). We can say, in fact, that it will occur in all cases in which the ion and atom are of the same species (including isotopes, see p. 39). Thus a helium atom can interact attractively with a positive helium ion to form a stable He_2^+ molecule with a dissociation energy of 2·8 eV. Just as for H_2^+, a repulsive interaction is also possible. Further examples are the homonuclear molecular ions of the heavier rare gases Ne_2^+, Ar_2^+, Kr_2^+ and Xe_2^+.

These considerations do not apply if the atom and ion are not of the same species. Thus positive ions of alkali metal atoms, in their ground states, all possess a closed shell structure and interact repulsively with neutral atoms which are not of the same species (see Chapter 11, p. 201).

Of all positive diatomic ions, the one with the highest dissociation energy is NO^+ (11·7 eV). For O_2^+, N_2^+ and CO^+ the dissociation energies are 6·48, 8·7 and 8·4 eV respectively.

Interactions between negative ions and neutral atoms follow very similar rules but are complicated by possibilities for autodetachment, consideration of which we shall defer to Chapter 7, p. 112.

4.6. *Molecular Vibration and Rotation*

Potential energy curves such as are illustrated in figures 4.1 and 4.2 strictly apply only when the nuclei are held at rest at each particular nuclear separation. However, even if the nuclei are allowed to move, the motion of the electrons, which are of relatively very small mass, rapidly readjusts to the changing nuclear separation. This means, as explained earlier, that to a very good approximation the relative motion of the nuclei is just that which would occur if they moved under the action of a potential energy given by the appropriate potential energy curve. Once this is realized, the usual methods of wave mechanics can be applied to determine the allowed energies and relative probability distributions associated with the nuclear motion. The total energy E of the molecule can then be written as

$$E = E_e + E_v + E_r,$$

where E_e, E_v, E_r are the respective electronic, vibrational and rotational energies.

Molecules

To a good approximation, for attractive potential energy curves, the motion can be separated into one of vibration along the line joining the nuclei and of rotation about an axis perpendicular to this line.

To determine the classical frequency of vibration about the equilibrium separation R_0 (see figure 4.1), we note that, if the separation changes from R_0 by a small amount δR, a restoring force of magnitude

$$\left(\frac{d^2V}{dR^2}\right)_{R=R_0} \delta R \qquad (4.1)$$

comes into play, V being the potential energy. Since this is proportional to δR, vibrations of small amplitude will be simple harmonic of frequency

$$v_0 = \frac{1}{2\pi}\left\{\frac{1}{M}\left(\frac{d^2V}{dR^2}\right)_{R=R_0}\right\}^{1/2}. \qquad (4.2)$$

Here M is the reduced mass of the nuclear motion.

Referring to (2.15), we see that the allowed energies E_0 of nuclear vibration will be given, according to the simple harmonic approximation, by

$$E_v = (v + \tfrac{1}{2})hv_0, \, v = 0, 1, 2, \ldots \qquad (4.3)$$

$\tfrac{1}{2}hv_0$ being the zero point energy (see p. 61).

For the simplest diatomic molecules hv_0 has the values given in table 4.1. It will be seen that the energy of the fundamental vibrational quantum is small compared with typical electronic excitation energies. In general, it becomes smaller the higher the reduced mass of the system.

The approximation (4.3) becomes less and less accurate as v increases. In particular, if the vibrational energy were to exceed D, the dissociation energy corresponding to the electronic state concerned (see figure 4.1), the molecule would dissociate. In other words, if the energy of relative nuclear motion exceeds D, there is a continuous spectrum of allowed energy values associated with dissociated states in which the nuclei ultimately separate to infinity. In fact, there exist only a finite number of allowed vibrational levels with energies less than D. Thus for H_2^+ there are 19 levels, some of which are shown in figure 12.1, p. 214. Typically the separation between the levels becomes smaller as the energy increases.

Because of the large masses of the nuclei, the vibrational motion is nearly classical, in which case the most probable nuclear separation is at either extreme of the classical amplitude. With the simple harmonic approximation, these

Molecule	H_2	D_2	O_2	N_2	F_2	Cl_2
hv_0/eV	0·54	0·38	0·20	0·29	0·11	0·07
$h^2/8\pi^2 MR_0^2/\text{eV}$	7.1×10^{-3}	3.6×10^{-3}	1.7×10^{-4}	2.8×10^{-4}	1.1×10^{-4}	2.9×10^{-5}

Table 4.1. Magnitudes of the fundamental vibrational quantum and rotational energy constant for certain molecules

Atomic and Molecular Collisions

points occur at separation $R_0 \pm a$. More generally, they occur at the so-called classical turning points R_1, R_2, which satisfy

$$V(R) = E_v,$$

where E_v is the allowed vibrational energy. In the classical motion, more time is spent near separations R_1 and R_2, while all separations between R_1 and R_2 are traversed during each vibration. Separations less than R_1 or greater than R_2 cannot be reached in the classical motion, so that it is very unlikely that in the vibrational state concerned the nuclei will be found at such separations at any time.

For the rotation of the nuclei about an axis through the centre of mass perpendicular to the line joining them, the allowed energy values are given to a first approximation by

$$E_r = \frac{h^2}{8\pi^2 M R_0^2} J(J+1), \quad J = 0, 1, 2\ldots$$

Values of the factor $h^2/8\pi^2 M R_0^2$ for some diatomic molecules are given in table 3.2. It will be seen that unless J is large compared with unity, the rotational energy is much smaller than the vibrational. In allowed rotational transitions J can change only by ± 1. The energy change in such a transition from a level J is given by

$$\Delta E_r = \frac{h^2}{4\pi^2 M R_0^2} \times (J \text{ or } J+1). \tag{4.4}$$

In fact the vibrational and rotational motions are not independent and, in higher approximations, the interaction between them must be allowed for. It will suffice, for most purposes which concern us in this book, to appreciate the relative magnitude of the contributions from electronic, nuclear vibrational and nuclear rotational motion to the energy of a molecule.

If the potential energy curve is of the repulsive, type, there will be no bound vibrational states, only a continuous spectrum of energy values for the relative nuclear motion (see Chapter 7, p. 113).

4.7. Behaviour of the Nuclei in a Transition between Electronic States— The Franck–Condon Principle

So far we have been considering the nuclear motion associated with a particular electronic state. How will the nuclei behave in a transition from one electronic state to another? Once again, the relatively large mass of the nuclei compared with that of the electrons is the determining factor—it is a very good approximation to suppose that the nuclei do not move during the transition. This is usually referred to as the Franck–Condon principle.

To see how this may be usefully applied, suppose that a transition takes

Molecules

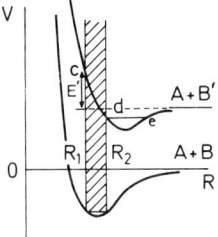

Figure 4.3.

place from the ground electronic state of a diatomic molecule, for which the potential energy curve has the form I in figure 4.2, to an excited state in which it has the form III. Initially the molecule will be in some particular vibrational state with energy E_v. In such a state the nuclear separation is most likely to be between R_1 or R_2 (figure 4.3), the limits of the classical amplitude of vibration in that state. According to the Franck–Condon principle, this will also be true immediately after the transition. It will then be most probable that the final state of the system, allowing for the nuclear motion, will be between the points c and d of figure 4.3. These are the points on the potential energy curve for the excited state for which the nuclear separations are R_1 and R_2 respectively.

This illustrates the way in which it is possible to determine the energies of the final molecular states allowing for the nuclear motion. The interpretation depends on the relation of these energies to the energy of the system when the constituent atoms are completely separated. Thus in Figure 4.3, the energy determined by the point c lies above the energy at infinite separation of an atom A in the ground state and one B in the excited state B'. On the other hand, at d the total energy is less than that at infinite separation. This means that the electronic transition from the particular initial state will lead either to (*a*) dissociation of the molecule into atoms A and B' with relative energy ranging from O to E' or (*b*) to a vibrationally excited molecule AB'. In the latter case the lowest vibrational level will lie close to *de* in figure 4.3.

Examples of the application of these considerations to the analysis of the possible consequences of electronic transitions in diatomic molecules are discussed in more detail in Chapter 7, p. 111 and p. 113.

CHAPTER 5
the scattering of slow electrons by atoms—the Ramsauer – Townsend effect

As explained earlier, the collisions of slow electrons with gas atoms were among the first impact processes involving atomic systems to be studied experimentally. The results obtained in the early 1920s were unexpected on the basis of classical mechanics and remain among the most clear-cut verifications of the wave aspect of electrons. In particular, it was found that the transparency of the heavier rare gases towards electrons did not decrease steadily as the electron velocity decreased but was a maximum at a certain velocity v_0. This effect, named after the two investigators, Ramsauer and Townsend, who discovered it almost simultaneously in two quite different types of experiment, can be understood quite easily in terms of the wave mechanical theory of scattering which we have sketched out in Chapter 2 p. 32, as we shall see below.

In this chapter we describe some of the key experiments which led to the discovery of the Ramsauer–Townsend effect and then explain how the results of these and related experiments can be interpreted.

Because the noble gases are the only permanent gases which are monatomic at ordinary temperatures, a great deal of experimental work has been concerned from the outset with studying electron collisions in these gases. The excitation energies are all quite high (the lowest, that of xenon, being 9.45 eV), so that electrons with energies less than 9 eV can only suffer elastic collisions in the noble gases. This simplifies the interpretation of the results. At the same time, for such slow electrons, the wavelength is of the order 10^{-9} m, comparable with or larger than atomic dimensions, so that we would expect wave effects to be of major importance.

The methods used to investigate electron scattering depend largely on the energy range to be studied. For energies greater than about 0·5 eV, it is possible to produce beams of electrons which are well collimated (i.e., sharply defined in direction) and possess very well defined energies (i.e., have high energy resolution). Most experiments therefore involve observation of the effects arising from passage of such a beam through a gas, or in crossing a second beam of neutral atoms or ions.

It is difficult to prepare such beams with energies less than 0·5 eV. To obtain information about collisions at such energies, it is necessary to study the motion

Scattering of Slow Electrons by Atoms

through the gas of a swarm of electrons with a considerable distribution of energy about a mean value, which may be only slightly larger than thermal (0·03 eV). Experiments of the latter type were carried out at an early stage in the study of ions in gases and, using modern techniques, are still proving of much importance. We shall begin by saying something about swarm experiments before going on to describe how beam experiments are carried out.

5.1. Diffusion of an Electron Swarm in a Gas Under the Action of an Electric Field

We consider the motion of a swarm of electrons through a gas at pressure p under the action of a uniform electric field of strength F.

As the electrons diffuse, they gain energy from the field but, in the steady state, this is balanced by the energy lost in collisions with the gas atoms. The swarm will then possess a definite mean *drift* velocity u in the direction of the field as well as a definite *mean kinetic energy* ε. It is not difficult to see how these quantities are related to the scattering cross-sections of the gas atoms for an electron.

To avoid unnecessary complication, we suppose that the mean free path (see Appendix 1) of the electron in the gas has a constant value l independent of the electron energy, and that in a collision with a gas atom the fraction of its energy which it loses is equal to a small constant λ. Then if c is the mean speed of random motion of the electron, the number of collisions made in traversing a distance x in the direction of the field will be cx/ul, since the actual path traversed will be of length $(c/u)x$. The energy lost in these collisions will be $\lambda \varepsilon cx/ul$ so, for a steady state,

$$\lambda \varepsilon cx/ul = Fex. \tag{5.1}$$

As $\varepsilon \simeq \tfrac{1}{2}mc^2$, where m is the electron mass, we have

$$\lambda c^3/u = 2Fel/m. \tag{5.2}$$

Again, if all directions of motion of an electron after a collision are equally probable, the mean distance traversed in the direction of the field in the interval δt between collisions will be $u\delta t$, so

$$\tfrac{1}{2}Fe(\delta t)^2/m = u\delta t. \tag{5.3}$$

Apart from a numerical factor of order unity, δt may be put equal to l/c, so

$$2uc \simeq Fel/m. \tag{5.4}$$

As l is inversely proportional to the gas pressure p at a fixed temperature T, both c and u are functions of F/p for fixed T. Also if c and u are measured for a fixed value of F/p, both l and λ may be obtained.

In practice, l will not be independent of the electron velocity, so that from c and u only certain mean values of l and λ over the velocity distribution of the electrons may be derived. However, for collisions with atoms, which will be elastic at the low energies concerned, λ may be calculated without difficulty, so that l may be derived in principle from measurements of either c or u alone. Thus, with the assumptions made in the above analysis, the fraction of its energy lost per impact by an electron that is scattered through an angle θ is given by (see (1.21) p. 8)

$$2(1 - \cos\theta)m/M, \qquad (5.5)$$

where m is the mass of an electron, M of a gas atom. If $I(\theta)$ is the differential cross-section per unit solid angle (see p. 10) for elastic scattering of an electron of the appropriate energy, the mean fractional energy loss per collision is given by

$$(2m/M)\int_0^\pi (1 - \cos\theta)I(\theta)\sin\theta\,d\theta \Big/ \int_0^\pi I(\theta)\sin\theta\,d\theta \qquad (5.6)$$

$$= (2m/M)Q_d/Q, \qquad (5.7)$$

where Q is the total elastic cross-section and

$$Q_d = 2\pi \int_0^\pi (1 - \cos\theta)I(\theta)\sin\theta\,d\theta, \qquad (5.8)$$

is known as the *momentum transfer* or *diffusion cross-section* (c.f. Chapter 8, p. 140). Like the total cross-section, Q_d will in general be a function of the electron energy.

The fractional amount of energy lost by an electron in traversing a small distance δx in a gas containing N atoms per unit volume will be $(2m/M)NQ_d\delta x$ just as if the total cross-section were replaced by Q_d and the fractional energy lost per collision by $2m/M$. The mean free path referred to in the analysis above is essentially a mean value of $1/NQ_d$ rather than $1/NQ$.

Thus, in terms of the approximate relations (5.1) and (5.2) we have, taking

$$l = 1/NQ_d, \lambda = 2m/M, \qquad (5.9)$$

$$Q_d = (eF/N)(8Mm)^{-1/2}u^{-2}. \qquad (5.10)$$

This can be taken as a relation between the diffusion cross-section, averaged over the velocity distribution of the electrons, and the drift velocity. Since the mean electron energy $\tfrac{1}{2}mc^2$ is given from (5.2) and (5.4) by

$$\tfrac{1}{2}mc^2 = Mu^2, \qquad (5.11)$$

it follows that the mean value of Q_d may be obtained as a function of the mean electron energy. A more detailed analysis may be carried out which enables the diffusion cross-section to be obtained as a function of electron velocity from observations of the drift velocity alone as a function of F/N.

In a molecular gas, it is no longer possible to take λ as $2m/M$, because the drifting electrons can lose energy by exciting molecular rotation and vibration.

However, if measurements can be carried out both of u and of c, λ may be determined, at least in principle. In practice, it is best to proceed by trial and error, using initial approximations for the inelastic as well as the elastic cross-sections and gradually improving them until good agreement is reached with the observed values of u and c as functions of F/N.

We shall limit ourselves here to describing the method used to measure drift velocities and some of the results obtained for Q_d from these measurements for monatomic gases.

5.2. The Measurement of Drift Velocities

Given present-day facilities in electronics, the precise measurement of drift velocities is straightforward. All that is essentially involved is the timing of the passage of a pulse of electrons between two points in a gas a known distance apart measured in the direction of the applied electric field.

In experiments carried out by Pack and Phelps in 1961, the pulse of electrons was produced by irradiating a suitable surface with a pulse of ultraviolet light. The passage of the pulse past two grids was then timed in the following way.

It was arranged that the fraction of electrons interrupted by the grid could be increased by applying a pulsed voltage between the grid wires. Measurements were then made of the electron current collected after passing through both grids, as a function of the time delay between the light pulse and the voltage pulse which was applied simultaneously to either grid. When this time delay is equal to the time taken for the electron pulse to drift from the photoelectric source to either grid, there will be a fall in the collected current, a typical example

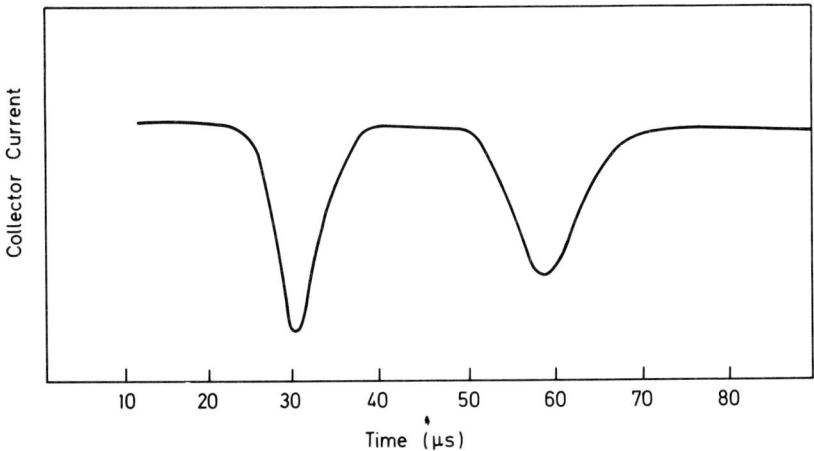

Figure 5.1. Characteristic form of the variation of the collector current with delay time between the light pulse generating the electrons and the voltage pulse applied to the grids, in the experiments of Pack and Phelps to measure electron drift velocity.

being as shown in figure 5.1. The time interval between successive minima in such observations is the time taken for the electrons to drift the distance between the grids.

This all seems deceptively simple, but the times involved are of the order of microseconds and the collector currents a few times 10^{-12} A. The grids were of gold-plated molybdenum wire 0·075 mm thick, 3·5 mm apart. Before the highly purified gas under study was introduced into the drift tube, the latter was evacuated to a pressure as low as 10^{-9} torr[†] (see Appendix 2).

Figure 5.2 shows drift velocities of electrons in helium measured by this technique at two temperatures. For $F/N < 10^{-22}$ V m² different results are obtained at the two temperatures. This is because, for these values of F/N, the mean energy of the electrons is comparable with the thermal energy of the gas molecules which was assumed to be negligible in the analysis above. Using the approximate formula (5.2), we obtain from these results for $F/N > 10^{-22}$ V m², the mean diffusion cross-section $\bar{Q}_d$ as a function of mean electron energy shown in figure 5.3. In the same diagram we also show Q_d as a function of

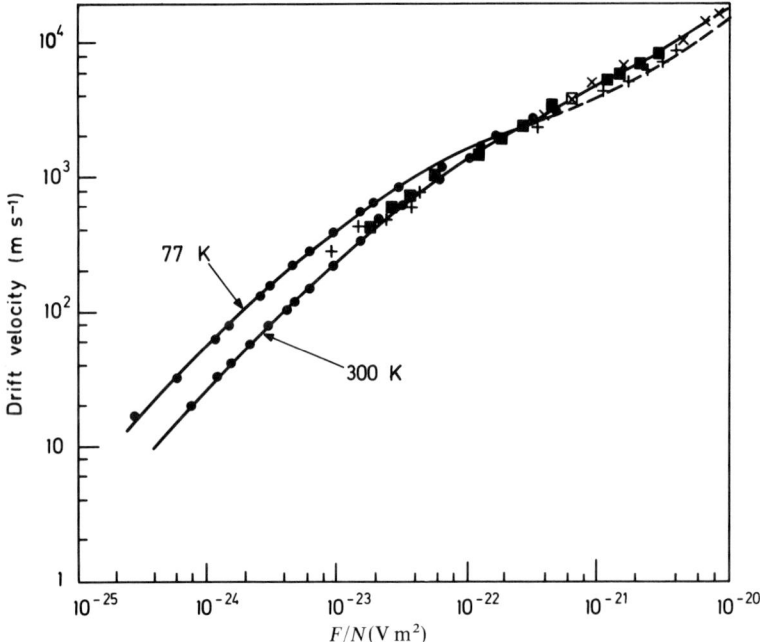

Figure 5.2. Drift velocities of electrons in helium measured at temperatures of 77 K and 300 K as functions of the ratio F/N of electric field strength F to gas concentration N. The points ●, ■, +, × show values obtained by different observers while the continuous curves represent the best fits to their data.

[†] The torr is the unit of pressure commonly used in atomic and molecular physics. It is the pressure corresponding to a column of mercury 1 mm high, and is equal to $1·333 \times 10^2$ newton m⁻².

Scattering of Slow Electrons by Atoms

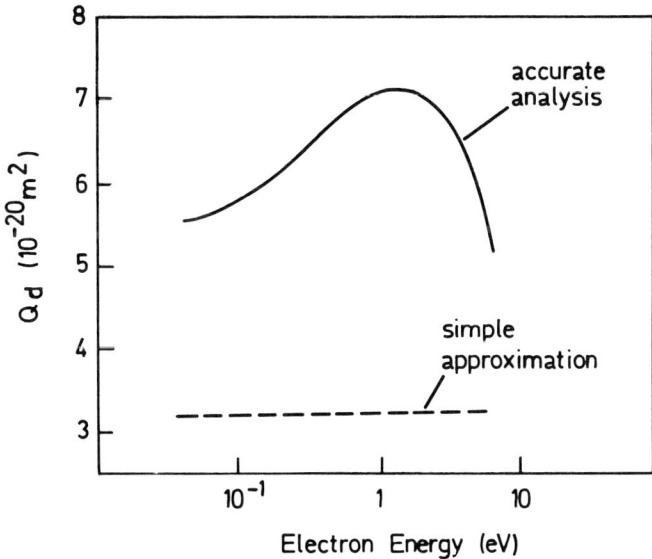

Figure 5.3. Diffusion cross-sections for electrons in helium derived from observed data. ---, using the simple approximation (5.10); ——, using an accurate analysis.

electron energy derived by a much more sophisticated and detailed analysis of the data. It will be seen that the rough approximations (5.2) and (5.4) give results which are correct within a factor of 2 or so.

It was from measurements of this kind that Townsend and Bailey, as early as 1922, found evidence that in argon the mean free path has a maximum at a mean electron energy near 0·5 eV (see inset of figure 5.5.) This effect would not be expected on classical theory, which would normally predict that the slower the electron, the more easily it would be deflected by the atomic field. It is referred to as the Ramsauer–Townsend effect because it was independently discovered a little earlier by Ramsauer using a beam technique which we will shortly describe. In section 5.5 we explain how the effect arises from the wave nature of the electron.

5.3. The Measurement of Total Cross-sections for Scattering of Electrons by Atoms using Beam Techniques—Ramsauer's Method

The method introduced by Ramsauer as early as 1921 is very direct and capable of giving quite accurate results. It takes advantage of the fact that electrons of given energy will pursue circular paths of definite radius (see Appendix 4) when moving in a plane perpendicular to the direction of a uniform and constant magnetic field. Electrons which suffer deflection or loss of energy, or both, in collision will, however, cease to follow the circular path which they pursued before the collision.

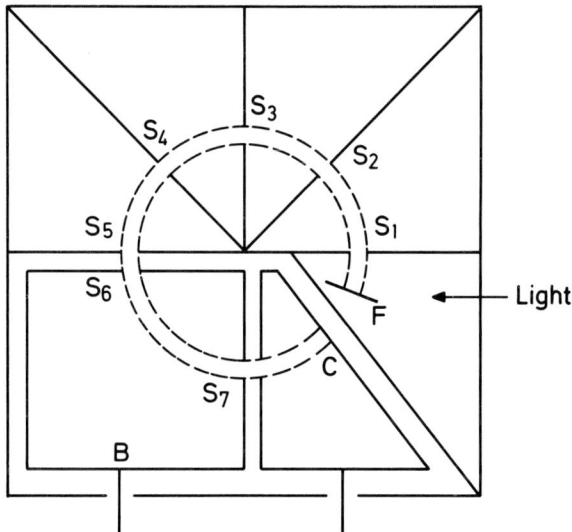

Figure 5.4. Illustrating the principle of Ramsauer's method for measuring cross-sections for collision of slow electrons of well defined energy with gas atoms.

In Ramsauer's experiments, a circular path of fixed radius was defined between quite narrow limits by a series of slits. A magnetic field could be applied perpendicular to the plane of the slits and its magnitude adjusted so that electrons of chosen energy, if they suffered no collisions, could follow the defined circular path from a source to a collector. The reduction of the collector current due to collisions when the gas pressure in the system was increased could be measured. Knowing the path length traversed by the electrons in the gas, the total cross-section could be derived.

In the original experiments, the source was a zinc plate from which electrons were ejected photoelectrically. Referring to figure 5.4, the electrons from the source F were accelerated to the desired energy before passing through the slit S_1. Those electrons which suffered no collisions described a circular path through the slits S_1–S_7 to enter the collector C. With a selected pressure p_1 of the gas under study in the apparatus, the currents i_1 to C alone and j_1 to the walls of the chamber B as well as to C were measured. If x is the length of the circular path between the slits S_6 and S_7,

$$i_1 = j_1 \exp(-\alpha p_1 x), \qquad (5.12)^\dagger$$

where α is the absorption coefficient at unit pressure. If i_2 and j_2 are the cor-

† This relation is derived as follows. In proceeding a distance δx along the circular path, the current i changes by an amount δi where

$$\delta i = -\alpha p_1 i \delta x,$$

responding currents when the pressure is p_2

$$i_2 = j_2 \exp(-\alpha p_2 x), \qquad (5.13)$$

so that

$$\alpha = \frac{1}{n(p_1 - p_2)} \ln(j_1 i_2 / j_2 i_1). \qquad (5.14)$$

Also if n is the number of gas atoms or molecules per unit volume at unit pressure

$$\alpha = nQ, \qquad (5.15)$$

where Q is the total collision cross-section.

Provided the slits are narrow enough, this method should give the true value of Q according to quantum theory. Electrons of 1 eV energy have a wavelength close to 10^{-9} m, which is comparable with or greater than atomic dimensions, so that their motion is very far from classical. This means, in general, that it is not necessary to have very high angular resolution in order to obtain a good approximation to Q (see p. 31)—the effective interaction with an atom falls off much faster than r^{-2} for large r.

5.4. Results for the Rare Gases

Figure 5.5 shows the total cross-section measured for the heavier gases argon, krypton and xenon. For all three gases, the cross-sections exhibit the same striking features, a maximum at an electron energy between 9 and 16 eV and a low minimum at an energy close to 1 eV. For comparison, we include on the same diagram the variation of the *mean* diffusion cross-section as a function of *root mean square* electron velocity observed by Townsend and Bailey for argon using the swarm technique. The minimum near 0·7 eV is clearly present and undoubtedly arises in the same way.

For helium no similar effects are observed, the cross-section falling off as the electron energy increases (see figure 5.6). Neon, on the other hand, while quite distinct, exhibits a cross-section which increases with electron energy to a flat maximum at an electron energy around 25 eV (see figure 5.6).

5.5. Interpretation of Results for the Rare Gases in Terms of Wave Mechanics

How can these remarkable results be interpreted? It has already been pointed

---

so

$$\frac{di}{dx} = -\alpha p i.$$

Integrating this equation $i = j_1 \exp(-\alpha p_1 x)$ where j_1 is the current when $x = 0$.

Atomic and Molecular Collisions

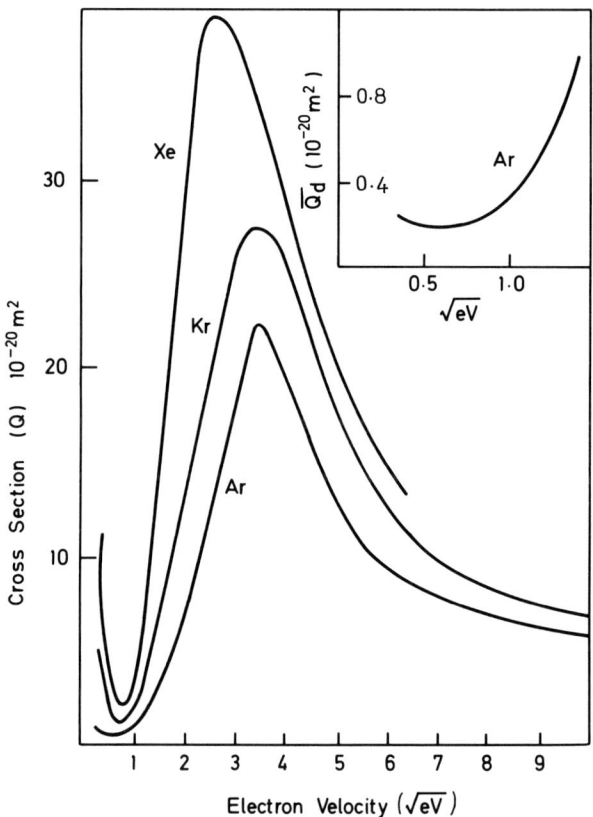

Figure 5.5. Variation with electron velocity of the total electron collision cross-section in Ar, Kr and Xe as measured in experiments of the Ramsauer type. The inset gives the variation of mean diffusion cross-section with root mean square electron velocity in argon, derived from the drift experiments of Townsend and Bailey. Note that 1 eV corresponds to an electron velocity of $0.6 \times 10^5 \text{ m s}^{-1}$.

out that the conditions are very far from classical so the wave theory of Chapter 2 p. 32 must be involved.

According to that theory, the total cross-section for scattering of particles of mass m and velocity v by an effective central potential $V(r)$ can be written in the form

$$Q = \sum_{l=0}^{\infty} q_l, \qquad (5.16)$$

where

$$q_l = \frac{\lambda^2}{\pi}(2l+1)\sin^2\eta_l, \qquad (5.17)$$

Scattering of Slow Electrons by Atoms

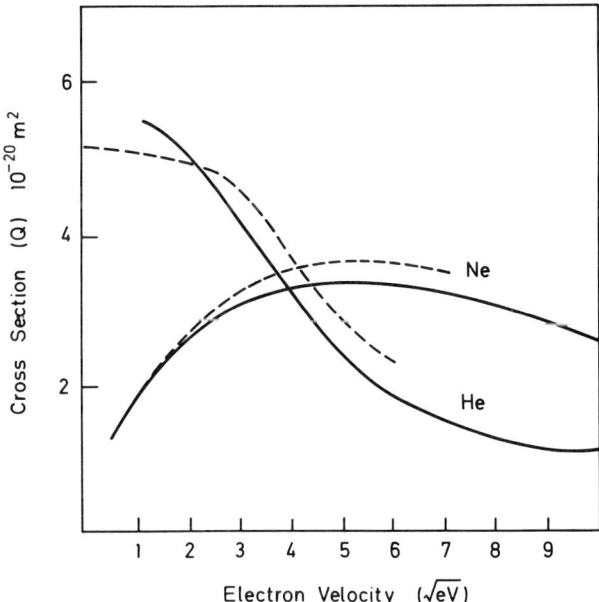

Figure 5.6. Variation with electron velocity of the total collision cross-section in He and Ne. ———, measured in experiments of the Ramsauer type; ----, calculated from the phase shifts shown in figure 5.9.

λ being the wavelength h/mv. The partial cross-sections q_l are the separate contributions from scattering of electrons with quantized angular momenta $\{l(l+1)\}^{1/2}h/2\pi$. In section 2.9 of Chapter 2, it was shown that the convergence of the series of partial cross-sections improves as the impact velocity decreases, so that, at very low velocities, only q_0 is important. There is no difficulty in understanding how q_0, while finite in the limit of vanishing velocity, may pass through a zero minimum at a low velocity.

Thus, suppose that η_0 tends to π as the electron velocity tends to zero. As v increases, η_0 may behave either as in figure 5.7(a) or figure 5.7(b). In the former case, it decreases steadily as v increases and the corresponding cross-section q_0 will vary as shown in figure 5.7(c). However, if η_0 at first rises above π to a maximum and then falls gradually thereafter as in figure 5.7(b) q_0 will vary as figure 5.7(d). In contrast to figure 5.7(c), it no longer falls gradually with increasing v but reaches a zero minimum at v_0.

Since at the Ramsauer–Townsend minimum the wavelength is close to 10^{-9} m, which is considerably greater than atomic dimensions, the arguments of Chapter 2 would suggest that the phase shifts η_l with $l > 0$ are likely to be small. It is therefore natural to ascribe the effect to the behaviour of the phase shift η_0. The fact that the phase shifts η_1, η_2 etc. will not be exactly zero however, means that the cross-section Q does not quite vanish at the minimum and this is not inconsistent with the observed results.

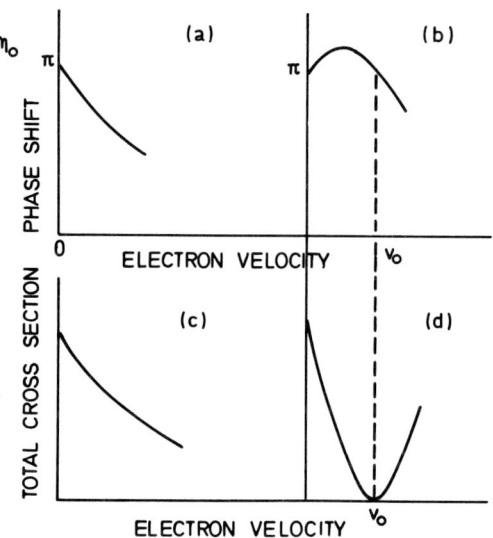

Figure 5.7.

The prominent maxima in the cross-sections for the heavier rare gases at energies between 9 and 16 eV must arise from the behaviour of higher-order phase shifts. To see that this is so, we note that the maximum possible value for q_l is $(\lambda^2/\pi)(2l+1)$. At 10 eV $\lambda = 4 \times 10^{-10}$ m so

$$q_l(\text{max}) = (2l+1) \times 7 \times 10^{-20} \text{ m}^2.$$

At 10 eV the cross-section Q has the values, in 10^{-20} m² of 22, 27 and 32 respectively. Since

$$\sum_{l=0}^{L} (2l+1) = (L+1)^2, \tag{5.19}$$

it follows that, if all partial cross-sections for $l < L$ say, are at their maximum value, then L must be at least 2. This result follows on the extreme assumption that all partial cross-sections with $l < L$ are at their maximum so that we must expect that some phase shifts with $l > 2$ will also be important near the cross-section maxima.

Figure 5.8 shows the phase shifts for argon calculated theoretically including the effects of exchange and of polarization of the atom by the impinging electron referred to in section 5.9 below. The η_0 phase shift behaves as in figure 5.7(b) so that the Ramsauer–Townsend effect appears. Using these phase shifts in the formula (5.17), we obtain the total elastic cross-section as a function of electron energy shown in figure 5.8(b) in comparison with observed results. The agreement is good and supports the interpretation in terms of the wave theory.

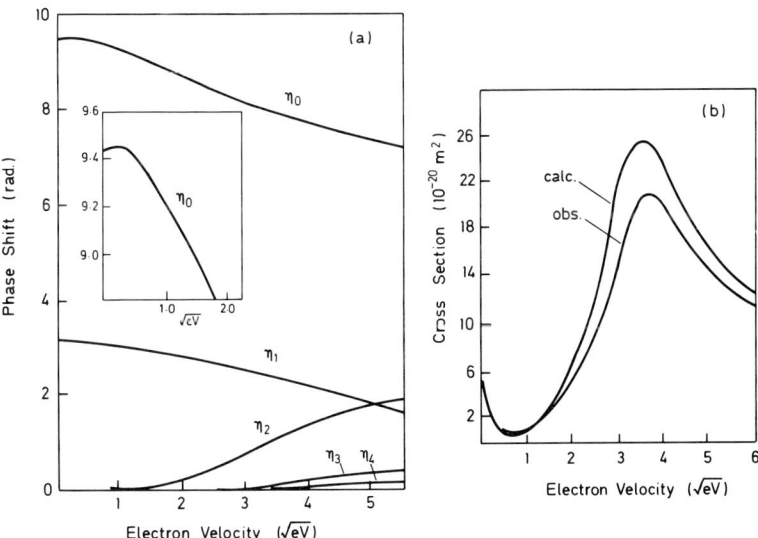

Figure 5.8. (a) Phase shifts $\eta_0, \eta_1, \eta_2, \eta_3$, and η_4 calculated theoretically for argon as a function of electron velocity.
(b) Comparison of the cross-sections calculated using these phase shifts with those observed.

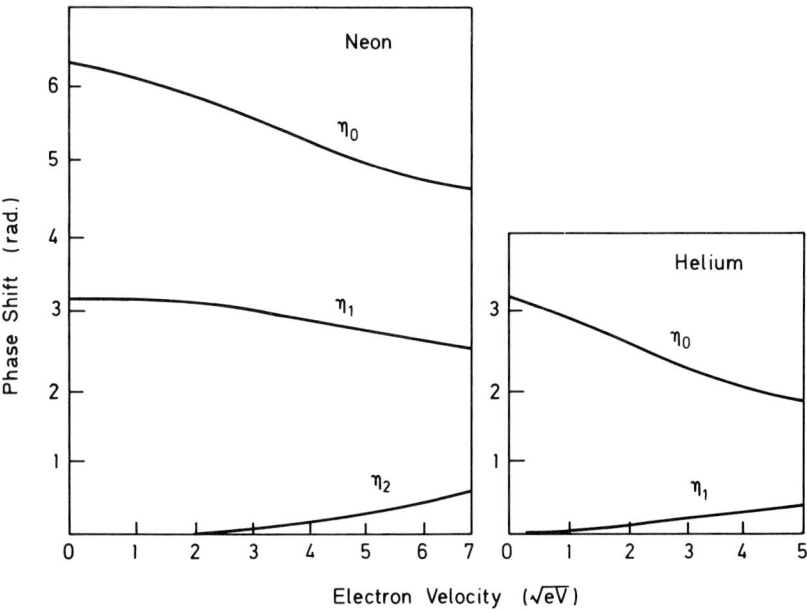

Figure 5.9. Phase shifts η_0, η_1, and η_2 for neon and η_0 for helium calculated theoretically as functions of electron velocity.

The maximum near 13 eV is largely due to the phase shift η_2 which passes through $\pi/2$ at that energy.

The rather similar behaviour of the cross-sections for krypton and xenon may be interpreted in a similar way. Thus as $v \to 0$ the phase shift η_0 tends to 4π for Kr and 5π for Xe but, in each case, it at first rises as v increases from zero to a maximum at an energy a little below 1 eV. The net effect on the cross-section is the same as for argon, the additional multiples of π being immaterial. Again, it is the phase shift η_2 which produces the maxima between 10 and 16 eV for the heavier rare gases just as for argon, the only difference being that at the maximum η_2 equals $3\pi/2$ for krypton and $5\pi/2$ for xenon, the extra multiples of π being once more immaterial.

For the lighter rare gases, helium and neon, the atomic field is so much weaker that η_2 never reaches a value of $\pi/2$ at any electron energy. η_0 tends to 0 for helium and to 2π for neon as $v \to 0$ but in the latter case it decreases as the velocity rises from zero, so there is no Ramsauer–Townsend effect. Figure 5.9 shows calculated phase shifts for each atom while in figure 5.6 comparison is made between total cross-sections calculated from the phase shifts and observed cross-sections. There is little doubt of the validity of the analysis.

So far we have been dealing only with the measured total cross-sections, but a detailed test of the validity of the interpretation of the results in terms of the wave theory is possible if data on the angular distributions of the scattered electrons at different electron energies are available; in other words, if differential as well as total cross-sections can be measured.

5.6. The Angular Distribution of the Scattered Electrons— Experimental Methods

The first measurements of the angular distribution of slow electrons scattered by helium, neon and argon atoms were published in 1931 by Bullard and Massey. The results, obtained by a method which we shall shortly describe, confirmed the wave interpretation but did not extend down to electron energies covering the neighbourhood of the Ramsauer–Townsend effect. A little later, Ramsauer and Kollath, using a different form of apparatus, were able to extend the observations down to electron energies as low as 0·6 eV and their data are of the form expected from the analysis we have described above.

Figure 5.10 illustrates the principle of the method used by Bullard and Massey and also by a number of later investigators. A beam of electrons from a suitable source F passes through the slit S into the gas contained in a region free from electrostatic fields. Electrons scattered from the area ABCD of the beam through angles in the small range between θ and $\theta + \delta\theta$ pass through the entrance slit L of a collector C. To ensure that only electrons of the required energy are collected, some energy-selecting device must be included in C. To measure the variation of the collected current with the angle θ, either the source

Scattering of Slow Electrons by Atoms

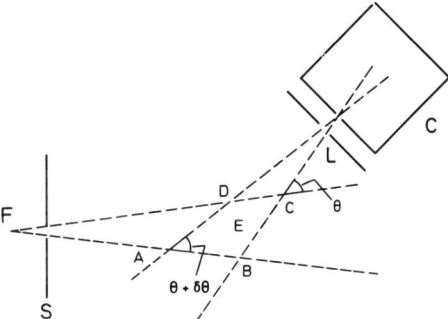

Figure 5.10. Illustrating the principle of the method used to observe the angular distributions of electrons scattered in gases.

F and slit S or the collector C may be rotated about an axis perpendicular to the plane of the paper at E. If the volume of the region from which the scattered current is observed were independent of θ, the current collected would be proportional to the intensity $I(\theta)$ scattered per unit solid angle. However, the scattering volume is proportional to $\operatorname{cosec} \theta$ so that the measured currents must be multiplied by $\sin \theta$ to obtain relative values of $I(\theta)$ at different scattering angles.

It is essential in experiments of this kind to check that the observed electrons have suffered single collisions only. This may be done by verifying that the scattered current is proportional to the gas pressure and also to the intensity of the beam issuing from the source slit S.

A further essential requirement is that no strong fields should be present due to charging up of the boundary walls of the scattering chamber. This was a serious matter in early experiments in which glass apparatus was used and indeed gave rise to spurious results. It was avoided by enclosing the scattering region in a cylindrical metal shield.

Finally, to reduce the level of impurities present which may not only give rise to scattering but also may affect the collecting power of slits, it is necessary to be able to outgas the apparatus thoroughly by baking (see Appendix 2).

Figure 5.11 shows diagrammatically the apparatus used by Bullard and Massey, which was similar to that used earlier by Arnot for the scattering of electrons by mercury vapour.

Electrons from a heated tungsten filament A were accelerated through a pair of slits B and C into the space enclosed within the cylindrical metal shield K, which contained gas at a pressure of some 10^{-3} torr. Those electrons scattered in the region D could be collected by a Faraday cylinder (see Appendix 6) E after passage through slits F and G in the enclosing metal case. The electron gun and the collector were both carried on the ground joint H, while the collector could be rotated around the axis by means of a second coaxial ground joint J. The geometry of the system was such that the angular range $\delta\theta$ of collection

Atomic and Molecular Collisions

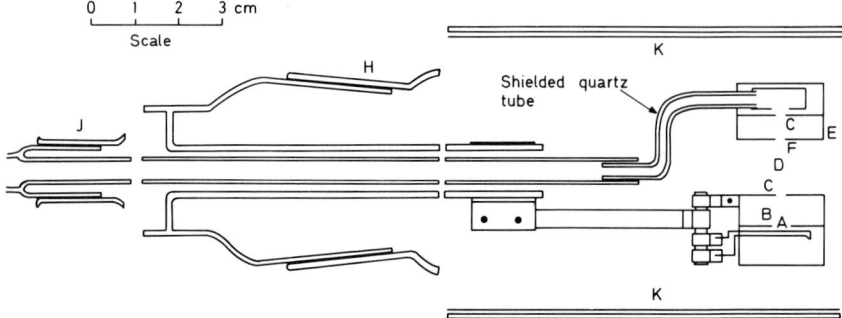

Figure 5.11. Diagrammatic sketch of the apparatus used by Bullard and Massey to observe the angular distributions of slow electrons scattered in gases.

was about 10°. Before experiments were carried out on any chosen gas, the apparatus was evacuated and the main tube baked at about 400°C.

In these experiments, the scattered electron currents of the order 10^{-12} A were measured by the rate of drift of a quadrant electrometer—electronic techniques had not yet come into use!

Bullard and Massey measured angular distributions of electrons with energies ranging from 4 to 40 eV. Arnot extended the measurements to higher energies, mainly through the use of a more elaborate collector which rejected electrons that had lost energy in inelastic collisions. Extension to energies as low as 0·6 eV was achieved by Ramsauer and Kollath with apparatus operating on a somewhat different but very basic principle. Figure 5.12 illustrates the general arrangement which they employed. Essentially, the electrons scattered

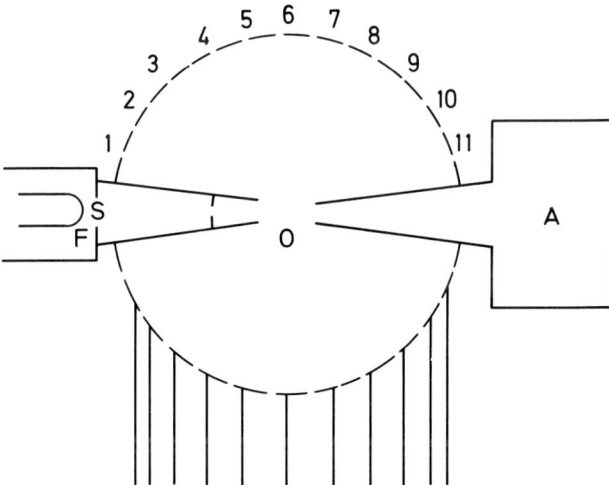

Figure 5.12. Illustrating the principle of the method used by Ramsauer and Kollath for observing the angular distributions of very slow electrons scattered in gases.

82

Scattering of Slow Electrons by Atoms

from a small, well defined region O of the gas were collected on a number of separate collectors 1 – 11 in the shape of parallel zones of a sphere centred on O. In figure 5.12, the electron source is the hot filament F, from which the electrons were collimated by passage through the slit S.

The energy and angular resolution as well as the beam geometry in these first experiments are poor by modern standards, but the results obtained at not too small angles of scattering have stood the test of time quite well. An illustration of a typical modern experiment, is described in the next chapter, p. 97.

5.7. Results of Angular Distribution Measurements

In figure 5.13 we illustrate the angular distributions of electrons of different energies scattered in argon as measured by Ramsauer and Kollath and by Bullard and Massey. Comparison is made with the distributions calculated from the formula (2.50) of Chapter 2 using the phase shifts η_0, η_1 and η_2 which yield the theoretical total cross-section curve shown in figure 5.8(b) that agrees well with the experimental data on that cross-section. There is good agreement as to the variation in form of the angular distributions as the electron energy varies. In the neighbourhood of the cross-section maximum, η_2 is the dominant

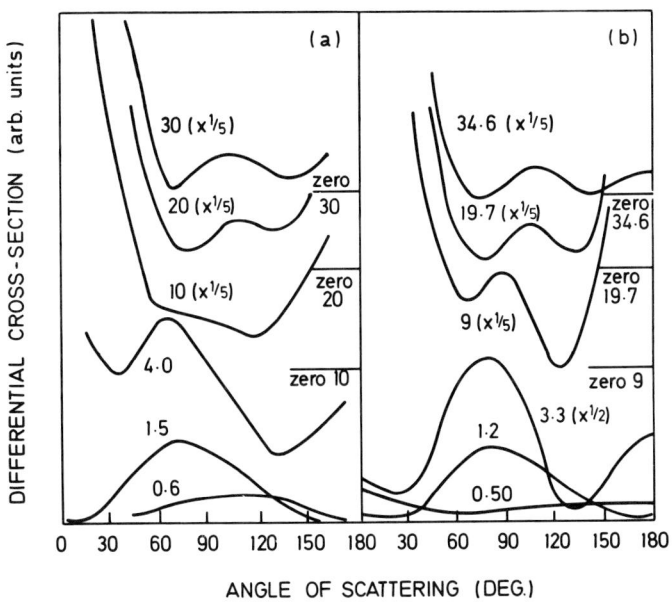

Figure 5.13. Differential cross-sections for scattering of electrons of different energies in argon (a) observed, (b) calculated,
The electron energy is indicated in eV on each cross-section curve.

phase, so that we would expect the angular distribution to be rather similar to that given by $\{P_2(\cos\theta)\}^2$, which is shown in figure 5.14. The general resemblance to the observed distributions over a considerable energy range is clear.

In the neighbourhood of the Ramsauer–Townsend minimum, the angular distribution is very variable. At these low electron energies, in general we would expect only the zero-order phase η_0 to be important, so that the angular distribution would be uniform. It is indeed true that η_0 is by far the largest phase, but it happens to be passing through an integral multiple of π, so that it has no

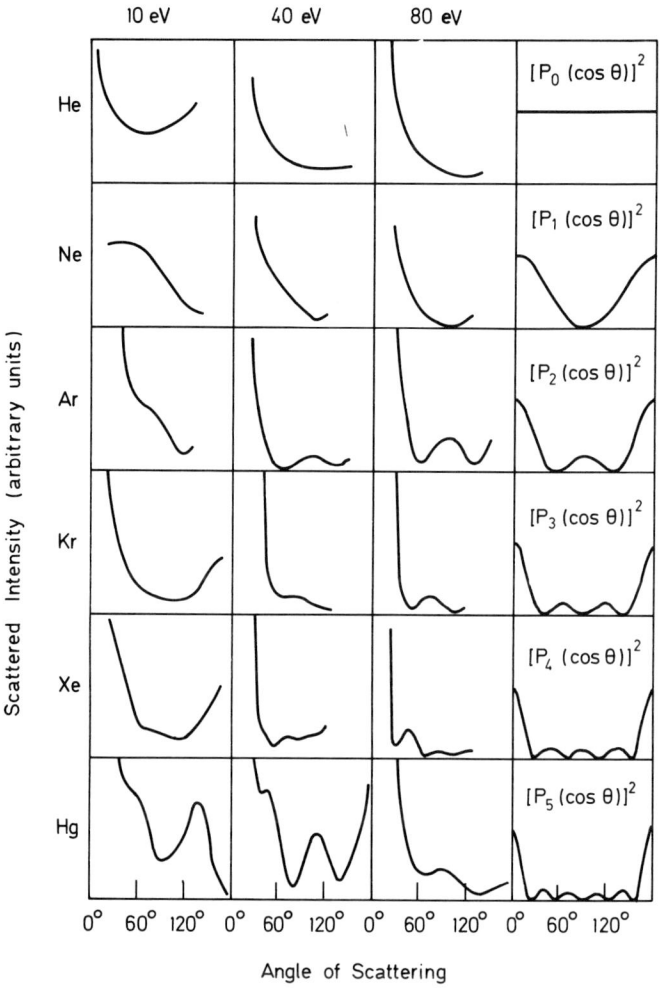

Figure 5.14. Observed angular distribution of electrons of 10, 40 and 80 eV energy scattered by different atoms. For comparison, the function $\{P_L(\cos\theta)\}^2$ is also shown in each case, η_L being the dominant phase shift at 80 eV.

influence and the angular distribution is determined by the very much smaller phase shifts η_1 and η_2.

There seems no doubt that the wave theory provides a very satisfactory detailed interpretation of the elastic scattering of electrons by argon atoms. Its success is, however, by no means confined to the case of argon. For helium and neon, even better agreement is obtained between the wave theory and experiment.

As an illustration of the way in which the angular distribution varies from atom to atom for fixed electron energy, we show in figure 5.14 the observed distribution for electrons of 80 eV energy compared with the distribution $\{P_2(\cos\theta)\}^2$ where η_L is the dominant phase shift. As the size of the atom increases L increases, and at the same time it must be remembered that the number of other phase shifts making significant contributions, and so blurring the distribution, will also increase. Nevertheless, on the whole, there is remarkably good correlation between the locations of maxima and minima in the observed distributions and those in the appropriate spherical harmonic distribution.

A striking example is shown in figure 5.15 of four cases in which, at the

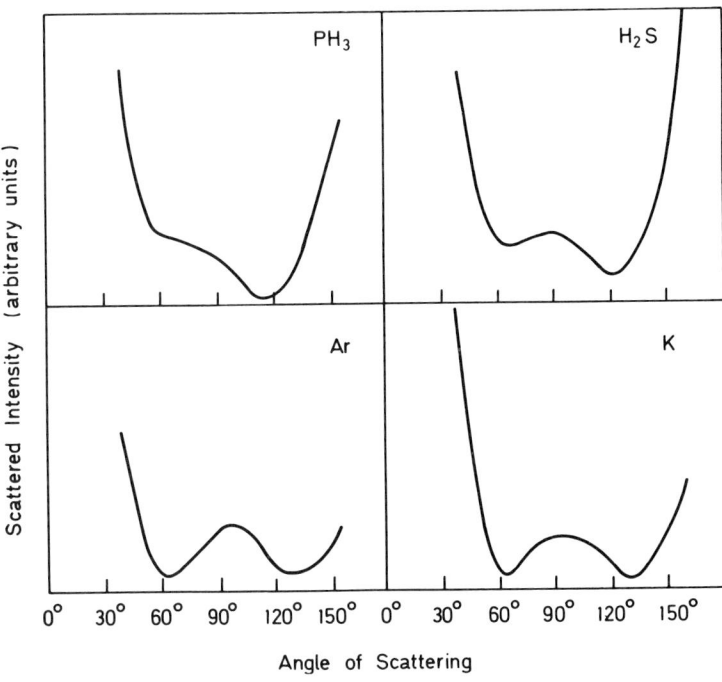

Figure 5.15. Observed angular distributions of electrons of 80 eV energy scattered by Ar, K, PH$_3$ and H$_2$S showing, by comparison with the form of $\{P_2(\cos\theta)\}^2$ shown in figure 5.14, that in all cases η_2 is the dominant phase shift.

same electron energy, 80 eV, η_2 is the dominant phase. These give observed distributions for elastic scattering by argon and potassium atoms and by phosphine (PH_3) and hydrogen sulphide (H_2S) molecules. In the latter two cases, the scattering will be largely due to the relatively heavy phosphorus and sulphur atoms respectively. In all four cases the resemblance to the distribution $\{P_2(\cos\theta)\}^2$ (see figure 5.14) is clear.

5.8. Interactions Effective in the Scattering of Slow Electrons by Atoms

Even though we can regard one of the colliding systems, the electron, as a structureless mass point, account must be taken of the electronic structure of the target atom even when elastic collisions alone are possible. In such cases, as mentioned earlier, the formulae (2.46) and (2.50) giving the cross-sections in terms of real phase shifts η_l are valid, but the theoretical calculation of the phase shifts in terms of the properties of the target atom is more complicated than for scattering by a structureless centre of force.

In most cases, apart from certain fine structure effects which we discuss in the next chapter, three physical factors must be taken into account in such calculations. These are:

(a) The mean electric field of the undisturbed atom.
(b) Electron exchange.
(c) The distortion of the atom by the incident electron with the consequent modification of the field which the atom exerts upon it.

The mean electric field of the undisturbed atom is attractive for electrons and gives rise to an interaction energy with an incident electron which falls off exponentially with distance r, at large r. Thus for a hydrogen atom, it is given by (3.14), and methods are available for calculating it with good accuracy for other atoms.

The main effect of the distortion of the atom is to introduce an additional effective interaction V_p which falls off as $-\frac{1}{2}\alpha e^2 r^{-4}$ at large r (see Chapter 3 p. 50), α being the polarizability of the atom.

Thus, if electron exchange is ignored, the background scattering is simply that calculated as described in Chapter 2, p. 32 for scattering of particles of mass m and given velocity v by a centre exerting an interaction energy $V_s + V_p$. However, even if the collision is elastic, the electron observed at infinity after the collision need not be the same as that which was incident—the incident electron may merely have changed places with one of the atomic electrons without any change of internal energy occurring.

The effect of this electron exchange is somewhat more complex, but may be represented by adding a further interaction V_e which is a function not only of r but also of the electron velocity v. Over a considerable range of low velocities, however, its effect may be simulated by a suitably chosen interaction V_e which is independent of v. This being so, it is possible to describe the elastic electron scattering in terms of the scattering by a structureless centre exerting an interaction $V_s + V_d + V_e'$ which falls off for large r as r^{-4}, so that a total elastic cross-section exists (see p. 31).

The calculated phase shifts shown in figures 5.8 and 5.9 were obtained with adequate allowance for all three factors listed above.

CHAPTER 6
electron scattering by atoms and molecules—inelastic scattering and resonance effects

6.1. Introduction

In the preceding chapter, we discussed collisions of electrons with atoms at such velocities that there was insufficient energy to excite even the lowest excited state of the atom. If the electron energy exceeds the excitation energy E_e (see p. 46) but is less than the ionization energy E_i, a bound state of an atom may be excited. As for elastic collisions, this may occur with or without electronic exchange. Indeed, for some cases excitation is possible only when such exchange occurs.

In collisions with a molecule, vibration and/or rotation may be excited with or without excitation of electronic states. Because of the small separations between allowed vibrational or rotational energy states, the minimum electron energy required to excite such states without accompanying electronic excitation is relatively quite small, being only a fraction of an electron volt (see Chapter 4, table 4.1). However, if an atom has a ground state with fine or hyperfine structure (see Chapter 3, p. 45), an upper fine structure or hyperfine structure level may be excited, in which case only a very small amount of energy need be transferred from the initial kinetic energy of the incident electron.

If the electron energy exceeds E_i, ionization of the atom or molecule may occur on impact. The energy of the ejected electron may have any value between 0 and $E_e - E_i$, and we may specify the relative probability that it will possess a kinetic energy between ε and $\varepsilon + d\varepsilon$ in terms of a differential cross-section $q_i(\varepsilon)d\varepsilon$ such that

$$\int_0^{E-E_i} q_i(\varepsilon)\,d\varepsilon = Q_i(E) \tag{6.1}$$

where $Q_i(E)$ is the cross-section for ionization of the particular atom or molecule by electrons of energy E.

If the incident energy E is high enough, more than one electron may be knocked off the atom or molecule in the collision. The process is then said to be one of *multiple ionization*.

A further possibility at sufficiently high energies is that while only one electron is ejected from the atom, it is not one of the most weakly bound electrons, but comes from an inner shell in which it is strongly bound. This is known as *inner shell ionization* and is important in the generation of X-rays. Thus one of the outer atomic electrons may drop into the vacancy left in the inner shell, thereby releasing a relatively large amount of energy in the form of radiation. Because of the high energy of the emitted photon, its frequency is very high.

A molecule AB may be dissociated by electron impact into separate atoms A, B which may or may not be in their ground states. This we may write symbolically as

$$AB + e \rightarrow A' + B'' + e, \qquad (6.2)$$

the ′ and ″ indicating the possibility that the atoms are produced in excited states. In general, the electron energy required to produce dissociation is considerably higher than the dissociation energy of the molecule, even when the atoms A and B are formed in their ground states. This is because the process AB (see Chapter 4, p. 60).

If the atom or molecule is not neutral but positively charged, the same possibilities arise, though with the following modifications. The interaction between an electron and a positive ion falls off with distance as slowly as $1/r$ for large distances r, so that no total cross-section in the sense discussed in Chapter 2, p. 32, exists. The finite limit of the scattering at zero angle is associated with elastic collisions, whereas the effective cross-sections for other types of collision are in general finite. Even for the elastic scattering, the differential cross-section per unit solid angle may be defined for angles of scattering greater than zero.

6.2. Direct and Resonant Scattering

We have explained in Chapter 2, p. 57, that a number of atoms may form stable negative ions by attachment of an electron but that, in general, only one truly bound state exists for this electron. Nevertheless, it was pointed out that atoms in excited states can attach electrons to form negative ions which have energies less than that of the parent excited state. If E_e is the excitation energy, relative to the ground state, the energy of the ion relative to the ground state will be $E_e - E'_a$ where E'_a can be regarded as the electron affinity of the excited atom (see Chapter 3, p. 57). This will, in general, be positive because E'_a will normally be much less than E_e. This means that, while the negative ion formed by capture to the excited state is stable with respect to that state, it is unstable with respect to the *ground* state and so must eventually break up to release the electron once again, leaving the atom in its ground state. Break-up of this kind is known as *autodetachment* (Chapter 3, p. 58). For our present purposes, the most important aspect of this process is that the lifetime against autodetachment is normally long compared with the time taken for a colliding electron to pass over a distance of atomic dimensions—say 10^{-10} m. An electron of energy 10 eV has a velocity near 10^6 m s^{-1}, so will traverse this distance in 10^{-16} s. The lifetime of a system of electron plus excited atom towards autodetachment may be 100 or more times longer.

Because of the possibility of autodetachment, the energy of the system of electron plus atom will be uncertain by an amount

$$\Delta E \simeq h/2\pi\tau, \qquad (6.3)$$

where τ is the lifetime towards autodetachment. Thus with $\tau = 10^{-14}$ s (see Chapter 3, p. 56).

$$\Delta E \simeq \frac{10^{-34}}{10^{-14}} \text{J}$$

$$= \frac{10^{-13}}{1 \cdot 6 \times 10^{-12}} \text{eV}$$

$$= 0 \cdot 06 \text{ eV}.$$

The energy of the electron ejected in autodetachment will lie in this range about $E_e - E'_a$. It also follows that there will be a relatively high probability that an electron, incident on the neutral atom in its ground state with this energy, will be temporarily captured to form the negative ion unstable towards autodetachment. If this happens, the electron will be held up in the neighbourhood of the atom for a time of the order of the lifetime towards autodetachment.

We may therefore expect that the scattering cross-sections will be modified by these effects, but only over narrow energy ranges about the energies $E_e - E'_a$, i.e., at energies not far below the threshold energies for exciting different bound states of the atoms.

Thus, for example, the elastic cross-section as a function of electron energy can be analysed into a slowly-varying background on which is superimposed a fine structure (see for example figure 6.6). The background is due to direct scattering, in which the electron spends a time in the neighbourhood of the atom which is comparable with the undisturbed passage time. The fine structure, localized in narrow energy ranges, arises because of the high probability of capture for times long compared with the passage time, when the electron energy is such as to make possible the formation of a temporary negative ion, i.e., the formation of a relatively long-lived *collision complex*. Because of the sharpness of the energy dependence, the fine structure effects are often referred to as arising from *resonance scattering*.

In general, the collision complex responsible for the resonance will possess a definite quantized angular momentum, so that it can only be formed when an incident electron of prescribed angular momentum quantum number L collides with a neutral atom in its ground state. The *amplitude* of the wave scattered through an angle θ by the resonance process will therefore be proportional

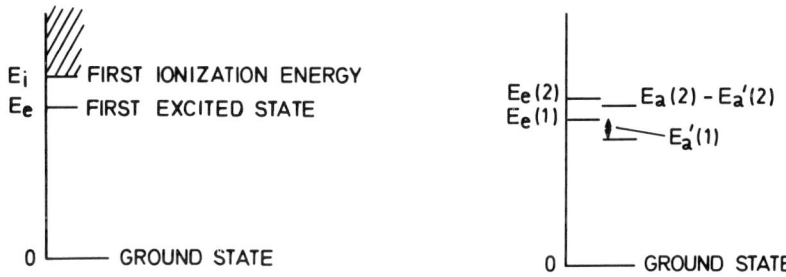

Figure 6.1.

to $P_L(\cos\theta)$. Although it must be combined with the amplitude due to direct scattering, close to an energy resonance the latter will be relatively small. The angular distribution should then vary as $\{P_L(\cos\theta)\}^2$ and may change dramatically on passing away from the resonance on either side (see for example p. 100).

Resonance effects of this kind are not confined to elastic collisions. Suppose, for example, that the complex formed has an energy $E_r(2) = E_e(2) - E_a(2)$ just below that $E_e(2)$ of the second excited state of the atom (see figure 6.1). If $E_r(2)$ is greater than the excitation energy $E_e(1)$ of the first excited state, the complex may break up either into (a) an atom in the ground state plus an electron of energy $E_r(2)$, or (b) an atom in the first excited state plus an electron of energy $E_r(2) - E_e(1)$. In the former case the ultimate process will be one of elastic scattering, but in the latter it will be one of inelastic scattering, in which the atom is left in the first excited state. The cross-section for the latter inelastic collisions will therefore show a fine structure about the incident electron energy $E_r(2)$, and so on.

The observation of resonance effects is possible only if the energy of the electron beams employed in the experiments can be confined within a range about the mean value comparable with ΔE, the energy uncertainty or *energy width* as it is often called, of the resonances. For this reason, resonances in scattering have only been observed systematically in recent years. Previously, measurements of cross-sections, such as those described in the previous chapter, gave information only about the gradually varying background. In this chapter we give some account of the techniques employed and the results obtained in recent experiments, which exhibit a rich variety of resonance effects.

6.3. Energy Analysers

Any apparatus designed for the study of resonance scattering must include a device for selecting from a beam electrons with energies falling within a very narrow band about some prescribed value. Such devices can not only provide incident electrons of very well defined energy but also select those scattered electrons which have energies falling within suitably narrow limits about a chosen value, i.e., they may function as energy analysers. Because of their vital importance for resonance scattering experiments, we shall discuss the principles involved in these devices before describing some typical modern experiments.

The first analysers depended on the fact that an electron of velocity v moving in a uniform, constant magnetic field of strength B normal to v experiences a force Bev in a direction perpendicular to both v and B. It follows that such an electron of mass m will describe a circular path of radius r given by

$$\frac{mv^2}{r} = Bev, \tag{6.4}$$

the centrifugal force along the radius being balanced by the magnetic force.

Consider now the situation illustrated in figure 6.2 in which the magnetic field has a plane boundary. Electrons entering the field through a fine slit at P with velocities normal to the field boundary will pursue semi-circular paths before passing out across the field boundary once more. Suppose that it is desired to select electrons of velocity v_0 say. Such electrons, entering normally, will pass across the field boundary at a point Q_0 distant $2r_0$ from P where

$$r_0 = mv_0/Be. \tag{6.5}$$

Electrons entering normally with some other velocity v will leave the field at

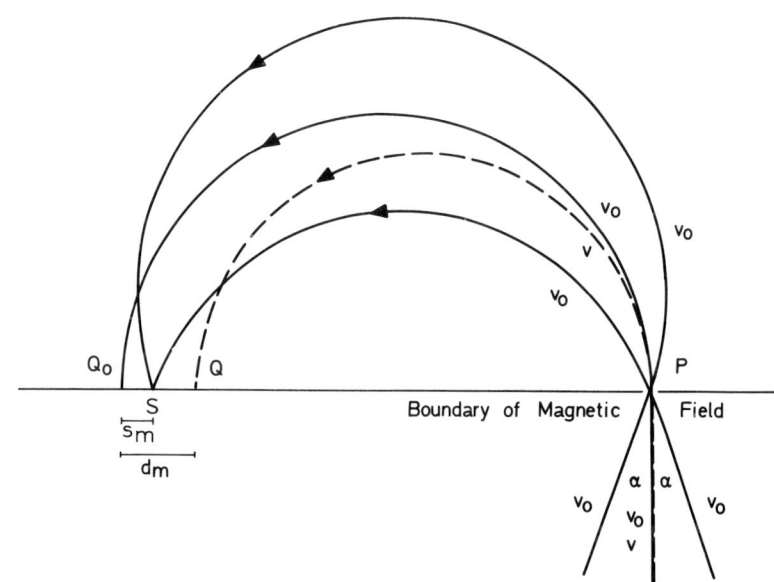

Figure 6.2. The principle of the magnetic energy analyser.

a point Q distant $2r$ from P where

$$r - r_0 = \frac{m}{Be}(v - v_0). \tag{6.6}$$

Velocity selection can be achieved by only allowing electrons to leave the field through a slit at Q_0. The resolving power of the analyser is then given by

$$d_m = r - r_0 = 2\beta r_0, \tag{6.7}$$

where $\beta = (v - v_0)/v$.

We must, however, allow for the fact that electrons entering at P will not all be moving in directions normal to the field boundary. The paths of electrons entering with velocity v_0 at angles $\pm \alpha$ from the normal take the form shown in figure 6.2. It will be seen that the two paths intersect again at the point S on the field boundary, i.e., are focused there. It may be shown by elementary geometry that the focus S is at a distance s from the point Q_0, where

$$s = \alpha^2 r_0, \tag{6.8}$$

terms involving higher powers of α being neglected.

If electrons leave through a slit at S just wide enough to pass electrons of the chosen velocity which enter at P in a cone of semi-angle α, the slit width must therefore be $\alpha^2 r_0$. Such a slit will also pass electrons entering normally at P with velocity differing by Δv from v where, using (6.7)

$$(2\Delta v/v)r_0 = \alpha^2 r_0. \tag{6.9}$$

So with this arrangement the velocity selection is given by

$$\Delta v/v = \tfrac{1}{2}\alpha^2.$$

Because the selection is made at the focus, as defined above, the ratio $\Delta v/v$ is of order α^2. At other points on the circular path, it would be of order α and so much larger.

Velocity selection may also be achieved by the use of electrostatic fields. In one such analyser, the electrons move between concentric cylinders of radii r_1, r_2 maintained at potentials V_1, V_2 respectively. The electrostatic field at a point between the cylinders at a distance r from the common axis is given by

$$X = A/r \tag{6.10}$$

where $A = (V_2 - V_1)/\ln(r_1/r_2)$. Referring to figure 6.3, suppose electrons enter at P midway between the cylinders at a radius r_0. Those entering normally with velocity v_0 given by

$$\tfrac{1}{2}mv_0^2 = Ae/r_0 \tag{6.11}$$

will describe circular paths. Analysis of the dynamics of the electron motion shows that electrons of velocity v_0 entering within the cone of semi-angle α will be focused again, not after traversing a semi-circle as in the magnetic field case discussed above, but at a point S after traversing a sector of angle 127° 17′. As for the magnetic case, the electrons pass out of the analyser through a slit at S of width sufficient to accept electrons of the chosen velocity entering in directions making angles up to $\pm \alpha$ with the normal. This width is given by

$$s_e = 4\alpha^2 r_0/3. \tag{6.12}$$

The resolution corresponding to (6.7) is given by

$$d_e = (2\beta - 4\beta^2)r_0, \tag{6.13}$$

and

$$\Delta v/v = 2\alpha^2/3. \tag{6.14}$$

For both the selectors we have described, the motion is assumed to be two-dimensional, so that the entrance and exit slits are long in the direction perpendicular to the plane of motion. Selection from a small circular hole instead of a slit may be carried out using a hemispherical electrostatic analyser in which the electrons move in the field between concentric spheres of radii r_1, r_2 held at potentials V_1, V_2 respectively. The field at a point between the spheres

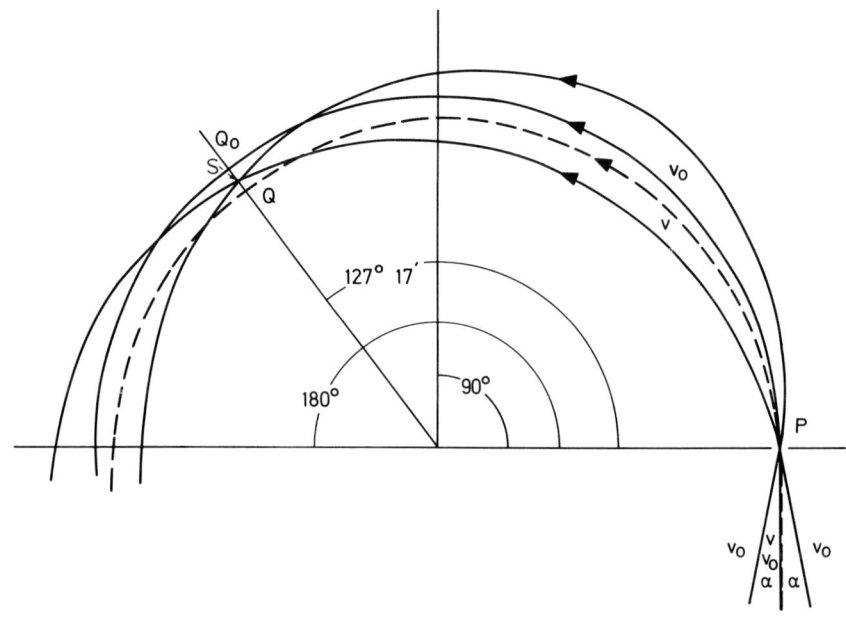

Figure 6.3. The principle of the electrostatic energy analyser.

at radius r is given by
$$X = B/r^2, \tag{6.15}$$
where $B = \left(\dfrac{V_2 - V_1}{R_2 - R_1}\right) R_1 R_2$ so that, for a circular path from a point midway between the plates at radius r_0,
$$\tfrac{1}{2} m v_0^2 = Be/r_0^2. \tag{6.16}$$
Focusing occurs after traversal of a semi-circle and it may be shown that
$$\Delta v/v = \tfrac{1}{2}\alpha^2, \tag{6.17}$$
where $\pm \alpha$ is the extreme angular divergence, in any plane, of the electron beam entering the analyser.

The approximate practical realization of the theoretical performance of these selectors requires great attention to the elimination of secondary effects of various kinds such as secondary electron emission from the electrodes in the electrostatic devices or the departure of the magnetic field from the ideal form with a sharp boundary shown in figure 6.2. Confinement of the entering beam divergence to a small angular range $\pm \alpha$ also requires careful attention.

6.4. Experimental Methods for Studying Fine Structure in Electron Scattering

Given suitable energy selectors and analysers, fine structure can be studied in a number of different ways which may be summarized as follows:

(a) From measurements of the total cross-section as a function of electron energy.
(b) From observation, as a function of electron energy, of the electron current transmitted through the gas without deviation.
(c) From observation of the variation with electron energy of the current scattered into a fixed narrow angular range.

Most measurements which have been made either fall under (b) or (c), although Golden and his collaborators have adapted the Ramsauer technique described on p. 73 to obtain total cross-sections at closely defined energies.

As an example of a typical transmission apparatus of high precision, we take that of Simpson, Kuyatt and Mielczarek, which uses hemispherical analysis both to define the energy of the incident electrons and to select that of the scattered. Figure 6.4 illustrates the general arrangement. Electrons from the hot filament F are accelerated towards the anode A_1 by a potential V_0 close to 20 V. The electrodes A_3, A_4 and A_5 constitute a lens (see Appendix 5) that focuses the slit in A_1 on to the entrance of the hemispherical analyser D_1. By maintaining A_5 at a potential E_1 of 1·35 V, the beam is decelerated before passing into D_1 through which it is deflected by 180° by means of the deflecting

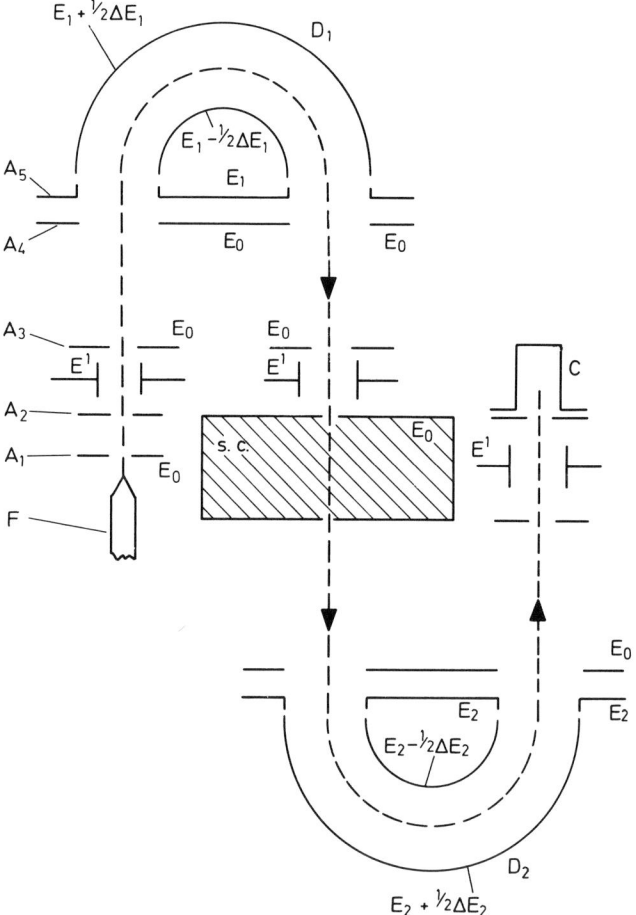

Figure 6.4. The arrangement of apparatus used by Simpson, Kuyatt and Mielczarek to observe resonance effects in the transmission of electrons of well-defined energy through gases.

voltage ΔE_1. It is then accelerated once more to pass through the scattering chamber SC, maintained at the potential E_0. Again, before passage through the second similar hemispherical analyser D_2 maintained at a potential E_2, it is decelerated and then finally accelerated once more into the collector C.

The scattering chamber is a stainless steel cylinder closed at the ends by molybdenum discs containing apertures through which the beam passes. If necessary, the beam may be centred by application of a suitable electrostatic field between the electrodes E′. The whole assembly is mounted within a stainless steel vacuum envelope. After baking, pressures as low as 10^{-9} torr can be obtained with a mercury diffusion pump and liquid air trap (see Appendix 2). Under these conditions, the energy spread of the electron beam, measured by that at which the intensity has fallen to half its peak value, is 0·038 eV and the collected current 3×10^{-10} A. Great care is taken to shield the whole apparatus magnetically. Electrons that have been scattered through an angle greater than 0·02 radian or have lost an energy greater than 0·05 eV are not collected.

With this apparatus, no attempt is made to measure absolute cross-sections because accurate measurements of the pressure in the scattering chamber are not possible. On the other hand, it is very suitable for the detection of fine structure in the variation of the transmitted current with electron energy.

Other transmission experiments have been carried out using cylindrical or other analysers and we shall refer to results obtained in these experiments as well as by the use of the equipment described above.

As an example of a typical experiment to observe the variation with electron energy of the current scattered into a fixed narrow angular range, we choose that used first by Andrick and Ehrhardt. This has been employed successfully in a number of experiments and has the advantage that measurements may be made with it over a wide range of angles. It can be regarded, in fact, as a modern version of the earlier experiments on the angular distribution of scattered electrons (see Chapter 5, p. 80).

The principle involved is essentially the same as in these earlier experiments, except that the electron beam is scattered by an atomic beam (see Chapter 9, p. 180), which it crosses at right angles, instead of in a static volume of gas. Furthermore, the definition of electron energy and also of the beam geometry is much more precise. For the former, 127° cylindrical analysers are used, while suitable electron lens systems (see Appendix 5) are introduced for beam definition. Figure 6.5 illustrates the general arrangement.

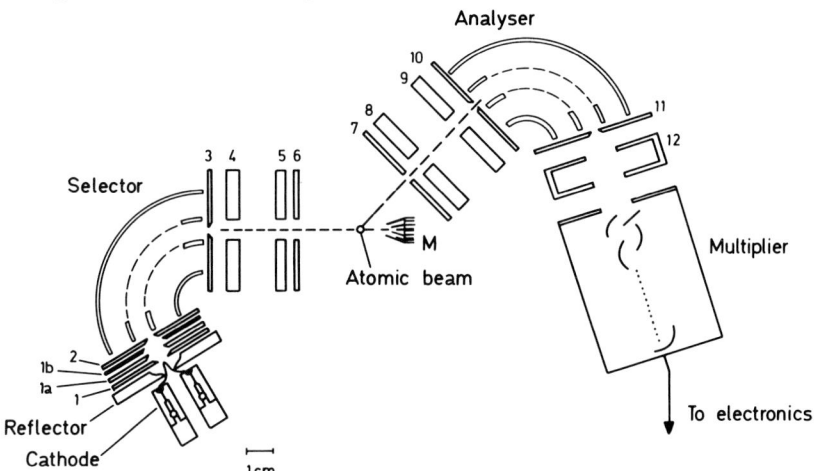

Figure 6.5. The arrangement of apparatus used by Andrick and Ehrhardt for observing resonance effects at different angles of scattering, in gases, of electrons of well defined energy.

Atomic and Molecular Collisions

Electrons from a hairpin-shaped hot cathode are focused on the entrance slit 2 of a 127° analyser by means of the lens electrodes, 1, 1a and 1b. After energy selection a beam with an energy spread at half width of about 0·05 eV is accelerated or decelerated to the desired energy (between 0·3 and 30 eV) and focused on the atom beam by means of the lens electrodes 4 and 5. The total current in the electron beam may be measured by a collector M and this may be used to adjust the voltages on the lens electrodes 4 and 5 to optimum conditions.

Electrons scattered at a fixed angle from the atomic beam are focused by means of the lens electrodes 8 and 9 on the entrance slit 10 of a second 127° analyser. The emergent electrons are then focused by the lens 12 onto the cathode of an electron multiplier which measures their intensity (see Appendix 6).

6.5. Typical Measurements of Fine Structure (Resonance) Effects in Elastic Scattering

The first resonance effect detected was that in helium at an electron energy of 19·3 eV (0·4 eV below the excitation energy). This was observed by Schulz in the variation of the intensity of scattering at 72° as a function of electron energy and his results are shown in figure 6.6(a). The occurrence of the resonance is marked by a sudden sharp fall in the scattered intensity followed by a recovery to a value in excess of the background value. Figure 6.6(b) shows the same resonance observed by Simpson in the variation of transmitted electron current with electron energy. These measurements were made with the apparatus of figure 6.4. Finally we show in figure 6.6(c) the variation of the total cross-section with electron energy observed by Golden and Bandel using a modern version of the Ramsauer method.

As described on p. 90, the resonance effect arises through the formation by electrons, with energies in a narrow range, of doubly excited negative ions with lifetime considerably greater than the normal time taken for an electron to traverse the atomic field. If the width ΔE of the resonance energy range can be measured, the lifetime may be determined as $h/2\pi\Delta E$. Such measurements are difficult because ΔE is usually of the same order or less than the so-called instrumental width. This arises from the spread of energy in the incident electron beam and other factors due to imperfect geometrical or energy resolution in the apparatus. In typical experiments with high instrumental performance, the instrumental width is of the order of 0·05 to 0·10 eV, while the best estimates from modern experiments give ΔE as close to 0·008 eV. This corresponds to a mean lifetime of the transient helium negative ions of 8×10^{-14} s.

The angular momentum quantum number L of the electron in the transient negative ion may be obtained, as described on p. 91, from observations of the resonance effect at different scattering angles—the intensity of the resonance scattering will vary as $\{P_L(\cos\theta)\}^2$. Figure 6.7 shows observations made by

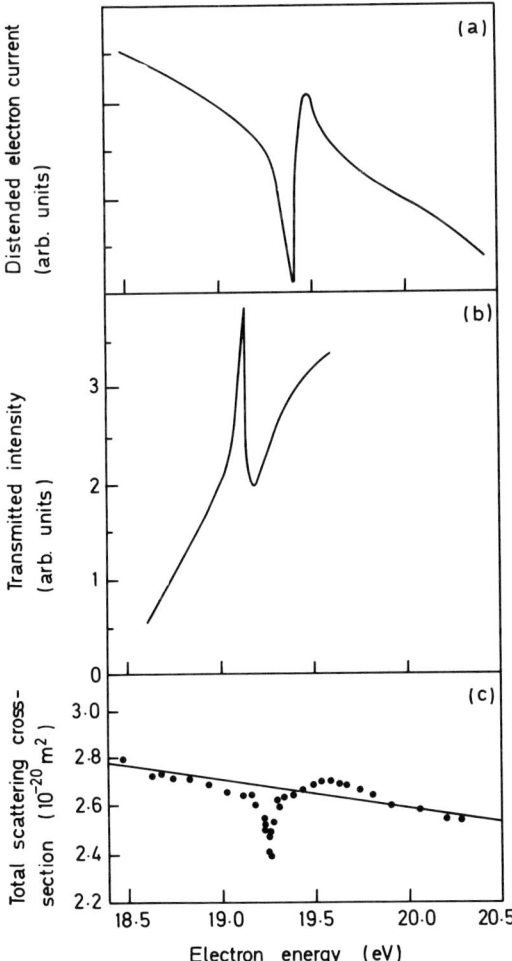

Figure 6.6. Observational manifestations of the resonance effect which occurs in the scattering of electrons of energy 19·3 eV by helium atoms.
(a) Variation with electron energy of the intensity of scattering of electrons at 72° as observed by Schulz—the first observation of a sharp resonance in electron scattering. The scattered current scale has been expanded to bring out the effect. The maximum depression is about 14% of the background value.
(b) Variation with electron energy of the transmission of electrons through helium as observed by Simpson.
(c) Variation with electron energy of the total cross-section for elastic scattering of electrons by helium atoms, as observed by Golden and Bandel. ●, experimental points.

Andrick and Ehrhardt using the apparatus described above (see figure 6.5).

From these data, the intensities of the background and resonance scattering may be separately determined as functions of the scattering angle θ, giving the results shown in figure 6.8. Allowing for the considerable spread of the values

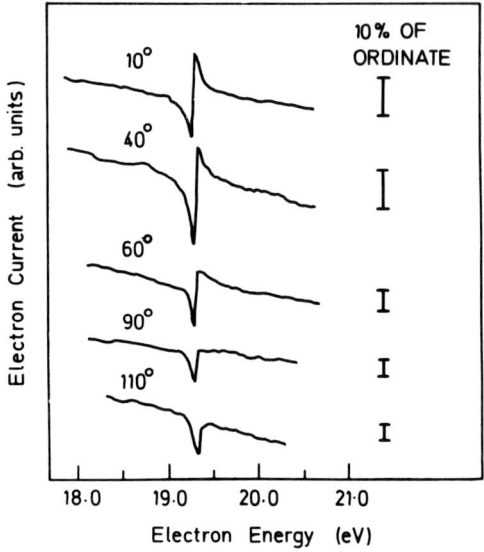

Figure 6.7. Observed variation with incident electron energy of the intensity of electrons scattered elastically in helium, at different angles of scattering. The range of the intensity variations may be judged from the vertical lines on the right hand side which indicate 10% of the total intensity near the resonance.

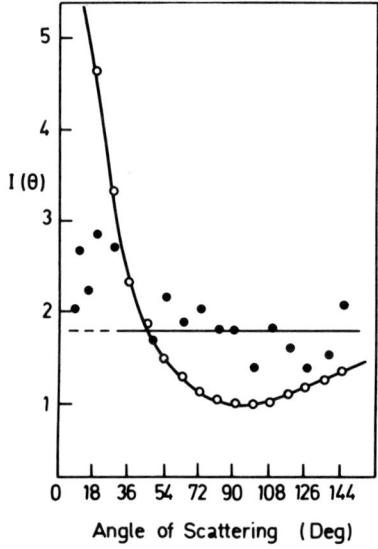

Figure 6.8. Angular distribution of the resonant and background elastic scattering per unit solid angle $I(\theta)$, of electrons in helium, derived from observations of the type shown in figure 6.7. ●●●, resonant; –O–O–O–, background.

Electron Scattering by Atoms and Molecules

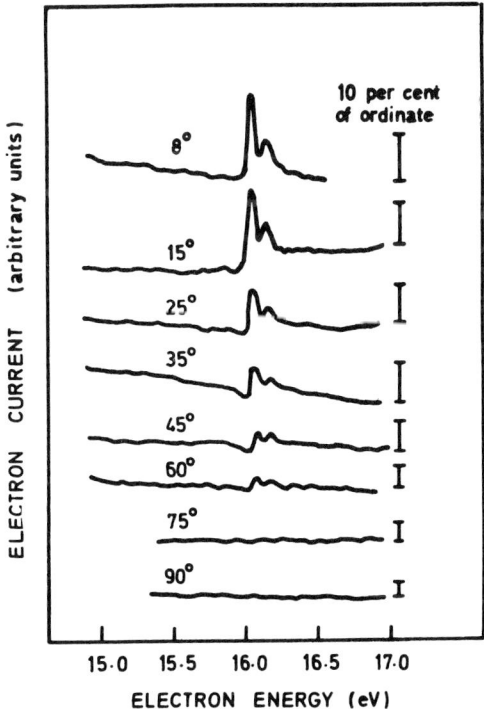

Figure 6.9. Observed variation with incident electron energy of the intensity of electrons scattered elastically in neon at different angles of scattering. The range of the intensity variations may be judged from the vertical lines on the right hand side which indicate 10% of the total intensity near the resonance.

obtained for the resonance scattering, there is evidence that it varies little with θ, in contrast with the background scattering. This is what would be expected if the electron is captured into a state of zero angular momentum, $P_0(\cos\theta)$ being a constant.

In contrast we show in figure 6.9 corresponding results for the resonance in neon at electron energies between 16·0 and 16·2 eV also observed by Andrick and Ehrhardt. Actually two very close resonances are involved, but it will be seen that both fade out at a scattering angle of 90°. This is because, for both, the quantum number L is not 0 but 1 so the intensity varies as $\{P_1(\cos\theta)\}^2$, i.e., $\cos^2\theta$, which vanishes at 90°.

A great number of observations of this kind have been carried out for scattering by the atoms of the rare gases, of mercury and of hydrogen.

6.6. Measurement of Inelastic Cross-Sections—Differential Cross-Sections

Differential cross-sections for collisions in which an incident electron loses a specified amount of energy in producing excitation or ionization of the atom

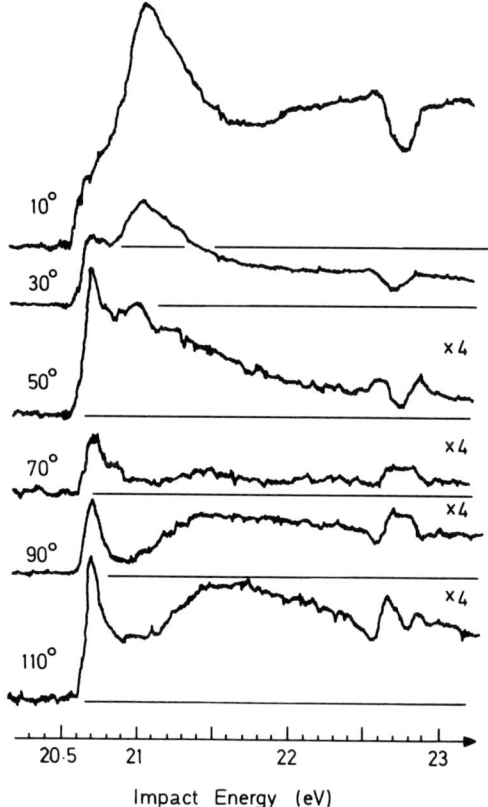

Figure 6.10. Variation with incident electron energy of the differential cross-section per unit solid angle for excitation of the 2^1S metastable state of helium by electron impact, as observed by Andrick and Ehrhardt.

or molecule may be studied with equipment such as that of Andrick and Ehrhardt (figure 6.5). This is because energy analysis of the scattered electrons is available.

Figure 6.10 shows results obtained in this way for collisions in which the metastable singlet state of helium is excited, requiring 20·6 eV energy. The observed variation of the differential cross-section for this process is shown as a function of incident electron energy for a number of scattering angles. These data are typical of the information forthcoming from experiments of this type.

6.7. Cross-sections for Ionization

The measurement of total cross-sections requires somewhat different techniques. We first consider total cross-sections for ionization, taking advantage of the characteristic feature that a slow positive ion results from the impact in such a case. Essentially what is done is to fire a beam of electrons of selected energy through the gas under investigation and collect all the positive ions produced from a given length of path l. Then if i is the incident beam current and i_+ the current of slow positive ions produced in the gas containing n atoms per unit of volume

$$i_+/i = nQ_i l,$$

where Q_i is the ionization cross-section. This is provided that conditions are such that an electron makes at most *one* collision in travelling the distance l, i.e., we must have $nQ_i l \ll 1$. i_+ is measured as follows. Electrons either scattered from the beam or produced by ionization are confined to the neighbourhood of the beam by means of a magnetic field along the direction of the beam. This field, of order 500 G, is not strong enough to confine the positive ions which are collected on a plane plate parallel to the beam, maintained at a potential negative by some volts with respect to that of a similar parallel plate on the other side of the beam.

The first experiments of this kind were carried out by Tate, Smith (P.T.) and Jones (T.) at the beginning of the 1930s. Figure 6.11 shows the arrangement used in modern experiments by Rapp, Englander-Golden and Briglia, which could also be adapted to the measurement of electron attachment cross-sections (see Chapter 7, p. 126). The electron beam from the cathode K passed into the collision chamber C after acceleration to the required energy and collimation through the electrodes 1, 2 and 3. The beam was collected on the plate ECP, which was maintained at a positive potential screened from the collision chamber by the region ECS. A magnetic field of 500 G or so was applied along the beam direction.

Positive ions produced by the beam were collected on the plate IC, which was maintained at a negative potential relative to the parallel plate JC on the other side of the beam. To ensure uniformity of the transverse electric field over the entire length of the plates both IC and JC were flanked by two guard electrodes, G and H respectively, G being at the same potential as IC and H as JC.

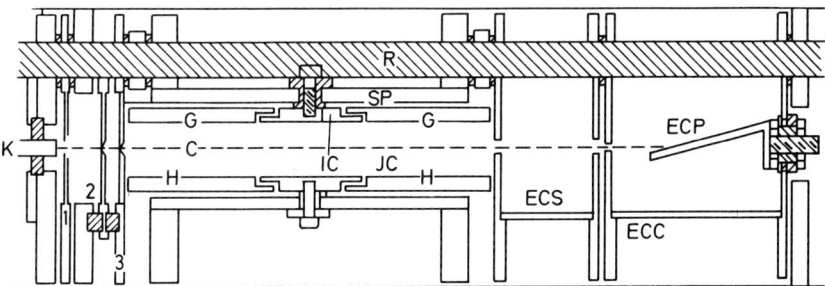

Figure 6.11. The arrangement of apparatus used by Rapp and Englander-Golden for the measurement of cross-sections for ionization of atoms by electrons.

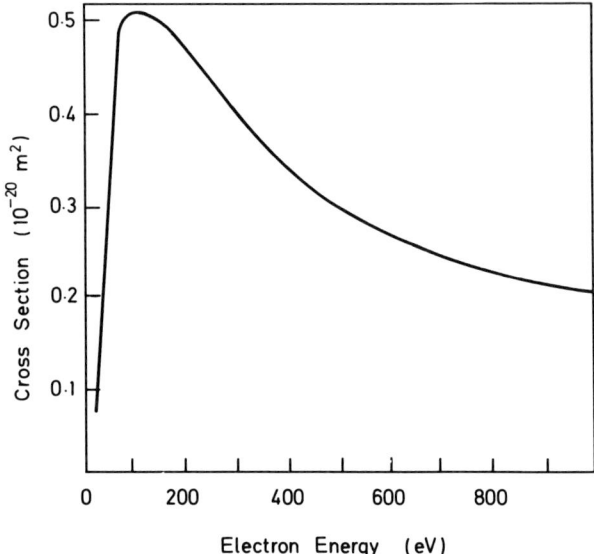

Figure 6.12. Observed cross-section for ionization of helium atoms by electron impact, as a function of electron energy.

The voltage difference V_p between the parallel plates must be large enough to collect all the ions formed by the beam over the length of the plate IC. For ionization of atomic gases, V_p need only be a few volts because the ions are formed with very small kinetic energy. This may not be so for molecular gases, and V_p may have to be somewhat larger to saturate the ion current. Figure 6.12 shows the ionization cross-section of helium as a function of electron energy, measured with this apparatus. The variation shown is typical, rising sharply from zero at threshold to a maximum at a few times the threshold energy and then falling off gradually as the electron energy E increases. At high energies the variation is as $E^{-1} \ln E$. Detailed study of the cross-section at energies not far above threshold reveals structure due to autoionization (see Chapter 3, p. 85).

Methods are available for the measurement of cross-sections for further ionization of positive ions and for detachment of electrons from negative ions, but the description of these methods, which employ crossed electron and ion beams, is beyond the scope of the present book. It is also possible to measure cross-sections for ionization of atoms such as H and O which do not exist uncombined in the gas at ordinary temperature. Again crossed beam techniques must be used.

6.8. Cross-Sections for Excitation of Bound States

Turning now to collisions in which the excited state of the atom is a bound state, the total cross-section for the process may be measured optically if the state concerned is not metastable. Let n_j be the number of excited atoms produced in the state j per unit volume by an electron beam of given velocity v. The rate of production of these atoms by electron impact per unit volume per second will be $nQ_j v n_e$, where Q_j is the total cross-section for collisions leading to excitation of the state j, n the number of normal atoms and n_e that of the electrons, per unit volume. Loss of excited atoms will occur through emission of radiation in transitions to lower states, the rate of which per unit volume per second may be written as $A_j n_j$, where A_j is the probability per second of the radiative transitions from the jth level (see Chapter 3, p. 46). Allowance must also be made for the fact that a contribution to the population of atoms in the state j will be made by radiative transitions from the various upper states i which are also excited by electron impact. This will be at a rate $\sum_i A_{ij} n_i$ per unit volume per second, A_{ij} being the probability per second of a radiative transition from the ith to the jth level.

In equilibrium we then have

$$nQ_j v n_e - A_j n_j + \sum A_{ij} n_i = 0. \tag{6.18}$$

The total number of quanta of the radiation emitted per second per unit length of the electron beam with frequency corresponding to a transition from the state j to a lower state k will be

$$J_{jk} = A_{jk} n_j S, \tag{6.19}$$

where S is the cross-sectional area of the beam. Using (6.18), this becomes

$$J_{jk} = (A_{jk}/A_j)\left(nQ_j \frac{I}{e} + \sum_i J_{ij}\right), \tag{6.20}$$

where

$$A_{jk}/A_j = J_{jk}/\sum_k J_{jk} \tag{6.21}$$

and

$$I = vSen_e, \tag{6.22}$$

is the current strength in the beam. It follows from (6.20) that Q_j may be derived from measurements of J_{jk} and J_{ij} for all states k and i respectively below and above the state j.

In practice it is often difficult to measure all the J_{jk} and J_{ij} involved but sufficient may be obtained to give a good approximation to Q_j. In some cases the data may be supplemented by theory.

The variation of J_{jk} with electron energy is usually referred to as the *optical excitation function* for the spectrum line arising from the $j \to k$ transition. From J_{jk} an effective cross section Q_{jk} for excitation of his line may be obtained by taking

$$J_{jk} = nQ_{jk} I/e. \tag{6.23}$$

It then follows from (6.20) that

$$Q_{jk}\left(1 - \frac{\sum_i J_{ij}}{\sum_k J_{jk}}\right) = \frac{A_{jk}}{A_j} Q_j. \qquad (6.24)$$

A great deal of care is required in obtaining reliable measurements in this way, but the technique is now refined to such an extent that autodetachment resonance structure can be detected clearly in observed excitation cross-sections. Figure 6.13 shows the forms of the cross-sections as functions of electron energy for excitation of different types of excited states of helium as would be measured with inadequate energy resolution to observe the fine structure. In all cases, the cross-sections vanish at the threshold, but the rate at which they

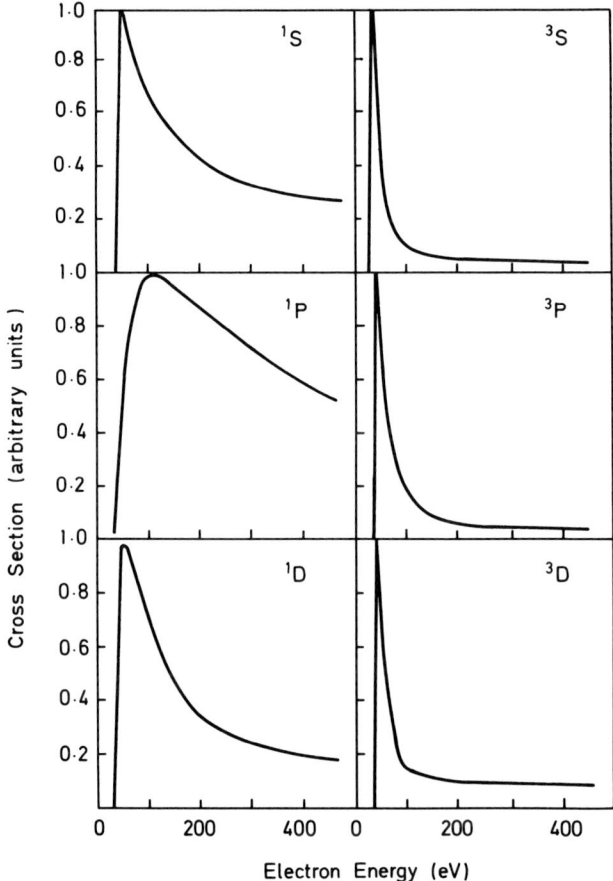

Figure 6.13. Forms of cross-sections for excitation of different types of states of helium atoms by electron impact, fine structure effects being smoothed out. The helium states concerned are shown in the top right hand corner in each case.

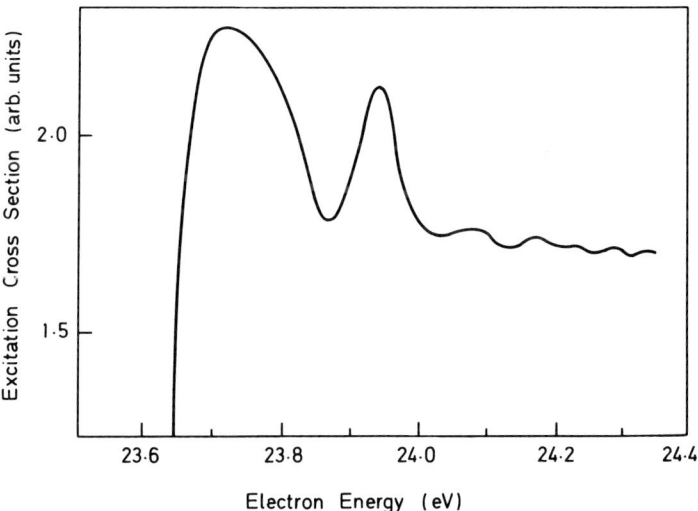

Figure 6.14. Fine structure effects in the cross-section for excitation of the helium spectrum line at 471·3 nm by slow electrons of well defined energy, as observed by Heddle, Keesing and Kurepa. Note the expanded electron energy scale.

rise to a maximum and then decline varies in different characteristic ways for the different types of state.

Figure 6.14 shows a typical recent example of an observed optical excitation function, close to the threshold, for the line of helium at 471·3 mm. The structure is due to excitation of doubly excited states which autodetach (see p. 58) to leave the atom in the upper state from which the spectrum line is produced.

The optical method is not available for excitation of metastable states. However, it is possible in many cases to observe metastable atoms through the secondary electron emission they produce on impact with metal surfaces or through the ionization they produce on impact with certain gaseous atoms or molecules (see p. 152). This may be made the basis of a technique for at least measuring relative excitation cross-sections for metastable atom production.

CHAPTER 7
collisions involving electron capture—recombination and attachment

In the preceding two chapters we have discussed collisions of electrons with atoms or molecules in which scattering, either elastic or inelastic, takes place but a free electron emerges from the collision[†]. We are now considering cases in which a captured electron is not so replaced. They include *recombination*, in which an electron is captured by a positive ion to form one or more neutral systems, and *attachment*, in which the capture occurs to a neutral atom or molecule to produce a negative ion. An understanding of the factors determining the rates of these processes is essential for the development of the theory of the formation and properties of the ionospheres not only of the Earth but of the planets. Recombination also plays an important role in determining the behaviour of the outer atmosphere of the Sun, the solar corona, as well as hot dense plasmas important for research on controlled nuclear fusion. It appears also to be important in determining the composition of molecular fragments in interstellar space. Attachment may be used to monitor very small concentrations of halogen-containing pesticides, a matter of much importance for environmental studies. Again, attachment plays a central role in determining the dielectric strength of such substances as sulphur hexafluoride.

In this chapter, we consider in some detail the important processes of recombination of electrons to positive ions and of attachment to neutral atoms and molecules, how they are studied experimentally and how information about the measured rates may be used in the detection of traces of halogen-containing pesticides. Applications to the interpretation of the behaviour of the Earth's ionosphere, the solar corona and to molecule formation in interstellar space are discussed in Chapter 14.

7.1. Recombination and Attachment Coefficients

It is convenient to consider these two processes together, although the experimental techniques for studying them are very different. Both processes involve a transition of an electron from a free state of motion with kinetic energy E to a bound state, with release of energy. The various mechanisms

[†] It is true that we have already discussed electron exchange in which the incident electron is captured but replaced by an emerging atomic electron.

which can lead to recombination or attachment differ in the means by which this energy is carried away.

The rates at which the processes occur may be expressed conveniently in terms of rate coefficients α, η respectively. Thus α is such that the chance per unit time that an electron in a medium containing n^+ positive ions per unit volume will recombine is given by αn^+. In other words, if n_e is the number of free electrons per unit volume then, due to recombination,

$$\frac{dn_e}{dt} = -\alpha n_e n^+. \qquad (7.1)$$

α has a dimensions $[L]^3 [T]^{-1}$. In most practical circumstances, the electrons will have a velocity distribution with a root mean square velocity $\bar{v}$. We may then define a mean recombination cross-section $\bar{Q}_r$ by

$$\alpha = \bar{v}\bar{Q}_r. \qquad (7.2)$$

Similar considerations apply to the attachment coefficient η. Thus, due to attachment of electrons to neutral molecules which are in a concentration n per unit volume,

$$\frac{dn_e}{dt} = -\eta n_e n. \qquad (7.3)$$

Measurements of attachment rates with electrons of well-defined velocity may be made much more readily than for recombination, so the attachment cross-section $Q_a(v)$ for electrons of velocity v may be obtained.

For an electron to be permanently captured either by a positive ion to form a neutral atom or molecule, or by a neutral atom or molecule to form a negative ion, a release of energy must occur. Thus in the captured state the electron possesses less total energy than when free and the surplus energy must be disposed of. We consider in detail two important cases here. The simplest one in principle is when the energy is carried away directly as electromagnetic radiation—*radiative capture*. The second is when the surplus energy produces dissociation of a molecule leading to *dissociative recombination* when the molecule is a positive ion, and to *dissociative attachment* when it is uncharged. There are other possibilities, but we shall not consider them in detail here (see however (7.20)). One, *dielectronic recombination*, we shall discuss in connection with the theory of the solar corona (Chapter 14, p. 275).

7.2. Radiative Recombination and Attachment

These processes may be represented symbolically by

$$A^+ + e \rightarrow A' + h\nu, \qquad (7.4a)$$

$$B + e \rightarrow B^- + h\nu, \qquad (7.4b)$$

Atomic and Molecular Collisions

In (7.4a) an electron of energy E is captured by a positive ion A^+ to form a neutral atom A. This may or may not be in an excited state so is denoted by A'. If E_n is the energy required to remove an electron from the excited atom we have

$$hv = E + E_n. \quad (7.5a)$$

Similarly for the radiative attachment process (7.4b), if E_a is the electron affinity of B, then

$$hv = E + E_a. \quad (7.5b)$$

In contrast to (7.4a), B^- will normally only have one stable bound state (see Chapter 3, p. 57).

To obtain an estimate of the order of magnitude of the rate coefficients, we recall that the mean time which elapses before an electron, in an excited state in atom, makes a radiative transition to a lower state is of the order 10^{-8} s. An electron of a few electron-volts energy will spend a time of the order 10^{-15} s in the neighbourhood of the atom. In that short period, it will only have a chance of the order $10^{-15}/10^{-8} = 10^{-7}$, of making a radiative transition to a lower state. The effective cross-section for radiative attachment is therefore likely to be only of the order of 10^{-7} of atomic dimensions, i.e., about 10^{-26} m², for electrons of a few eV energy. For less energetic electrons, it will be somewhat larger, because of the longer time spent in the neighbourhood of the atom.

Since electrons of 1 eV energy have a velocity of 6×10^5 m s^{-1}, the attachment coefficient for such electrons will be of order 10^{-20} m³ s^{-1}. Figure 7.1 shows the variation with electron velocity of the attachment coefficient for atomic oxygen, derived from experimental data on the rate of the inverse process of photodetachment (see Chapter 13, p. 244). It will be seen that these results are consistent with the conclusions derived from the simple argument outlined above.

For radiative recombination, account must be taken of the fact that, unlike

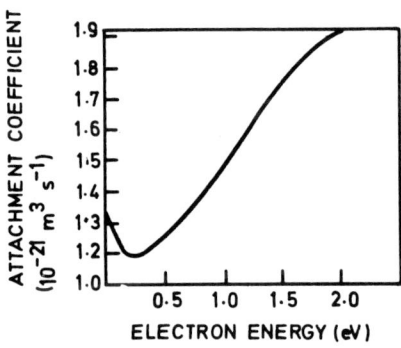

Figure 7.1. Radiative attachment coefficient for electrons to oxygen atoms, as a function of electron energy.

Collisions Involving Electron Capture

a negative ion, a neutral atom or molecule has an infinite number of bound states, so that the radiative recombination rate is the sum of the rates for capture into different final states. The rate coefficient, though small, is considerably higher than for radiative attachment. Thus for recombination of electrons at room temperature with O^+ ions, the value calculated by the methods of quantum theory is $3\cdot7 \times 10^{-18} \text{ m}^3 \text{ s}^{-1}$.

7.3. Dissociative Recombination and Attachment

For collisions with molecules, an alternative way in which the energy released by recombination or attachment may leave the system is through dissociation of the molecule.

To consider the possibilities which may arise in this way we discuss first the case of recombination

$$AB^+ + e \rightarrow A' + B'' \tag{7.6}$$

in which the atoms A and B may or may not be in excited states. To indicate this they are denoted by A' and B'' respectively.

In figure 7.2, we show the potential energy curve I (see Chapter 4, p. 61) for the ground state of the molecular ion AB^+. Because this must be a stable molecular state, the curve will have a minimum at a nuclear separation R_0. This minimum will lie below the energy of the separated atom A and ion B^+ by an amount D, which is the dissociation energy of AB^+.

Consider now the potential energy curves for states of the neutral molecule AB. Suppose that one of these (curve II of figure 7.2) which corresponds to a repulsive state (see Chapter 4, p. 61), dissociating at large nuclear separations

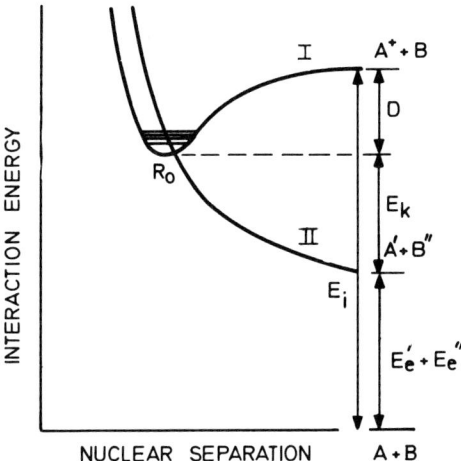

Figure 7.2. The energy relations in dissociative recombination.

Atomic and Molecular Collisions

into atoms A' and B", intersects the curve I near R_0. It follows that an electron with very small kinetic energy may be captured on collision with the ion, forming a neutral system AB in the repulsive state corresponding to the curve II. Once this occurs, the atoms A and B will separate with relative kinetic energy given by E_k in figure 7.2. Thus

$$E_k = E_i - D - E'_e - E''_e, \qquad (7.7)$$

E_i being the ionization energy of A, E'_e, E''_e the respective energies of excitation of A and B. However, it does not mean that capture of a slow electron will result in all cases in the production of neutral atoms. There is always the chance that the capture process will be reversed through autoionization before the atoms can separate very far.

Nevertheless, the chance of reversal will decrease as the separation between the atoms increases, until it becomes negligible. If now τ is the mean lifetime of the complex before autoionization occurs and θ the time before the atoms separate so far as to render negligible the chance of autoionization, the chance that the initial capture will result in production of two separated neutral atoms will be $\tau/(\tau + \theta)$.

For a typical autoionization process, τ is of order 10^{-14} s or more. In a typical case of oxygen atoms separating with a kinetic energy of 1 eV, the relative velocity of separation is 5×10^3 m s^{-1}, so that the time required to separate 5×10^{-11} m, which is about $\frac{1}{2}R_0$, will be 10^{-14} s. It seems then that $\tau/(\tau + \theta)$ will be not much less than unity. The cross-section for the dissociative recombination process will therefore be of the order Q_c, the cross-section for a collision in which the initial capture occurs. Since Q_c will be of the order πR_0^2, $\simeq 10^{-19}$ m^2, and the electron velocity before capture about 10^5 m s^{-1}, the recombination coefficient due to this process will be of the order 10^{-14} m^3 s^{-1}, between 10^3 and 10^4 times as large as for radiative recombination.

A very similar discussion is appropriate for dissociative attachment, but there are some important differences. Perhaps the most important is that the electron affinity E_a of an atom is so much less than the ionization energy E_i (see Chapter 3, p. 46). E_a will usually be less than D and the relation of the potential energy curves will be as shown in figure 7.3. Intersection between the repulsive curve for AB$^-$ and that for the ground state of AB will now occur at nuclear separations much larger than R_0. Since the separation in the actual molecules will be close to R_0, there is little chance (Chapter 4, p. 67) that an electron of very low kinetic energy can be captured. Instead, capture can take place only to molecules with separation between $R_0 \pm \frac{1}{2}a$, where a is the amplitude of vibration. At the extreme points, the energy of the repulsive curve of AB$^-$ lies at energies E_1, E_2 respectively above that of AB. It follows that, for an electron to be captured with appreciable probability, it must have an energy between E_1 and E_2, giving rise to an atom A and negative ion B$^-$ which separate with relative kinetic energies between E_3 and E_4 (see Figure 7.3(a)).

Collisions Involving Electron Capture

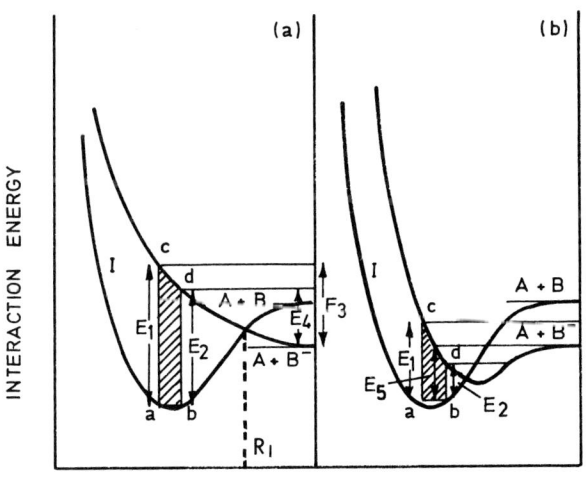

Figure 7.3. The energy relations in dissociative attachment.

Because the atom and ion must separate to a distance greater than R_1 in figure 7.3(a) before autodetachment, leading to release of the captured electron, is no longer possible, the likelihood that the initial capture of the electrons will finally result in attachment is much smaller than in the case of dissociative recombination discussed above. In fact the peak cross-section for dissociative attachment of this type is usually no larger than 10^{-23} m^2, giving rise to a rate coefficient as low as 10^{-17} m^3 s^{-1}.

For dissociative attachment, a further possibility arises which is illustrated in figure 7.3(b). In this case, the initial capture takes place into a state of the molecule AB$^-$ which has a minimum. Electrons with energy between E_1 and E_2 can now be captured, but only when the energy is greater than E_5 can the capture lead to dissociative attachment. Otherwise the final state to which the electron is captured lies below the limit for dissociation into A and B$^-$. The atom and ions cannot separate indefinitely, but oscillate relative to each other until autodetachment occurs and the electron is released again to return to infinity. This case is distinguished from the one illustrated in figure 7.3(a) by the fact that the energy of relative motion of the separated atom and ion will range from zero to a value not greatly in excess of $E_1 - E_5$ (see figure 7.3(b)) instead of between two finite values.

The kinetic energy of the ions produced is the fraction

$$\gamma = M_n / M_n M_i,$$

of the total kinetic energy E_k, M_n, M_i being the masses of the neutral atom and negative ion respectively. In both the cases illustrated in figure 7.3 we note the important relation

$$E_k = E_e - D + E_a, \qquad (7.8)$$

where D is the dissociation energy of the molecule AB and E_a the electron affinity B. Thus, if E_i is the kinetic energy of the ion

$$E_i = \gamma\{E_e - D + E_a\}. \qquad (7.9)$$

It follows that, if E_i can be measured for a given E_e, $D - E_a$ can be obtained. Usually D will be known from chemical data, so that the less readily determined electron affinity of the atom B can be derived.

In practice, allowance must be made for the thermal motion of the gas molecules. Because of this, the ion energy for given E_e will not be given exactly by (7.9), but there will be a distribution of ion energy depending on the gas temperature. The most probable ion energy, however, is still given by (7.9).

It is assumed in deriving (7.9) that the atom A is formed in the ground state. If, however, it is formed in an excited state with excitation energy E_x then (7.9) is replaced by

$$E_i = \gamma\{E_e - D + E_a - E_x\}. \qquad (7.10)$$

For simplicity, we have confined the discussion of these dissociation processes to collisions with diatomic molecules. Many more possibilities arise when the molecules are polyatomic so that the corresponding rate coefficient may be very large. In particular, attachment of electrons with very low kinetic energy may occur with rate coefficients as high as 10^{-13} to 10^{-14} cm^3 s^{-1}. Use may be made of the high rates for some molecules to develop very sensitive methods indeed for detecting these molecules in very small concentrations, as described in more detail on page 135.

We next describe the experimental methods which have been used to measure the rates of recombination and attachment. After discussing some of the typical results obtained, we conclude with a brief account of the application of electron attachment to the detection of minute concentrations of DDT and other pesticides. Applications to the Earth's atmosphere are discussed in Chapter 14.

7.4. *The Experimental Study of Dissociative Recombination*

Laboratory measurement of recombination rates is difficult because neither of the reacting species, the electrons nor the positive ions, can be maintained in bulk concentration to provide a high density of target material for a beam experiment.

Almost all experiments have been based on the observation of the decay of electron concentration with time in a discharge afterglow. When an electric discharge is generated in a gas a considerable concentration of free electrons and of positive ions is built up, together with a variety of excited species of

Collisions Involving Electron Capture

neutral atoms and molecules and, in certain cases, also negative ions. Once the exciting power is cut off, an appreciable time, of the order of milliseconds, elapses before the gas returns to its normal unexcited condition. This period is referred to as the *afterglow*, because light continues to be emitted, albeit with rapidly falling intensity, during this stage.

There are three principal sources of electron loss in the afterglow—recombination with positive ions, diffusion to the walls of the containing vessel and attachment to form negative ions if an attaching gas is present. Assuming that means are available for measuring the rate of decrease of electron concentration with time, how can the three loss processes be separately distinguished and their rates measured?

If n_e is the concentration of free electrons at an early time in the afterglow and α the coefficient for recombination to the positive ions present, then, if there were no other loss process,

$$\frac{dn_e}{dt} = -\alpha n_e n^+. \tag{7.11}$$

Moreover, in a plasma such as exists in the active discharge or the afterglow, n_e must equal n^+. This follows because, if in any region n_e were in excess, the net negative charge due to this would draw in positive ions until the balance were redressed and vice versa. We therefore have

$$\frac{dn_e}{dt} = -\alpha n_e^2, \tag{7.12}$$

or

$$\frac{dt}{dn_e} = -\frac{1}{\alpha n_e^2}.$$

This is readily integrated to give

$$t = \frac{1}{\alpha n_e} + \text{constant}. \tag{7.13}$$

If at the initial time, which we may take as $t = 0$, n_e has the value $n_e(0)$, the constant in (7.13) must be $-1/\alpha n_e(0)$, so

$$t = \frac{1}{\alpha}\left\{\frac{1}{n_e} - \frac{1}{n_e(0)}\right\}. \tag{7.14}$$

Thus the characteristic feature of loss by recombination is that the reciprocal of the electron concentration varies linearly with the time. If this is so, the slope of a plot of $1/n_e$ against t will have a slope equal to α. It is to be noted also that if the recombination mechanism is either radiative or dissociative, α will be independent of the gas pressure.

Loss by diffusion to the walls is a little more complicated to analyse, but it results in an exponential decay of the electron concentration with time in the afterglow

$$n_e = n_e(0) \exp\{-\beta t\}, \qquad (7.15)$$

where β is proportional to the effective diffusion coefficient and hence is inversely proportional to the gas pressure p. β also depends on the geometry of the container.

Finally, if electron loss by attachment predominates then

$$\frac{dn_e}{dt} = -\eta n_e n, \qquad (7.16)$$

where η is the attachment coefficient to the neutral molecules which are in concentration n. (7.16) may be integrated in a similar way to (7.12) to give

$$t = -\eta \ln n_e + \text{constant},$$

so

$$n_e = n_e(0) \exp\{-\eta n t\}. \qquad (7.17)$$

This is of the same exponential form as for diffusion loss, but now the exponent ηn is proportional to the pressure, in contrast to the diffusion case.

If conditions of pressure, container geometry etc. are suitably adjusted so that it is found that $1/n_e$ is a linear function of t over a large enough range of $1/n_e$, then it can be assumed that recombination is the dominant rate of loss and α determined. The proviso 'over a large enough range of $1/n_e$' is an important one in practice because, over a limited range, a linear variation with t may be simulated by diffusion loss. In a cylindrical container, n_e must range through a factor of four or more over the region in which $1/n_e$ varies linearly with t to within 2%. Failure to meet this condition vitiated the results of a number of early experiments.

Even if conditions may be obtained in which recombination dominates, a number of other problems must be overcome before reliable and useful data may be obtained. For application to the interpretation of other physical phenomena, it is necessary to have data on recombination to specific positive ions. If, for example, we study an afterglow in argon, we cannot assume that the positive ions will be mainly Ar^+. In fact, under afterglow conditions in pure argon, the ions are normally Ar_2^+. This makes a great deal of difference because, whereas Ar_2^+ may recombine through the dissociative process, the much slower radiative process is alone possible for Ar^+. Again, the rate of dissociative recombination to O_2^+ is required for ionospheric applications (see Chapter 14, p. 264), but can we be sure that the principal positive ions in an afterglow in oxygen will be O_2^+? Experience shows that we certainly cannot, and special conditions must be realized to avoid predominance of polyatomic ions O_3^+ and O_4^+.

Collisions Involving Electron Capture

One way of reducing the importance of these clustered ions is to work with the gas at a low pressure, because the rate at which clustering occurs will be proportional to at least the first power of the pressure. However, if this is done, the importance of diffusion loss is increased, so that it is rarely possible to achieve conditions in which recombination predominates while clustering is negligible.

For this reason, experiments are carried out using a buffer gas. This is a gas such as helium or neon with high ionization energy. If, for example, a discharge is produced in helium containing a small concentration of argon, the positive helium ions rapidly transfer their charge to the argon to produce Ar^+,

$$He^+ + Ar \rightarrow He + Ar^+.$$

This is because energy is released in such a process, the ionization energy of Ar (15·76 eV) being less than that of He (24·58 eV). By maintaining the helium pressure high enough, diffusion loss is rendered negligible, while at the same time the partial pressure of argon may be maintained so low that diatomic ions Ar_2^+ are not formed in appreciable concentration. $HeAr^+$ ions are very much less stable, if at all, and do not complicate the situation. In this way, an afterglow may be obtained in which the ions are almost exclusively monatomic Ar^+.

The same technique may be applied with molecular gases such as O_2 and N_2 so as to obtain afterglows in which O_2^+ and N_2^+ respectively are the dominant species. However, the situation is not so simple as for the monatomic gases because the molecular ions formed by charge transfer

$$He^+ + N_2 \rightarrow He + N_2^+$$
$$He^+ + O_2 \rightarrow He + O_2^+$$

may possess excess vibration. In that case, the experimental results refer to recombination to such ions, whereas, in applications, ions in their lowest vibrational state may alone be involved. For O_2 there is the further problem that the O_2^+ ions resulting through charge transfer from He^+ may be in an excited electronic state. It was found empirically that the best results for O_2^+ were obtained with an $Ne-O_2$ mixture containing a partial pressure of krypton (see p. 123).

Because of the complexity of the processes which might be occurring in an afterglow, it is important to have an independent means of checking that the dominant positive ions are indeed those for which the recombination coefficient is to be measured. In many experiments, therefore, the positive ions were sampled during the afterglow period through a fine hole in the wall of the containing vessel. Positive ions drifting through the hole entered a mass analyser (Appendix 7) which operated with high time resolution so the time variation of the sampled current of each kind of positive ion observed could be measured and compared with that of the electrons (see figures 7.6 and 7.8).

Early experiments, carried out before World War 2, gave inconclusive results, as the complexity of ionic reactions in the afterglow was not realized. However, soon after the war, advantage was taken of the major developments which had occurred in microwave techniques due to the wartime importance of radar. Coupled with the development of high time-resolution electronics, this made it possible to study the time variation of various species in the afterglow much more thoroughly and reliably than hitherto. One of the major new features introduced was the technique for monitoring the electron concentration.

The natural frequency v_0 of electrical oscillation of a cavity with walls of some insulating material such as quartz, is changed when a mean concentration $\bar{n}_e$ of free electrons is produced within it, by an amount

$$\Delta v_0 = A\bar{n}_e/v_0, \tag{7.18}$$

where A is a constant depending on the distribution of electron density and microwave electric field within the cavity. If Δv_0 can be measured as a function of time in the afterglow, we have a means of determining the time variation of $\bar{n}_n$.

A convenient method for determining the resonant frequency at any time is to measure the intensity of reflection of microwaves of adjustable frequency by the cavity plus afterglow. When the frequency of the microwaves is equal to the resonance frequency, the reflected intensity is a minimum or, in other words, the microwave power transmitted through the cavity is a maximum.

To apply this technique conveniently, the discharge in the cavity is initiated by application of a powerful microwave pulse. This may be repeated at regular intervals, long enough for the afterglow to decay completely after one pulse before the next begins. For monitoring the mean electron concentration, a probing microwave signal of low power is used, so as not to disturb the afterglow. With a fixed frequency of this signal, the time in the afterglow at which its transmission through the afterglow is a maximum, is observed. This gives the resonance frequency and hence the electron concentration at that time. The probing frequency is then changed and the transmission remeasured in a subsequent afterglow period so that, because of the reproducibility of the afterglow conditions after successive pulses, it is possible to obtain $\bar{n}_e$ as a function of time in the afterglow.

This is not exactly what is required, because in the analysis outlined above we have assumed that n_e is uniform throughout the afterglow. In practice, any very marked departures from uniformity will be smoothed out rapidly by diffusion so that, with an initiating pulse such that the applied electric field does not vary much across the cavity, near-uniform conditions are usually obtained very rapidly.

In typical cases, the pulse initiating the discharge in the cavity lasts for a time which may be adjusted from 10 μs to a few ms and repeated between 10 and 100 times per second. However, when working with gases which attach electrons to form negative ions, the repetition rate must be drastically reduced so as to avoid serious accumulation of negative ions—the afterglow conditions are such that such ions tend to remain within the volume of the cavity rather than diffusing to the walls.

In most applications, it is desirable to have information about the way the recombination coefficient varies with the temperature, over a considerable range. It is possible with suitable design to be able to heat the experimental cavity plus gas to temperatures of 500 K or so, but in order to obtain measurements up to 3000 K or so it is necessary to use shock heating and study the decay of an afterglow produced from a discharge in the heated gas travelling behind a shock wave.

By means of microwave heating, it is possible to raise the temperature T_e of the electrons without raising that, T_g, of the gas. It is to be understood in this context that the electron temperature defines the mean energy of the electrons as $\tfrac{3}{2}kT_e$, where k is the Boltzmann constant (see Appendix 1). Thus, if a microwave electric field of strength F and frequency v is applied to the cavity containing the afterglow, the electron temperature is raised to T_e, where

$$T_e = T_g + \tfrac{1}{6}M(eF/2\pi mv)^2, \qquad (7.19)$$

M being the mass of a gas atom and m that of an electron.

Figure 7.4 shows the general arrangement of a typical experiment to measure the recombination coefficient in a neon afterglow. Provision is made for three

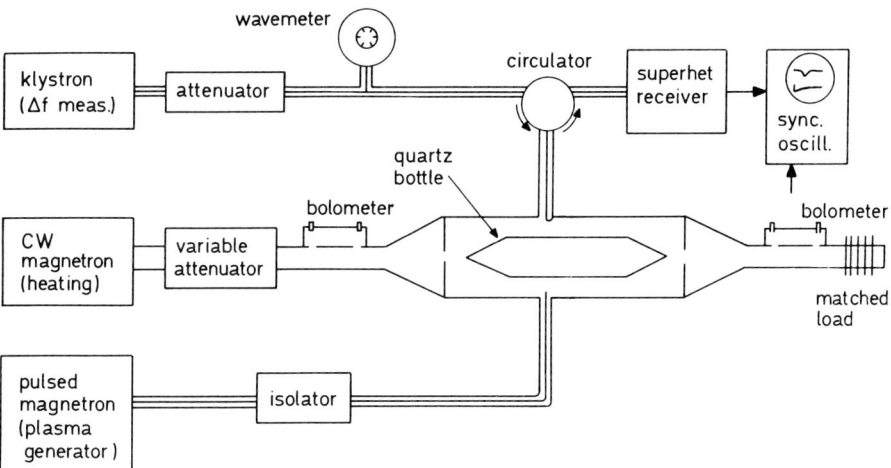

Figure 7.4. The general arrangement of a recombination experiment. The pulsed magnetron (plasma generator) supplies the power to produce breakdown of the gas in the quartz bottle while the klystron (Δf measuring) supplies the signal for observing the reflectivity of the plasma at different stages of the afterglow. Finally the continuous wave (CW) magnetron provides the power for raising the electron temperature in the plasma.

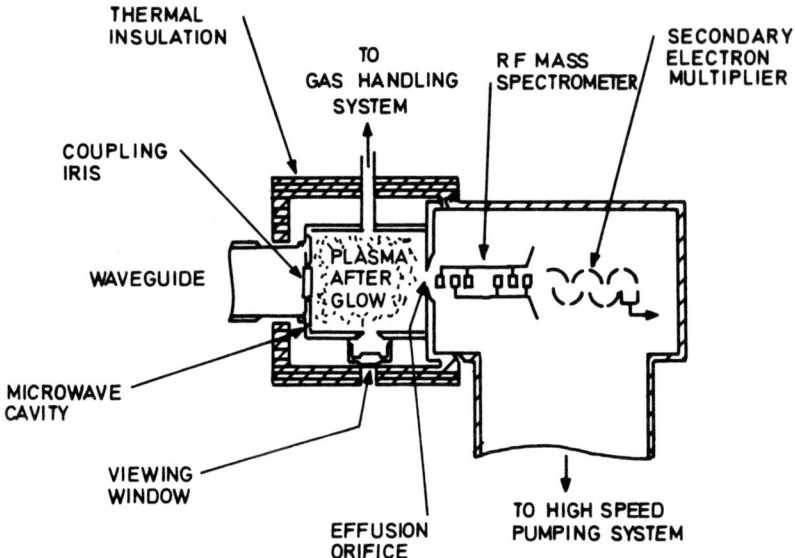

Figure 7.5. Typical arrangement of appratus in an afterglow experiment on recombination, using a radio-frequency mass spectrometer to analyse the positive ion composition as a function of time in the afterglow.

microwave power sources, the strongest generating the breakdown pulses, the weakest providing the probing signals and the intermediate one the electron heating. In this experiment, no mass analyser was included. Figure 7.5 shows the general arrangement of an experiment in which time-resolved mass analysis is carried out using a radio-frequency mass spectrometer (see Appendix 7) through which the current transmitted was detected by an electron multiplier. The orifice in the cavity wall through which the ions passed out to the spectrometer was about 0·25 mm in diameter.

7.5. Results of Recombination Measurements—The Rare Gases

A great deal of work has been carried out in a helium afterglow but, as it happens that helium is the only rare gas in which dissociative recombination is unimportant, we shall consider it briefly later.

The most extensive experiments have been carried out with neon and argon. Thus, at a quite early date, in 1951, Biondi found that, while the measured recombination coefficient in pure argon at room temperature was quite high (7×10^{-13} m^3 s^{-1}), for helium containing 0·1% of argon it was too small to be measured. This is what would be expected if Ar_2^+ ions are responsible for the results in pure argon.

Experiments using time-resolved mass analysis confirmed that, in afterglows in pure neon and argon, the positive ions are diatomic Ne_2^+ and Ar_2^+ respective-

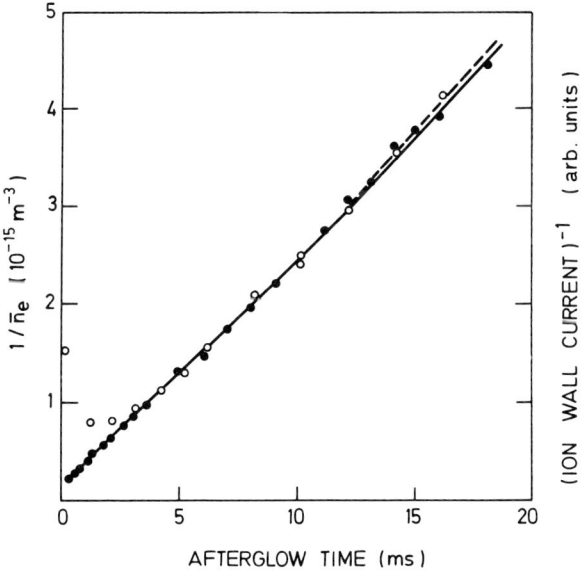

Figure 7.6. Comparison of the variation of the reciprocals of the mean electron concentration $\bar{n}_e$ and of the wall current j^+ of $Ne_2{}^+$ ions, with time in an afterglow in neon, as observed by Kasner.

●, $1/n_e$; ○, $1/j^+$.

ly. Moreover, the measured current of the diatomic ions decayed at the same rate as the electrons, as seen in typical results for neon shown in figure 7.6.

If the process of recombination is dissociative, as described above, excited neon atoms should be produced in the recombination process with kinetic energies considerably greater than the mean value at the gas temperature (see (7.7)). If these excited atoms radiate their surplus energy before there is time for it to be lost in collisions with rare gas atoms, there should be an appreciable Doppler shift of frequency due to their excess velocity. When various factors are taken into account, such as averaging over the directions of motion of the emitting atoms, the variation of the intensity of an emitted 'line' with wave number which would be expected is shown in figure 7.7 for a typical transition in neon. In practice, account must be taken of the thermal motion of the atoms and of the possibility that the excited atoms concerned might be formed from some process other than dissociative recombination. Figure 7.7(a) then shows the type of 'line' shape expected when these factors are included. This is to be compared with the shape for the line of wavelength 585·2 nm emitted from the neon afterglow as observed in a remarkable experiment by Frommhold and Biondi in 1969 (see figure 7.7(b)). The strong similarity with the predicted shape provides strong evidence for the dissociative mechanism of recombination.

Similar evidence was obtained for argon but not for helium. It seems that,

Atomic and Molecular Collisions

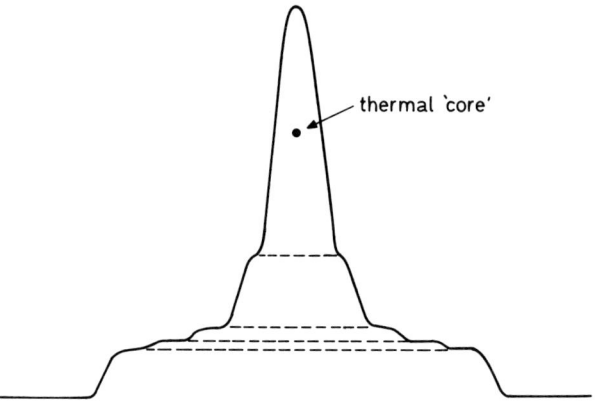

Figure 7.7. (a) Expected shape of a spectrum line emitted from an afterglow by excited atoms which are produced by dissociative recombination. The central peak is due to excited atoms with a thermal velocity distribution while the stepped pedestal is due to fast excited atoms arising from recombination. The step structure arises because the velocity with which a particular excited atom is produced depends on the state of excitation of the other atom which is simultaneously formed. Each such velocity gives rise to a step in the pedestal.

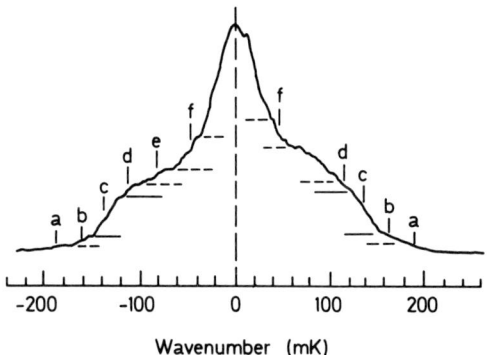

(b) Shape observed by Frommhold and Biondi for the spectrum line of neon at 585.2 nm emitted from a neon afterglow.
The horizontal scale is in reciprocal wavelength or wave number in units milli Kayser or 10^{-3} cm^{-1}.

for this gas, recombination does not proceed through the dissociative process. The observed value of α in the afteglow experiments is much smaller than for the other rare gases and it is likely that the basic process involved is one in which the surplus energy released by recombination is carried off by a second free electron

$$He^+ + e + e \rightarrow He' + e \qquad (7.20)$$

an example of a three-body process.

Collisions Involving Electron Capture

7.6. Results of Recombination Measurements—Molecular Gases

For applications, the values of α for N_2^+, O_2^+ and NO^+ are of special interest. Figure 7.8 illustrates results obtained by Kasner and Biondi for an afterglow at room temperature in neon, at a pressure of 19 torr, containing a partial pressure of $2\cdot 5 \times 10^{-4}$ torr of nitrogen. Under these conditions, the N_2^+ ion is dominant though N_3^+ and N_4^+ are observable. The rate of fall of n_e with time in the afterglow follows closely that for N_2^+ at afterglow times greater than 3 ms. From these data, α was found to be $2\cdot 7 \times 10^{-13}$ m^3 s^{-1} but, as mentioned on p. 117, there is some uncertainty about the degree of vibrational excitation of the N_2^+ concerned. For O_2^+, for reasons explained on p. 117, it is even more difficult to ensure that the O_2^+ ions are in their lowest electronic and vibrational states. Values of α at room temperature close to 3×10^{-13} m^3 s^{-1} have been obtained. For NO^+ the results are less ambiguous and give α as $4\cdot 1 \times 10^{-13}$ m^3 s^{-1}. The variation of these coefficients with T_e has been observed, as also for N_2^+, but the dependence on T_g is still not well established.

We shall refer to these observed results in Chapter 14, where they are considered in relation to electron loss in the Earth's ionosphere.

The subject of recombination is a very extensive one and we have concentrated on one or two aspects. In particular, in plasmas which are much denser than those typical of afterglow experiments, recombination essentially involves the process (7.20). It would be out of place here to discuss the extensive experi-

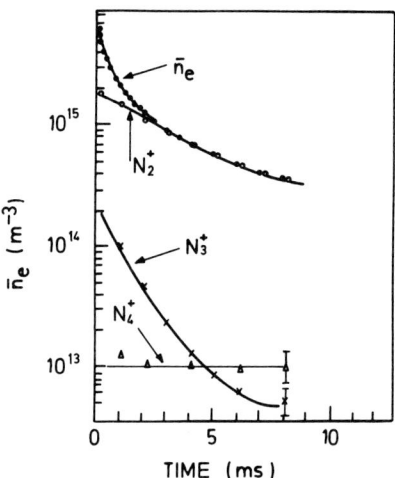

Figure 7.8. Comparison of the decay rates of the mean electron concentration n_e and the wall currents of different positive nitrogen ions, as indicated, in an afterglow in a mixture of neon and nitrogen, as observed by Kasner and Biondi. The scale for the ion currents is adjusted so that for N_2^+ agrees with $\bar{n}_e$ at times later than 4ms.

7.7. The Experimental Measurement of Attachment Rates

The techniques employed for the experimental investigation of attachment rates depend on the energies of the electrons involved. Unlike recombination, beam methods may be used to measure rates for electrons with energies above 0·5 eV or so. Using these methods, the variation of the attachment cross-section with electron energy may be measured, as well as the nature, kinetic energies and angular distributions of the ions produced. For attachment at much lower energies, it is necessary to work with a swarm of electrons possessing an energy distribution about some mean value. In such experiments, the mean attachment rate is measured as a function of mean electron energy.

The afterglow technique is one example of how the latter type of experiment may be carried out for electrons of mean energy corresponding to room temperature. There are considerable difficulties in ensuring reliable and definite results, however, because of the variety of possible reactions which can occur. Nevertheless a number of experiments have been carried out in which electron loss has been identified as due to attachment.

Most swarm experiments have depended on studying the motion of a swarm of electrons drifting through a gas at pressure p under the influence of a constant electric field F. This motion has been discussed in Chapter 5, p. 69, for cases in which no attachment occurs and it was shown that the mean energy is a function of F/p at a fixed gas temperature. Let β be the chance that an electron of the swarm will attach to a gas molecule in drifting a distance δx in the direction of the electric field. This will change the electron current I by an amount δI, where

$$\delta I = -I\beta\delta x. \tag{7.21}$$

Hence, if I_0 is the current at some initial point and I that at a distance x from it downstream, along the direction of the field, we have, by integrating (7.21),

$$I = I_0 \exp(-\beta x). \tag{7.22}$$

Most swarm methods depend on the measurement of β as a function of F/p. To relate β to the mean attachment cross-section $\bar{Q}_a$, we note that, in passing a distance δx in the direction of the field, an electron actually traverses a much greater distance due to its random motion. Thus if u is the drift velocity (see Chapter 5, p. 69) and c the random velocity, the actual path traversed will be $(c/u)\delta x$. It follows that, if n is the number of neutral molecules per unit volume,

$$\beta = n\bar{Q}_a c/u. \tag{7.23}$$

The velocities c and u may be measured by standard techniques (see p. 71), so that, from measurement of β, $\bar{Q}_a$ may be derived.

Collisions Involving Electron Capture

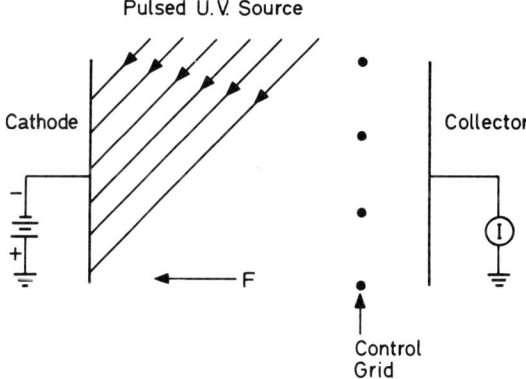

Figure 7.9. Illustrating the principle of the pulse method for measuring attachment rates.

Several different methods for measuring β have been devised and used with success. We select for description a pulse method due to Doehring and developed by Chanin, Phelps and Biondi.

The principle of this method is as follows (see figure 7.9). A pulse of electrons, released from a suitable cathode surface by a pulse of ultraviolet light, is allowed to drift through the gas under study, under the action of a uniform electric field F, to be collected at an anode at a distance d from the cathode. During their passage, some of the electrons will form negative ions by attachment to the gas molecules. These ions, because of their much greater mass, will drift about 1000 times more slowly than the free electrons. This means that, after an electron pulse is collected at the anode, a further delayed current will arrive due to the ions. Figure 7.10 illustrates the typical form of the signal received at the anode in the experiments of Chanin, Phelps and Biondi, showing this effect.

If u_i is the drift velocity of the negative ions, the ion current received at the anode at a time t after the pulse will have been produced at a distance $u_i t$ from it. At this distance, the electron concentration in the pulse will have been reduced from its initial value $n_e(0)$ to $n_e(x)$, where

$$n_e(x) = n_e(0) \exp(-\beta x), \qquad (7.24)$$

where x, the distance from the cathode, is equal to $d - u_i t$. Thus

$$n_e(x) = n_e(0) \exp\{-\beta d + \beta u_i t\}. \qquad (7.25)$$

Since the ions observed at time t are formed from these electrons, it follows that the ion current will vary as $\exp(\beta u_i t)$ provided t is not greater than the time taken for a negative ion to drift the whole distance d. Thus the shape of the delayed pulse due to the negative ions should rise exponentially with delay time up to a certain time t_c, after which it should drop rapidly to zero. From the

Atomic and Molecular Collisions

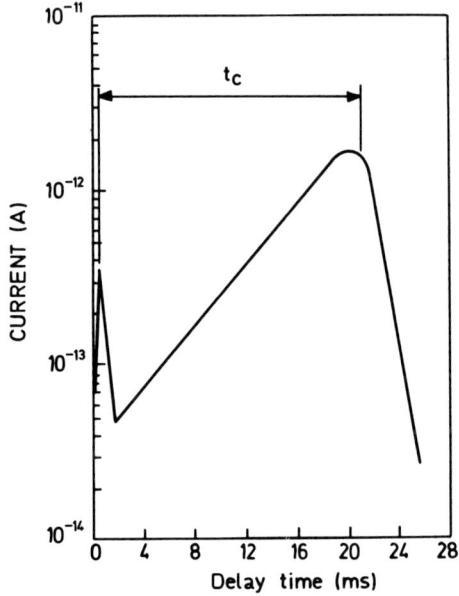

Figure 7.10. Typical form of the signal received at the anode in pulse experiments for measuring attachment rates.

exponential rise, βu_i may be obtained, while, from t_c, u_i may be derived separately.

It is assumed in this analysis that, once a negative ion is formed, it has a negligible chance of losing its captured electron before it reaches the anode. The technique has been extended to take account of electron detachment when it is significant.

We now turn to consider the way in which measurements have been carried out using electron beams. The total attachment cross-section may be obtained as a function of electron energy using methods similar to those described in Chapter 6, p. 103 for measuring total ionization cross-sections for electron impact.

Thus, referring to figure 6.11, if the electron energy is less than the threshold for ionization but negative ions may be formed by dissociative attachment at a lower energy, then the negative current received by the positive plate electrode JC will be a measure of the attachment cross-section. However, the currents will normally be much smaller than in ionization experiments because the dissociative attachment cross-section is usually much less than that for ionization, except very close to the threshold. Also, the negative ions formed in dissociative attachment may possess a kinetic energy of some electron-volts so that the potential difference necessary between the collecting plates to saturate the collection of the ions will need to be correspondingly large.

Most beam experiments provide information about the variation of the attachment cross-section with electron energy rather than its absolute value. In recent experiments of this kind, however, provision has been made for the measurement of the energy distributions of the negative ions produced by electrons of a particular kinetic energy. As described on p. 114, the most probable ion energy, allowing for the thermal distribution of velocities of the target molecules, is given by

$$\gamma\{E_e - D + E_a - E_x\}, \qquad (7.26)$$

where γ is the ratio $M_n/(M_n + M_i)$ of the mass of the neutral atoms produced to that of the target molecule, E_e is the electron energy, D the dissociation energy of AB, E_a the electron affinity of B and E_x the excitation energy, if any, of the product atoms.

Figure 7.11 illustrates the arrangement in a typical modern experiment on these lines, that used by Chantry and Schulz. The chamber CC contains the gas under investigation at a suitable pressure. This must be low enough to ensure not only that electrons in the beam B fired through the chamber make at most one collision in passage, but also that the product ions and atoms do not interact further with the gas before passing out of the chamber. Negative ions formed are allowed to escape through an aperture A in the side of the chamber. To increase the efficiency of collection of the ions, the plate R is maintained at a negative potential with respect to the electron beam so as to repel ions towards the aperture. To avoid penetration of electric fields from the subsequent energy-analysing system, the aperture is covered by a grid G.

The energy analysis employs what is known as a Wien filter (Appendix 4). Between the plates F_A and F_B, symmetrically placed with respect to the aperture A, a uniform electric field F is maintained while the region is bathed in a uniform magnetic field B at right angles to F and to the plane of the paper. Only ions entering at A in a direction parallel to the plates with velocity $v = F/B$ will pass undeflected through the system. To correct for small deviations in

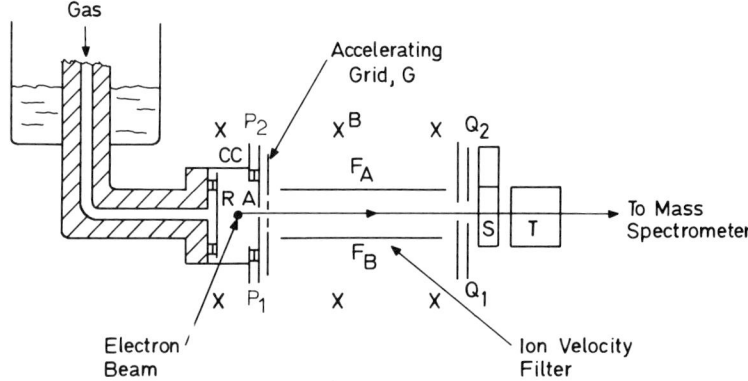

Figure 7.11 Arrangement of apparatus used by Chantry and Schulz for studying dissociative attachment.

the motion of the ions due to the magnetic field encountered before entering the filter, a small electric field is applied between the plates P_1 and P_2 and adjusted to minimize the current passing through the filter. A similar pair of plates Q_1 and Q_2 is provided at the output aperture of the filter. The ions then pass through the electrode system S which cuts off ions travelling in directions making angles greater than about 8° with the incident direction in the plane perpendicular to the paper. Finally, the ions issuing from S are accelerated to a suitable energy for analysis in a mass spectrometer.

7.8. Some Results of Attachment Experiments—Dissociative Attachment in O_2

Figure 7.12 shows the cross-section for production of negative ions in O_2 as a function of electron energy as observed by Rapp and Briglia with the apparatus described in Chapter 6, p. 103. It will be seen that the energy variation is of a quite different form from that for ionization (Chapter 6, figure 6.12), resembling a resonance peak, which it essentially is. The maximum cross-section is only $1 \cdot 4 \times 10^{-22}$ m², less than 1/100 of that for ionization.

Using the equipment described on p. 127 (see figure 7.11), Chantry and Schulz have measured the energy distributions of the negative ions produced in O_2. Measurements were made both at room temperature and with the inlet gas cooled in liquid nitrogen. The results illustrated in figure 7.13 show that the distributions are narrower at the lower temperature—a major contribution

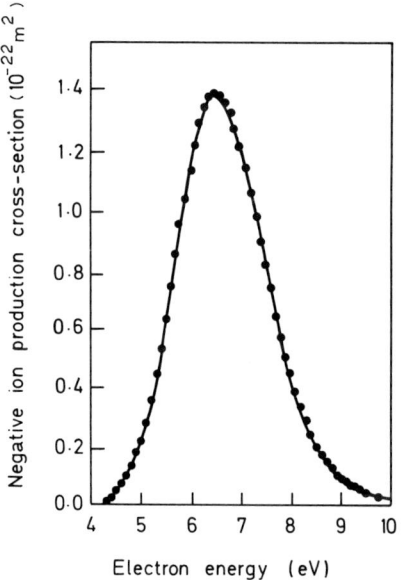

Figure 7.12. Cross-section for production of negative ions by dissociative attachment of electrons to O_2 molecules, as observed by Rapp and Briglia.

Collisions Involving Electron Capture

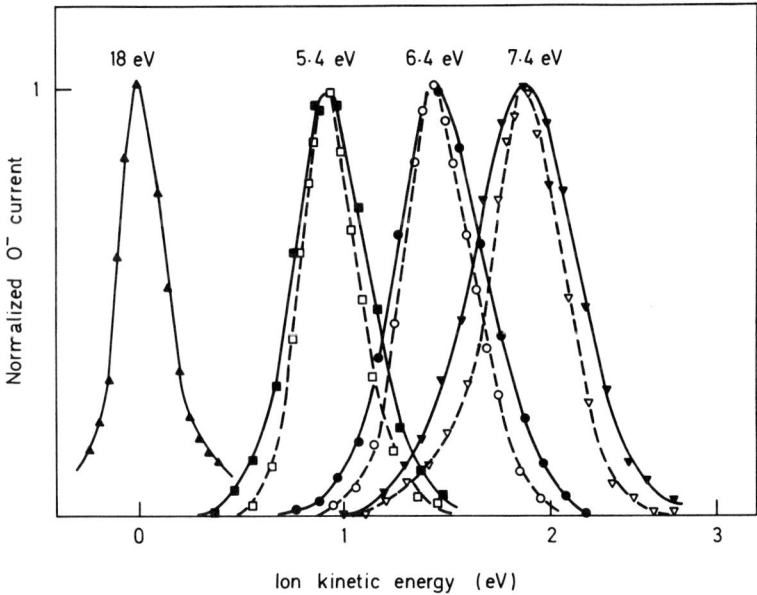

Figure 7.13. Energy distributions of negative ions produced in O_2 by electrons of different energies as indicated above each distribution.
—▲—, —■—, —●—, —▼—, observed at room temperature;
—△—, —□—, —○—, —▽—, observed with inlet gas cooled to liquid nitrogen temperature.

to the energy spread of the ions comes from the temperature motion of the target molecules. Figure 7.14 shows a plot of the most probable ion energy as a function of electron energy. This is seen to be a straight line of slope $\frac{1}{2}$ as predicted by (7.9), γ being $\frac{1}{2}$ in this case (see (7.7)). At zero electron energy, it extrapolates to 3·6 eV, which, according to (7.9), should be the difference $D(O_2) - E_a(O)$ between the dissociation energy of O_2 and the electron affinity of O. From spectroscopic measurements $D(O_2) = 5·09$ eV and $E_a(O)$, from photodetachment observations (see Chapter 13, p. 252), is 1·46 eV, so that the difference is 3·63 eV, which agrees very well.

Attachment occurs also at much lower impact energies as observed by swarm techniques. The reaction involved is a three-body one

$$O_2 + e + M \rightarrow O_2^- + M, \qquad (7.27)$$

where M is a second molecule which, in pure oxygen, is one of O_2. It leads to the production of molecular negative ions O_2^- instead of the atomic ions O^- which result from dissociative attachment. Because a second molecule is required, the rate of the process, in pure O_2, is proportional to the square of the O_2 concentration. Referring to (7.23) the mean attachment cross-section for a drifting swarm of electrons is given by

$$\bar{Q}_a = \beta u/nc, \qquad (7.28)$$

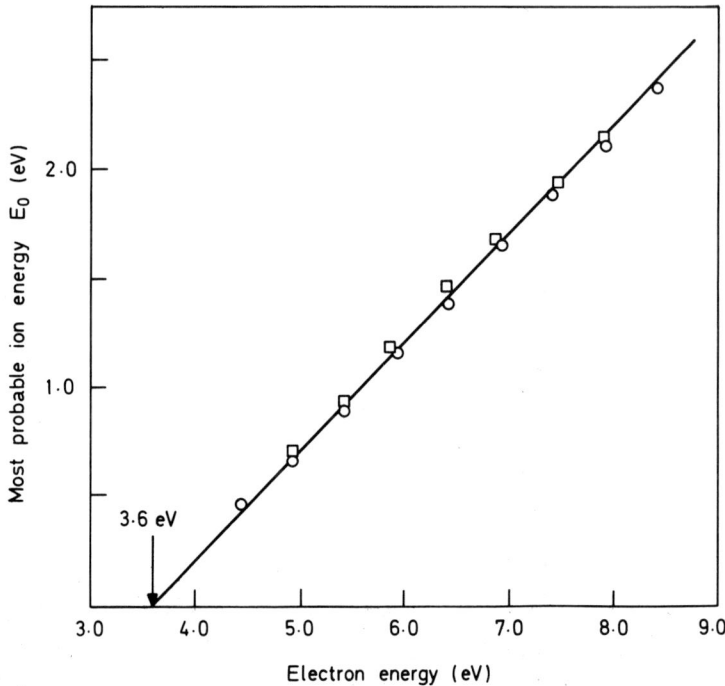

Figure 7.14. Most probable observed negative ion energy in O_2 plotted as a function of the energy of the electrons producing the ions. □, ○, observed values in two separate sets of observations by Chantry and Schulz.

where β is the attachment probability per unit path length in the direction of the electric field F, n is the molecule concentration, u and c the drift and random velocities. At a given temperature, n is proportional to the gas pressure p, while u/c is a function of the ratio F/p. A plot of β/p against F/p will therefore represent the variation of Q_a with F/p. If Q_a is independent of p, the same plot should be obtained for all p. Figure 7.15 shows such plots obtained by Chanin, Phelps and Biondi from their observations in O_2, using the pulse method described on p. 125. It will be seen that, for low values of F/p, the plots are not coincident but parallel and, at a given F/p proportional to p. These separate plots merge into a single plot at sufficiently high F/p. This behaviour is due to the fact that, at low F/p and hence low mean electron energy, attachment occurs through (7.27) $\bar{Q}_a$ proportional to p but as the mean energy increases two-body dissociative attachment, with Q_a independent of p, gradually dominates.

The reaction (7.27) is important in the lower ionosphere and in other conditions in which slow electrons are concerned. At 300 K, the three-body attachment rate coefficient in pure O_2 is $2 \cdot 7 \times 10^{-36}$ m^3 s^{-1}. If water vapour is present, the molecule M may be H_2O, in which case the rate coefficient is increased to 2×10^{-35} m^3 s^{-1}. This is probably the reason why negative ions are readily

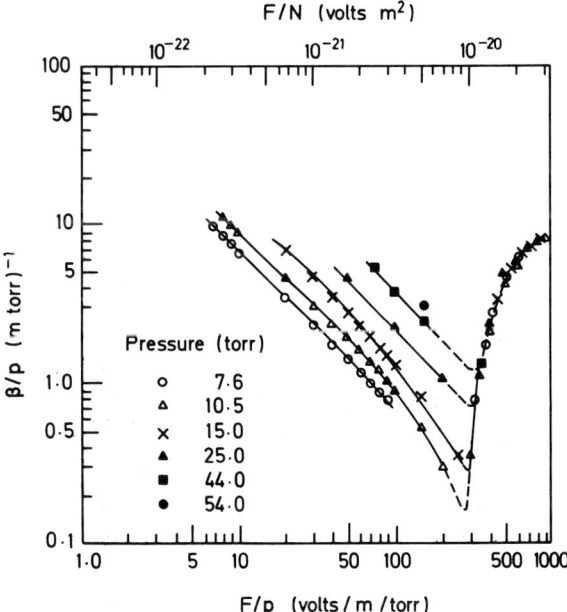

Figure 7.15. Observed variation of β/p with F/p for electrons drifting in oxygen at a pressure p at room temperature under the action of a uniform electric field F, β being the attachment probability per unit path length in the direction of the field. At high F/p the excitation is independent of the pressure p, but at lower values of F/p, β increases with p. This shows that a three-body process is involved under these conditions.

formed in moist air, rather than by any direct attachment of electrons to H_2O molecules.

7.9. Dissociative Attachment in CO

The attachment cross-section for CO, as observed by Rapp and Briglia, is shown in figure 7.16. It is of similar form to that for O_2, though the peak cross-section is somewhat smaller. On the same diagram, the variation of the cross-section with electron energy observed by Stamatovic and Schulz in later experiments is also shown, the peak being adjusted to agree with that measured by Rapp and Briglia. In these experiments, the energy spread in the electron beam was considerably smaller and detailed structure is apparent in the cross-section curve. Thus there is evidence of some new process beginning at an electron energy close to 11 eV.

Measurements of the energy distribution of the ions, made by Chantry, confirmed that the main peak is due to the process

$$CO + e \rightarrow C + O^-, \tag{7.29}$$

in which the neutral atom and O^- ion are formed in their normal states. Also,

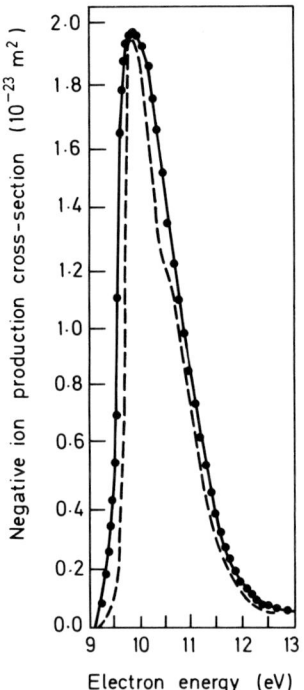

Figure 7.16. Cross-sections for production of negative ions by dissociative attachment of electrons to CO molecules.
—●—, absolute values observed by Rapp and Briglia;
---, observed by Stamatovic and Schulz. The absolute value has been adjusted to agree with that measured by Rapp and Briglia at the maximum.

the process with threshold near 11 eV was identified as one in which the carbon atom was formed in the first excited state with energy 1·25 eV. The peak cross-section for this was found to be $9·5 \times 10^{-23}$ m^2.

As C$^-$ ions are stable, the process

$$CO + e \rightarrow C^- + O \qquad (7.30)$$

must also be possible. C$^-$ ions were first observed to arise from dissociative attachment in CO, by Lagergren in 1955. It was not until 15 years later that Stamatovic and Schulz were able to measure the cross-section, which is as low as 6×10^{-25} m^2.

7.10. Attachment to Polyatomic Molecules—Nitrous Oxide (N$_2$O)

We mention here the results of measurements made for attachment to N$_2$O because they are remarkable in exhibiting a very strong dependence on temperature. Furthermore, the attachment cross-section for N$_2$O at 1000 K for very low energy electrons is greater than 10^{-19} m^2, very large compared with the

Collisions Involving Electron Capture

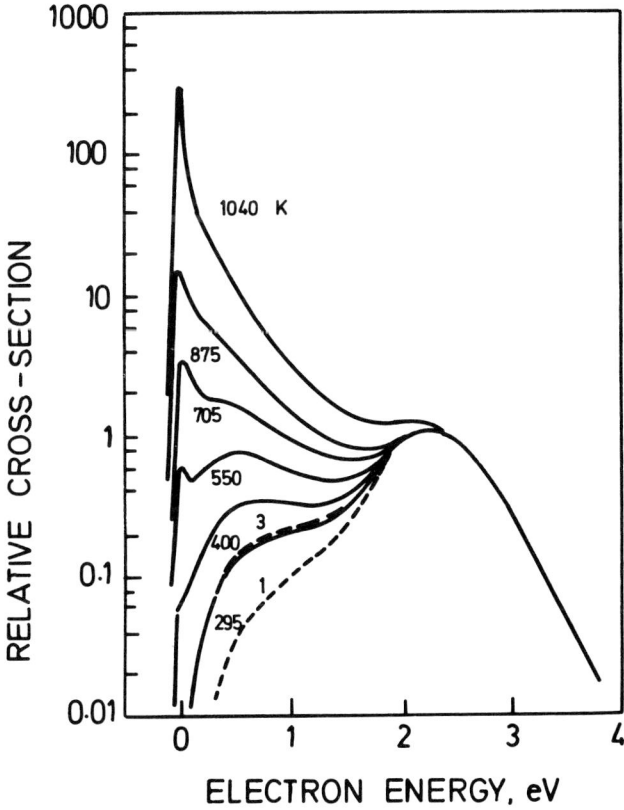

Figure 7.17. Cross-sections for attachment of electrons in N_2O, measured at different temperatures as a function of electron energy.

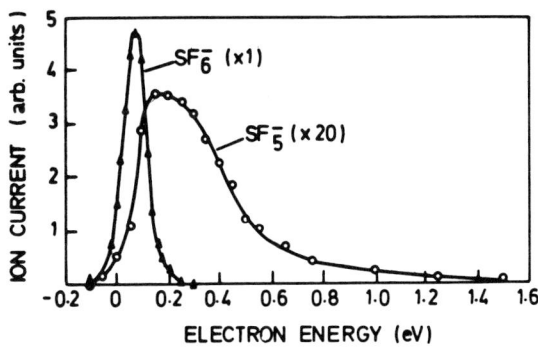

Figure 7.18. Currents of SF_6^- and SF_5^- ions produced by electron attachment in SF_6, as a function of electron energy. The ion currents are proportional to the cross-sections for formation of the respective ions.

peak values for the diatomic molecules O_2 and CO and for many other molecules.

Figure 7.17 shows the variation of the total negative ion currents with electron energy in N_2O at different temperatures, observed by Chantry. The great sensitivity to temperature is evident and is almost certainly due to the fact that attachment occurs much more readily to vibrationally excited N_2O than to the unexcited molecule.

7.11. Attachment to Sulphur Hexafluoride (SF_6)

This molecule is highly symmetrical and relatively inert chemically. While it undergoes the dissociative attachment reaction

$$SF_6 + e \to SF_5^- + F, \tag{7.31}$$

with a cross-section varying with electron energy as shown in figure 7.18, at very low electron energies, the undissociated ion SF_6^- is formed (see figure 7.18) with a high effective cross-section. Direct attachment of an electron will form a complex SF_6^- which, if left alone, would eventually break up by autodetachment, releasing the electron and restoring the neutral molecule. However, it appears that the lifetime of the complex initially formed is quite long, so, under experimental conditions, there is a high probability that the complex loses sufficient energy in collisions to become stabilized.

From swarm experiments, the attachment rate coefficient for thermal electrons at room temperature is found to be close to $2 \times 10^{-13} \, m^3 \, s^{-1}$ in a wide variety of different mixtures of SF_6 with other gases. This corresponds to a mean attachment cross-section as large as $1.4 \times 10^{-18} \, m^2$. Estimates made of the lifetime of the complex initially formed indicate that it is longer than 10 μs.

The resonance width corresponding to such a long lifetime is less than 10^{-10} eV, so capture will occur only if the electrons have a very sharply defined energy. This means that, in actual experiments, the observed variation of SF_6^- production with electron energy will simply be determined by the shape of the electron energy distribution. Use may be made of this result to measure the distribution when it is required in certain experiments, simply by introducing an admixture of SF_6 and observing the SF_6^- production as a function of electron energy. Because of its chemical inertness, the added SF_6 leads to no complications.

The high attachment rate for slow electrons in SF_6 makes it very useful as a tracer (see p. 138).

A further use of SF_6 is in the calibration of an electron energy scale, because the resonance peak for SF_6^- production occurs very close to zero energy.

Finally, the ability of SF_6 to attach slow electrons rapidly may be made use of to remove such electrons if required in some experiments. Because of this SF_6 is often referred to as a slow electron 'scavenger'.

Molecule	CH_3Cl	CH_2Cl_2	$CHCl_3$	CCl_4	CF_3Cl	$CFCl_3$	CHF_2Cl	$CHFCl_2$	CF_4
Attachment rate	5.4×10^{-4}	1.4×10^{-4}	3.4×10^{-2}	2.4	4×10^{-3}	0.98	5.4×10^{-4}	1.1×10^{-2}	7×10^{-6}
Molecule	CH_3Br	CH_2Br_2	CF_3Br	CF_2Br_2	CH_3I				
Attachment rate	2.9×10^{-2}	0.28	0.11	2.3	0.22				

Table 7.1. Observed attachment rates in 10^{-13} m^3 s^{-1} of thermal electrons to various polyatomic molecules.

7.12. Attachment to Other Polyatomic Molecules

The rate of attachment of electrons with near-thermal energies varies greatly from molecule to molecule. However, it is usually large for molecules containing two or more halogen atoms. For an effective attachment cross-section of 10^{-18} m^2, which is much larger than the average, the attachment rate for thermal electrons, possessing a root mean square velocity of about 10^5 m s^{-1}, is 10^{-13} m^3 s^{-1}. Table 7.1 gives observed attachment rates to a number of polyatomic molecules containing one or more halogen atoms. It will be seen that for several molecules the observed rate is large. In all cases, these are molecules containing more than one halogen atom, though it does not follow that the presence of more than one such atom guarantees a high attachment rate. We shall see below how use may be made of high attachment rates to detect trace concentrations of certain molecules.

7.13. The Electron Attachment Detector

Suppose that a stream of molecules is being carried through a cell containing a concentration of n_e electrons per unit volume. If k_a is the rate coefficient for attachment of electrons to the molecules, the chance that any molecule will undergo attachment in its passage through the cell will be given by

$$k_a n_e t_r, \tag{7.32}$$

where t_r is the time taken for a molecule to pass through the cell. For typical flow rates and a cell with dimensions of a few cm, t_r is of order 1 s. It follows that, if n_e is of the order 10^{13} m^{-3}, the chance of attachment will be between 10^{-2} and unity for k_a between 10^{-15} and 10^{-13} m^3 s^{-1}.

Again, if n_m is the concentration of the molecules per unit volume, the chance that an electron will undergo attachment will be

$$k_a n_m t_e, \tag{7.33}$$

where t_e is the time an electron spends in the cell before collection at a suitable electrode. Even with n_m as low as 10^{15} m^{-3}, corresponding to a partial pressure of only 4×10^{-8} torr at ordinary temperatures, the chance is between t_e and $100 \, t_e$ for k_a between 10^{-15} and 10^{-13} m^3 s^{-1}. Even if t_e is as small as 100 μs, it is practicable, with modern techniques, to detect the small change in electron current across the cell.

Atomic and Molecular Collisions

It is clearly possible to use these results as a basis for a sensitive detector of certain molecules present in only very minute quantities in a gas which is inactive as regards attachment. The first such devices were introduced by Lovelock and Lipsky in 1960. They used a cell containing two electrodes, with dimensions shown in figure 7.19, and produced the free electrons by ionization of the main gas by electrons from a beta radioactive source. Measurements were made of the variation of the current between the electrodes with the voltage applied between them—the so-called current–voltage characteristic. With pure nitrogen flowing between the plates, the current rapidly saturates as the voltage increases, but when attaching impurities are present, saturation is much more gradual, as seen from figure 7.20. The voltage required to produce a current which is a chosen fraction f, say, of the saturation current will depend on the nature and concentration of the attaching substances present.

With this very simple device, the presence of relative concentrations of chlorinated pesticides, such as DDT, in treated crops was established by Goodman, Goulden and Reynolds in 1961. This involved using gas chromatography, a technique for separating out the various components of a gas or vapour by flowing the mixture through a column containing stationary solid or liquid material which reversibly adsorbs the gaseous components. Because of this each component is retarded in its passage through the column by an amount depending on its affinity for the stationary material. Hence when the mixture is admitted to the column, the different components will emerge at different times. The emergence of a component which attaches electrons

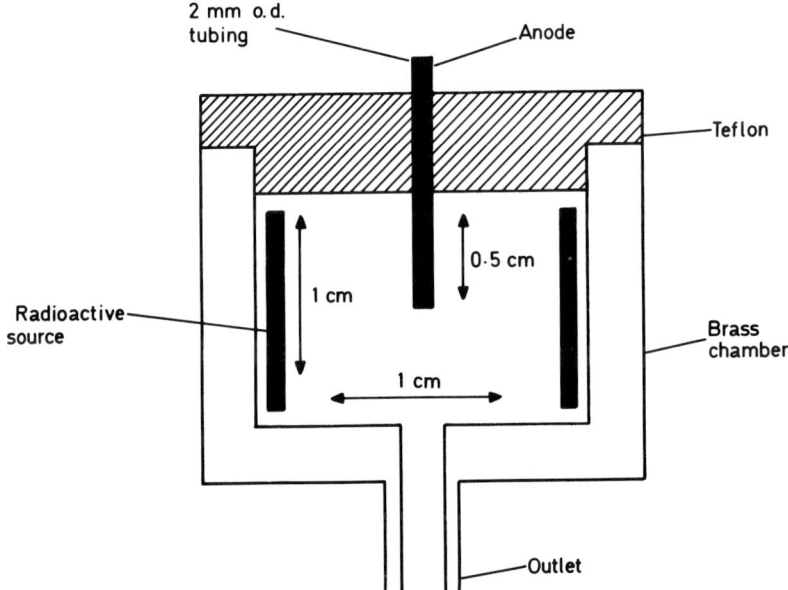

Figure 7.19. Sectional drawing of the electron capture detector used by Lovelock and Lipsky.

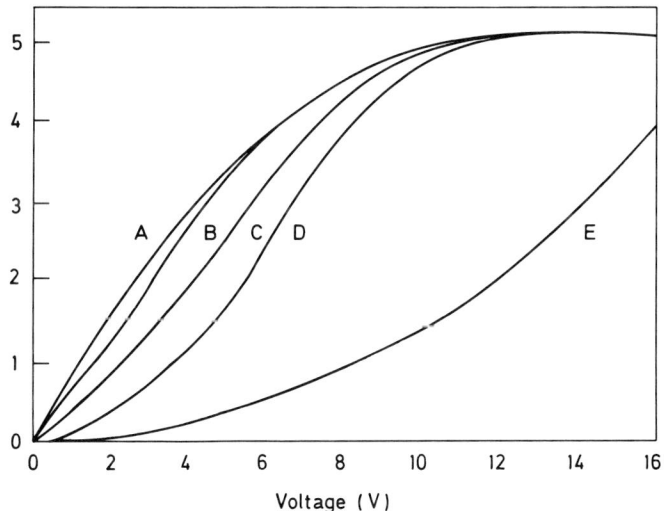

Figure 7.20. Typical voltage–current characteristics of the detector shown in figure 7.19. Curve A is obtained with pure nitrogen, which does not attach electrons flowing between the plates. B, C, D, E are obtained when various organic molecules which can attach electrons are present.

rapidly is then observed, at its characteristic emergence time, with an attachment cell. Figure 7.21 shows typical observed results of this type for pesticides in different crops.

The technique has been considerably refined and now uses pulsed rather than d.c. current measurements. In 1971, Lovelock, Maggs and Adlard established the use of the detector for quantitative measurement of the concentration of attaching molecules. Thus if U is the flow rate of the main, carrier, gas, the proportion of molecules lost by attachment in passage through the detector will be given by

$$\alpha = k_a n_e / (k_a n_e + U). \tag{7.34}$$

For SF_6, k_a is known and n_e may be determined from the detector current, so that α may be calculated from the flow rate. To compare with observation, the gas flows through two identical detectors in series. If S_1 and S_2 are the signals in the respective detectors, measured in terms of the loss of electrons by attachment, when N_c molecules pass through the system, then

$$S_1 = \alpha N_c, \quad S_2 = \alpha(N_c - \alpha N_c), \tag{7.35}$$

giving

$$\alpha = 1 - S_2/S_1, \quad N_c = S_1^2/(S_1 - S_2). \tag{7.36}$$

Good agreement was obtained between the value of α obtained in this way and that measured by (7.34).

Atomic and Molecular Collisions

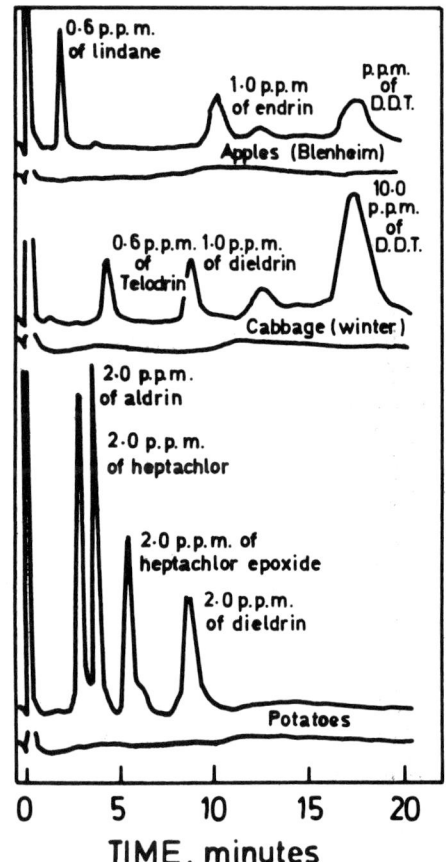

Figure 7.21. Signals in an electron attachment detector obtained after passage of toxicant-treated and control crops of apples, cabbage and potatoes through a gas chromatograph system (from Goodwin, Goulden and Reynolds). The peaks in the records are due to the toxicant pesticides indicated (such as lindane, aldrin, DDT etc). The concentrations of the toxicants are given in parts per million (ppm). FSD denotes full-scale deflection in the recording system.

Because of its high attachment rate, SF_6 is a very good tracer to use with attachment detectors. It should, for example, be possible to follow the passage of an air sample through the atmosphere across the Atlantic by introducing SF_6 at the beginning and measuring its concentration thenceforward at other points of observation.

While the attachment detector will not operate for molecules which do not attach readily, it is possible, by chemical means, to add halogen atoms to such molecules so as to produce strongly attaching molecules which can be observed. On the other hand, the selective response of the attachment detector is often very valuable in identifying a particular compound among a complicated mixture.

CHAPTER 8

collisions between neutral atomic and molecular systems— a general survey

We now consider collisions between neutral systems, either atoms or molecules, both of which possess electronic structure. Much of the study of such collisions is carried out under so-called gas kinetic conditions in which the kinetic energies of the colliding systems are comparable with $\frac{3}{2} kT$ where k is the Boltzmann constant (see Appendix 1) and the temperature T is below 1000 K. In many cases also the energies are not sharply defined, both systems possessing an energy distribution. There is a very wide range of phenomena for which such collisions are important, as in the viscosity and thermal conductivity of gases, the rate of diffusion of one gas into another, the dispersion of high-frequency sound in gases as well as optical phenomena such as the quenching of resonance fluorescence, sensitized fluorescence and the behaviour of gaseous lasers. In recent years, it has become possible to prepare beams of neutral molecules with selected velocities, which can be used to study individual collisions with other molecules under much more closely prescribed conditions. This work, which is now in full swing, has provided a wealth of new information about collisions between neutral systems at low energies and about the detailed nature of the interaction energies involved.

It is possible also to carry out experiments in which the scattering of beams of neutral systems moving at much higher energies is studied. Although it is not possible to accelerate neutral molecules by electromagnetic fields, methods have nevertheless been developed to obtain high-energy beams.

We now consider in rather general terms the different kinds of collision which can occur, noting some of the salient features of these collisions as well as their significance in determining the nature of other physical phenomena. A detailed account will be reserved for elastic collisions between gas atoms for which much experimental and theoretical information exists. This will form the subject matter of Chapter 9 and 10.

8.1. Atom–Atom Collisions—Elastic Scattering

As long as we confine ourselves to collisions between monatomic systems, there is no difficulty in distinguishing elastic from other types of collision. In general, the energy required to produce excitation of either system will be of the order of several electron-volts, far in excess of that available under gas-kinetic conditions. As mentioned above, a great deal of experimental and theoretical work on elastic collisions between gas atoms is being carried out as a result of recent developments of technique.

At room temperature, the root mean square velocity of a neon atom is 6×10^2 m s^{-1}, so that its wavelength is 3×10^{-11} m. This is much shorter than the range of interaction between neon atoms. The same argument applies even more strongly for heavier atoms.

However, the interaction between gas atoms falls off as r^{-6} at large separations r (see Chapter 4, p. 60), so that, according to the arguments of Chapter 3, p. 31, the total elastic cross-section Q, defined in terms of the differential cross-section per unit solid angle $I(\theta)$ by

$$Q = 2\pi \int_0^\pi I(\theta) \sin \theta \, d\theta, \tag{8.1}$$

which is unbounded in classical theory, is finite and may be measured with apparatus of sufficiently high angular resolution. An account of the calculation and measurement of $I(\theta)$ and Q forms a large part of the following two chapters.

The viscosity of a gas is determined by the rate at which energy is transferred between colliding gas atoms. The greater the rate at which energy is equalized by sharing in collisions, the smaller is the viscosity. In a collision between two similar atoms, the energy is shared equally between them after impact if the angle of scattering in the centre of mass system (see Chapter 1, p. 5) is 90°. It follows that collisions in which $\theta \simeq 90°$ are the most effective in reducing the viscosity. Accordingly, we would expect the equivalent collision cross-section Q_η effective in determining the viscosity, to differ from Q in that greater weight is given to collisions in which θ is nearly 90°. Thus detailed analysis shows that

$$Q_\eta = 2\pi \int_0^\pi \sin^2 \theta \, I(\theta) \sin \theta \, d\theta. \tag{8.2}$$

The special importance of collisions in which $\theta \simeq 90°$ appears in the weight factor $\sin^2 \theta$, which does not appear in the formula for the total elastic cross-section.

Collisions most effective in reducing the rate of diffusion are those in which back scattering occurs, so that $\theta \simeq 180°$. Accordingly, it is found that the appropriate collision cross-section Q_d is given by

$$Q_d = 2\pi \int_0^\pi (1 - \cos \theta) I(\theta) \sin \theta \, d\theta, \tag{8.3}$$

with the weighting factor $(1 - \cos\theta)$ a maximum at $\theta = 180°$.

Because of the weighting factors, the contribution from small-angle collisions

Collisions between Neutral Systems

to Q_η and Q_d is much reduced compared with that to the total cross-section Q. It is only for these collisions that quantum effects are important. As a result Q_η and Q_d may be calculated with good accuracy by classical theory, except for helium at low temperatures (below 50 K). This is in contrast to the total cross-section Q, which only exists as a finite, definite quantity because of quantum theory.

Information about the interactions between gas atoms has been obtained by analysis of experimental data on viscosity, thermal conductivity and diffusion in terms of classical mechanics, for most gases. Quantum collision theory has to be used, however, for helium at low temperatures.

At impact energies corresponding to gas-kinetic conditions, a growing volume of measurements of $I(\theta)$ and Q have been made and analysed in terms of a semi-classical theory to obtain atomic interactions (see Chapter 10, p. 86). For much higher impact energies, extending up to 2000 eV or so, the angular resolution necessary to measure the total cross-section Q is difficult to attain, so that, over the angular range studied, $I(\theta)$ is as given by classical theory. In place of Q, a 'classical total cross-section' Q_c is introduced, defined by

$$Q_c = 2\pi \int_{\theta_0}^{\pi} I(\theta) \sin \theta \, d\theta, \tag{8.4}$$

where θ_0 is the smallest angle of scattering observable with the experimental equipment used. If θ_0 can be clearly specified, measurement of Q_c gives valuable results which can be analysed in terms of classical theory to obtain atomic interactions.

The extension of elastic scattering measurements to these relatively high impact energies requires a means of producing a beam of atoms of the required energy and of detecting both the primary and scattered beams. Fast beams of atoms A are produced by charge exchange from beams of A^+ ions which are readily accelerated to the required energies. Such an ion beam is passed through a gas of atoms B so that charge transfer (see Chapter 12, p. 219) may occur on collision i.e.

$$A^+ + B \to A + B^+. \tag{8.5}$$

The electron passes from B to A without any appreciable change in the magnitude and direction of the velocity of the A^+, so the ion beam is partially converted to a neutral atom beam. The latter may be separated by application of an electric or magnetic field which deviates the ions but not the neutral atoms.

Atoms possessing kinetic energies of some hundreds of eV produce secondary electrons on impact with suitable surfaces and this may be made the basis of detection, just as for fast positive ions. An alternative method, suitable when working with intense beams, is to collect the fast atoms in a cup and measure the heat input (see Appendix 6).

8.2. Collisions Involving Molecules—Excitation and Transfer of Vibration and Rotation

If one or both of the colliding systems is a molecule rather than an atom, a new possibility must be taken into account. Vibration and/or rotational motion in the molecule may be changed as a result of the collision. Since the quanta of vibrational energy are at most a few tenths of an electron-volt (see chapter 4, table 4.1), and those of rotation much smaller still, it is clear that account must be taken of such possibilities even under gas-kinetic conditions. If one of the systems is an atom, we need only consider inelastic collisions in which the molecular system gains or loses vibrational and/or rotational energy. However, if both systems are molecular, such energy may be transferred from one molecule to the other. If the molecules are similar, this transfer need not involve any change in the relative kinetic energy in the collision.

It is possible, from quite simple arguments, to derive the conditions under which these types of inelastic collisions are improbable, in other words the conditions under which

$$Q_v/Q \ll 1, \tag{8.6}$$

Q_v being the cross-section for a collision in which a change of vibrational energy occurs and Q the total collision cross-section for all types of collision.

Thus, consider a collision between two molecules A and B, the first of which is vibrationally excited. If the molecules approach very gradually, so that, in the time τ during which they are in effective interaction (the so-called *time of collision*), the atoms in A execute a great number of complete vibrations, then there will be plenty of time for these atoms to readjust to the slowly changing conditions. Under these circumstances, which are referred to as *nearly adiabatic*, the chance of any transition taking place in the vibrational motion in A will be small.

In classical terms this may be restated as follows. We consider the amplitude of oscillations that will be set up by applying a disturbing force to an oscillator of natural frequency v. This disturbance will vary with time t according to some function $F(t)$, but we may regard it as built up from harmonic components of all frequencies. In this analysis, the contribution from the harmonic component with frequency between μ and $\mu + d\mu$ will be some function f_μ. It is only those components of $F(t)$ which are in close resonance with the natural frequency v of the oscillator which will produce an appreciable disturbance in its motion. Thus the amplitude of the forced oscillations will be proportional to f_v. If $F(t)$ varies so slowly with t that in the time $1/v$ of one natural oscillation, there is very little fractional change of $F(t)$, f_v will be very small.

We write the time of collision in the form a/v, where v is the relative velocity of the colliding systems and a is the range of interaction between them. The chance of any vibrational transition will be small when

$$av/v \gg 1. \tag{8.7}$$

Collisions between Neutral Systems

This may be recast in a more convenient form by noting that, for simple harmonic vibrations of total energy E and frequency v, the amplitude d of vibration is given by

$$d \simeq \frac{1}{\pi v}(E/2M)^{1/2}, \tag{8.8}$$

where M is the reduced mass (Chapter 1, p. 6) of the vibrating molecule, so that, if $E = hv$,

$$d \simeq \frac{1}{n}(h/2\pi vM)^{1/2}. \tag{8.9}$$

Also, under gas-kinetic conditions at temperature T,

$$v^2 \simeq kT/M_1, \tag{8.10}$$

where M_1 is the mass of the lighter molecule concerned in the collision and k is the Boltzmann constant. On substitution in (8.7) we have, finally, the condition

$$\frac{a}{\pi d}\left(\frac{hvM_1}{kTM}\right)^{1/2} \gg 1, \tag{8.11}$$

which must be satisfied in order that the chance of excitation or deactivation of a vibrational quantum hv should be small.

For most molecules, hv will be greater than kT for transitions between the deepest vibrational levels for which $a \gg d$. Under these conditions, the chance of a vibrational transition on impact will be small. When $hv < kT$, the condition (8.11) may still be satisfied because $a \gg d$, but, for sufficiently small hv/kT, the probability of a transition on impact will be comparable with unity. The fact that (8.11) may be satisfied under suitable conditions is of importance in a number of physical phenomena, particularly the dispersion and absorption of high-frequency sound waves, which we discuss in more detail below. Before this, we must examine how the above argument may be applied to rotational transitions.

Referring to (4.4) of Chapter 4, the allowed rotational energy changes for a diatomic molecule of reduced mass M and equilibrium separation R_0 are given by

$$\Delta E_r = h^2 J/4\pi^2 MR_0^2. \tag{8.12}$$

Substituting this for the vibrational quantum hv in (8.7), we obtain the condition

$$ahJ/4\pi^2 R^2 Mv \gg 1. \tag{8.13}$$

In all cases, except for the lightest molecules at very low temperatures, the wavelength h/Mv is much less than a or R, which are of the same order of magnitude. The condition (8.13) is therefore rarely satisfied unless J is very

Atomic and Molecular Collisions

large, so that we would expect that rotational transitions may take place readily in gas-kinetic conditions. In fact, the left-hand side of (8.13) may even be so much smaller than unity that the opposite extreme conditions are approached—the collision takes place so fast that there is inadequate time for a transition to occur.

Because of the considerable probability that rotational transitions will take place in gas-kinetic collisions, involving only relatively small energy transfer to or from relative kinetic energy of translation, it is very difficult to distinguish experimentally truly elastic collisions involving no such transfer. A further complication in analysing data on collisions in which molecules are involved is that the interaction between the colliding systems will depend not only on the distance between their centres of mass but also on their relative orientations.

The interaction between molecules such as water (H_2O) or ammonia (NH_3), both of which posses permanent electric dipole moment, is of very long range. The relative motion is effectively classical, so that the impact parameter p is well defined. For large values of p, all the collisions are inelastic, involving rotational transitions in both molecules. The resulting total collision cross-sections may therefore be very large, though including only a very small contribution from elastic scattering.

8.3. Dispersion and Absorption of High-Frequency Sound

The velocity v of propagation of sound waves in an ideal gas at pressure p, density ρ and temperature T is given according to the well known Laplace formula by

$$v^2 = \gamma p/\rho, \tag{8.14}$$

where γ is the ratio of the specific heat c_p at constant pressure to that c_v at constant volume. As it stands, this formula shows no dependence of the velocity on the frequency, i.e., no dispersion. This is found to be correct for propagation through atomic gases, but high-frequency sound in molecular gases does show dispersion, as well as absorption, much greater than would be expected from the effects of viscosity, heat conduction and radiation.

The explanation of this anomalous behaviour depends on the persistence of vibrational excitation through many collisions in the molecular gas. In such a gas, the specific heat c_v includes contributions not only from the translational motion but also from the molecular vibration and rotation. Thus

$$c_v = c_v(\text{trans}) + c_v(\text{rot}) + c_v(\text{vib}). \tag{8.15}$$

During the passage of the sound wave, thermal changes will occur with a rapidity which increases with the frequency. The Laplace formula supposes that, nevertheless, the changes in the excitation of all the degrees of freedom

Collisions between Neutral Systems

concerned can follow without appreciable time lag. However, if the rate at which vibrational transitions can occur is not fast enough for this to be so, the contribution from $c_v(\text{vib})$ will decrease with increasing frequency at high frequencies, so c_v will depend on v. Since $\gamma = 1 + R/c_v$, where R is the gas constant, it follows that γ will depend on v and the propagation will exhibit dispersion.

In the limit of very high frequencies, $c_v(\text{vib})$ will vanish and the velocity of propagation v_∞ will be given by (8.14) with

$$\gamma = 1 + R/\{c_v(\text{trans}) + c_v(\text{rot})\}. \tag{8.16}$$

Detailed analysis shows that, when the acoustical frequency is $\omega/2\pi$, the velocity of propagation is given by

$$v^2 = (v_0^2\omega_i^2 + v_\infty^2\omega^2)/(\omega_i^2 + \omega^2), \tag{8.17}$$

where v_0 is the velocity of propagation at very low frequency and

$$\omega_i = \left\{1 + \frac{c_v(\text{vib})}{c_v(\text{trans}) + c_v(\text{rot})}\right\}\bigg/\tau. \tag{8.18}$$

τ is a time, known as the *relaxation time*, which is given by

$$\frac{1}{\tau} = \beta_{01} + \beta_{10}, \tag{8.19}$$

where β_{01} is the number of collisions made per second by a vibrationally excited molecule which lead to de-excitation and β_{10} the corresponding number made per second by neutral molecules which lead to vibrational excitation. At a temperature T, the two rates are related by

$$\beta_{01} = \beta_{10}\exp(-h\nu/kT), \tag{8.20}$$

where $h\nu$ is the vibrational quantum involved. Figure 8.1(a) illustrates the general shape of the variation of v^2 with the angular frequency ω.

Because of the vibrational lag, absorption also occurs, so that, if α/λ is the absorption coefficient for sound of wavelength λ,

$$\alpha = 2\pi(v_\infty^2 - v_0^2)\omega_i\omega/(v_0^2\omega_i^2 + v_\infty^2\omega^2). \tag{8.21}$$

The form of the variation of α with ω is shown in figure 8.1(b).

It follows from (8.17) and (8.21) that, if the velocity and/or absorption of sound waves is measured as a function of frequency, the relaxation time τ may be obtained in gases at different temperatures and pressures and hence the chance of a vibrational transition occurring on collision. From its definition, we expect that $1/\tau$ should be proportional to the gas pressure. Unfortunately when $h\nu/kT \gg 1$, the contribution to the specific heat from vibrational degrees of freedom is very small in any case, so that it is difficult to detect dispersion and absorption effects due to persistence of vibration. Nevertheless, a great

Atomic and Molecular Collisions

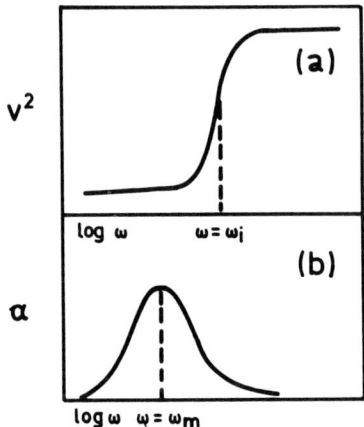

Figure 8.1. Form of the variation of the square of the velocity of propagation v and the absorption coefficient with the logarithm of the angular frequency ω for sound propagating in a molecular gas in which persistence of vibration occurs. In the notation of (8.21) $\omega_m = v_0 \omega / v_\infty$.

amount of information on the rates of vibrational deactivation on impact has been obtained for a wide variety of gases and vapours from observations of dispersion and absorption of ultrasonic waves. Before discussing some of these, we draw attention to a further technique which makes it possible to determine vibration relaxation times at temperatures above 1000 K—the upper limit for dispersion experiments is about 600 K.

8.4. Persistence of Vibration in Shock Wave Experiments

We have already referred to the use of shock waves to heat gases to temperatures above 1000 K.

Information about vibrational relaxation comes from observation of the thickness of the transition region in the gas immediately behind the shock front. The extent of this region is determined by the specific heat of the gas through which the shock is passing. In a molecular gas, because of persistence of vibration, the effective specific heat will vary with distance behind the shock wave, that is to say with the time since the shock front passed through. If the variation of density or temperature of the gas can be observed throughout the transition region, the vibrational relaxation time may be derived. Techniques for making such measurements have been successfully developed and have yielded valuable information on vibrational relaxation times at high temperatures.

8.5. Some Results of Vibrational Relaxation Measurements

Figure 8.2 shows a selection of results obtained for a number of molecular

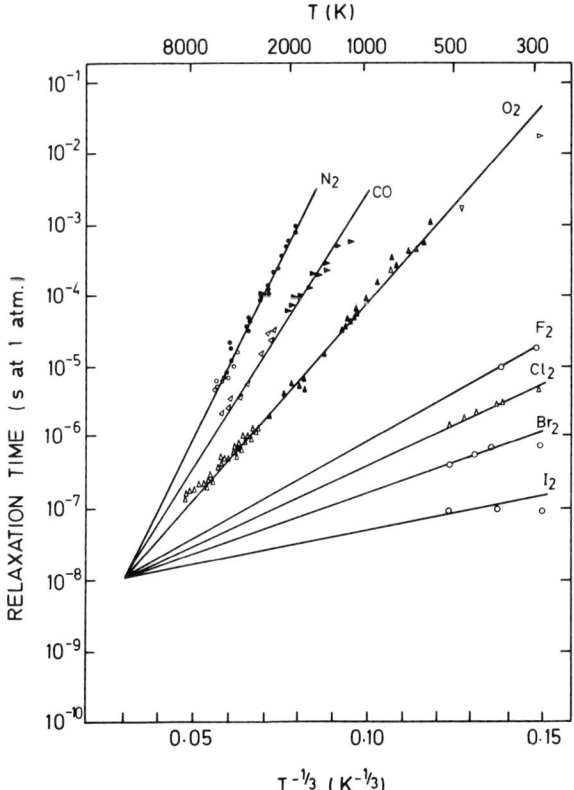

Figure 8.2. Relaxation times for vibrational deactivation in different molecular gases at a pressure of 1 atmosphere, observed as a function of temperature T. With times on a logarithmic scale, the observed points fall nearly on a straight line, for each gas, when the abscissa is taken to be $T^{-1/3}$. This is as expected from the theory.

gases. In all cases the relaxation time τ is given for a pressure of 1 atmosphere at 0°C. For N_2 and CO, hv is so large that the only data available have been obtained from shock wave measurements and the same is largely true for O_2. On the other hand, for the halogen molecules, ultrasonic methods have been used.

It is instructive to derive from the relaxation times the probability that a vibrating molecule will be deactivated in a collision. This may be done without difficulty using (8.19) and (8.20) and comparing β_{01} with the collision rate in the gas derived from gas kinetic studies. Typical values obtained are given in table 8.1.

It will be seen from this table above that in all cases vibration persists through a large number of collisions. As expected from the general account given on p. 143, the probability of deactivation increases rapidly with the temperature

Atomic and Molecular Collisions

$N_2(7.1)$		$O_2(4.9)$		$F_2(2.65)$		$Cl_2(1.7)$		$I_2(0.7)$	
T/K		T/K		T/K		T/K		T/K	
550	1.6×10^{-8}	300	10^{-8}	300	8×10^{-6}	300	2.6×10^{-5}	400	2.6×10^{-3}
800	1.6×10^{-7}	500	10^{-7}	375	1.5×10^{-5}	440	8×10^{-5}	500	4.2×10^{-3}
1200	4.7×10^{-7}	1000	4×10^{-6}			660	7.7×10^{-4}		
2500	1.2×10^{-5}	1900	1.5×10^{-4}			900	3.2×10^{-3}		
4200	1.5×10^{-4}	4000	5×10^{-3}			1500	1.1×10^{-3}		
4900	2.7×10^{-4}	6000	4×10^{-3}						

(Vibrational frequencies in 10^{13} s^{-1} are given in brackets for each molecule.)
Table 8.1. Typical values of the probability of vibrational deactivation per collision between similar diatomic molecules at different temperatures.

and, for a given temperature, with decreasing vibrational frequency. Similar data are available for mixtures of diatomic molecules.

For polyatomic gases, the relaxation times are, in general, shorter because the vibrational quanta are smaller. In fact, it is very important in carrying out experiments in diatomic gases to work under conditions of high purity, because only a small admixture of a polyatomic impurity may have a profound effect on the observed relaxation times.

8.6. Relaxation in H_2 and D_2—Rotational Persistence

For the lightest diatomic molecule H_2 the vibrational quanta are so large (0·26 eV) that it is very difficult to observe vibrational relaxation even with shock tubes. We shall return to this problem a little later.

Meanwhile, we return to a question which we have not yet considered—the possibility of rotational energy persistence in collision. In H_2 and D_2, such persistence can be observed in the absorption of ultrasonic waves and in the thickness of the transition region behind a shock front. From the observations, it is found that the probability of deactivation of an H_2 molecule from its first excited rotational state on collision is about 4×10^{-3} at room temperature, falling to about half this value at 90 K. For D_2 at room temperature, it is 5×10^{-3}. Under these conditions the quantity $ah/4\pi^2 R_0^2 Mv$ in (8.13) is not much greater than unity, so extreme rotational persistence is not expected. Detailed theoretical calculations yield results in reasonable agreement with the observed values.

Returning now to the question of vibrational excitation in H_2 and D_2, measurements have been made quite recently by a remarkable technique. Hydrogen in a suitable container was irradiated with an intense pulse of light from a ruby laser (see Chapter 13, p. 235), so that an appreciable fraction of the molecules were excited to the first vibrational state. The rate at which the excitation decayed was observed through the change Δn of the refractive index of the hydrogen with time since the existing pulse was cut off. Under the experimental conditions, Δn is proportional to the number $N(t)$ of molecules vibrationally deactivated since $t = 0$. Δn was monitored by observing the scattering of light from a helium–neon laser. If $\Delta n \ll 1$, the intensity of scattering is proportional to Δn^2 and hence to $\{N(t)\}^2$. τ is then obtained from the relation

$$N(t) \propto \{1 - \exp(t/\tau)\}. \tag{8.22}$$

In this way, Ducoing, Joffrin and Coffinet found for H_2 at room temperature that the probability of vibrational deactivation per collision is 6×10^{-8}, somewhat higher than might have been estimated by comparison with results for other diatomic molecules (see table 8.1).

8.7. Inelastic Collisions Involving Electronic Excitation

We now consider, very briefly, some aspects of a very extensive subject— collision processes between atomic or molecular systems in which electronic excitation is involved.

The simplest type of collision of this kind is one involving direct excitation or deactivation of one atom on collision with another. Under gas-kinetic conditions, as we have remarked earlier, there is insufficient energy for one atom to excite another on impact. However, there is no energy limitation for the inverse process

$$A' + B \rightarrow A + B, \tag{8.23}$$

in which an atom B in its ground state, colliding with an excited atom A′, deactivates it, so the excitation energy E_e is transferred to kinetic energy of relative motion. This is often referred to as a quenching collision, because the occurrence of such collisions reduces the probability that the excited atoms A will release their energy as radiation, i.e., the radiation is *quenched*.

In view of the quantum relation, we might expect that the arguments outlined on p. 143 for vibrational deactivation could be adapted to these collisions simply by replacing the vibrational frequency v by E_e/h. On this basis, we could expect the process (8.23) to be very improbable when

$$aE_e/hv \gg 1, \tag{8.24}$$

or, under gas-kinetic conditions at temperature T, when

$$\frac{aE_e}{h}\left(\frac{M_1}{kT}\right)^{1/2} \gg 1, \tag{8.25}$$

M_1 being the mass of the lighter atom. Since E_e will be of the order of at least a few eV, very much greater than the largest vibrational quanta, the condition (8.25) is even more strongly satisfied than for vibrational excitation. We would therefore expect that (8.23) will be extremely unlikely to occur under gas-kinetic conditions. However, we must be somewhat careful in applying (8.24) because, if the atoms attract each other, their relative velocity when in close collision may be much higher than that v at infinite separation. At the same time, the energy separation between the states of the molecules A'B and AB, which dissociate ultimately to $A' + B$ and $A + B$ respectively, may become much smaller at certain separations (see figure 8.3). (8.24) should be applied to the most favourable conditions, of smallest energy separation and largest relative velocities, and (8.25) may not then be so adequately satisfied.

Experimental results for atom–atom impacts do in fact confirm the general conclusion that quenching collisions of the type (8.23) are very improbable, but this is found not to apply in general when the quenching system is a molecule and not simply an atom.

The general principle of the most usual experimental method for investigating quenching rates is as follows. A cell containing the gas of atoms A is irradiated by a beam of light so that a considerable fraction of the atoms are raised to the state A' by light absorption. These atoms, if left alone, will return to the ground state, emitting radiation of the same wavelength. This is known as *resonance fluorescence*. The intensity of this fluorescent radiation at right angles to the exciting beam is measured when the gas consists purely of atoms A, and again when it is mixed with the quenching gas at a number of different partial pressures p.

Suppose that the exciting beam raises R atoms A to the excited state per second. Then in equilibrium, the number of excited atoms n_A present per unit volume will be given by

$$n_A/\tau_r = R, \qquad (8.26)$$

where τ_r is the mean lifetime of an excited atom before emitting radiation. n_A/τ_r is then the rate of de-excitation per second provided no other source of de-excitation is present. When quenching collisions may occur with molecules B with mean cross-section $\bar{Q}_q$, excited atoms are destroyed through this process at a rate

$$n_A Z = n_A n_B \bar{Q}_q \bar{v}, \qquad (8.27)$$

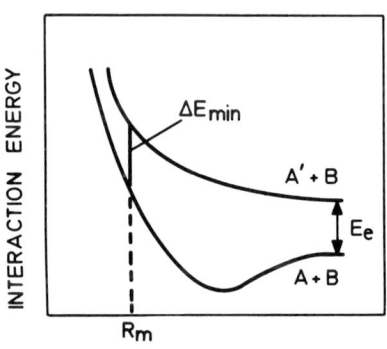

Figure 8.3.

Gas	N_2	O_2	H_2	CO	NO	CO_2	H_2O	N_2O	NO_2
$\bar{Q}_q \bar{v}$	$<2 \times 10^{-22}$	3.6×10^{-19}	2.8×10^{-22}	4.9×10^{-21}	8.0×10^{-17}	3.6×10^{-19}	7×10^{-17}	1.1×10^{-17}	5.0×10^{-16}

Table 8.2. Observed values of the rate coefficient $\bar{Q}_q \bar{v}$, in m³ s⁻¹, for quenching of O(¹S) at 300 K by different foreign gases.

where n_B is the number of molecules B per unit volume and $\bar{v}$ is the root mean square relative velocity of collision. We now have in place of (8.26)

$$n_A\left(Z + \frac{1}{\tau_r}\right) = R. \qquad (8.28)$$

Hence, in equilibrium, the concentration of excited atoms n_A is reduced by a factor

$$1 + \tau_r Z = 1 + \tau_r n_B \bar{Q}_q \bar{v}, \qquad (8.29)$$

The intensity of fluorescent radiation is proportional to n_A and so is reduced by the same factor. Hence by measuring the reduction factor at different partial pressures of B and hence different n_B the product $\tau_A \bar{Q}_q \bar{v}$ may be obtained. Since τ_r may be measured in other experiments, $\bar{Q}_q \bar{v}$, the rate coefficient for the quenching, may be obtained.

Experiments of this kind have even been carried out when the excited atoms are metastable states of atomic oxygen which have very long radiative lifetimes. Thus for the ¹S state, the upper state for emission of the green line which is such a permanent feature of the spectrum of the radiation emitted by the night sky and by the polar aurora, τ_r is 0.07 s. The ¹D state which is the upper state for the red oxygen line, also a prominent feature of the night sky and auroral emission, has an even larger radiative lifetime, 100 s. In the experiments referred to, the excited O atoms were produced by dissociation of suitable oxygen-containing molecules by ultraviolet radiation. Despite the very low intensity of the radiation emitted from the metastable atoms, measurements were made of the intensity as a function of foreign gas pressure and $\bar{Q}_q \bar{v}$ derived.

For the quenching of ¹S oxygen atoms at room temperature by foreign gas molecules, Felseth, Stuhl and Welge found the values of $\bar{Q}_q \bar{v}$ given in table 8.2. It will be seen that the rate coefficient varies greatly from molecule to molecule. For quenching by atoms the values are too small to be measured. This behaviour is typical of quenching reactions.

8.8. Excitation Transfer—Sensitized Fluorescence

An alternative possibility which may arise on collision between an excited atom A' and an atomic or molecular system B is that the excitation be transferred from A to B

$$A' + B \to A + B'. \qquad (8.30)$$

If $E_e(A)$ is the excitation energy of A' and $E_e(B)$ of B', such a collision will be possible if an amount of energy $E_e(A) - E_e(B)$ is transferred to kinetic energy of relative translation.

Processes of this kind give rise to the phenomenon of *sensitized fluorescence*. Thus, when a mixture of species A and B is irradiated by radiation of the correct frequency to excite the atom A to A' then, in addition to the resonance fluorescent radiation emitted by the atoms A', radiation may also be observed which arises from excited states of B.

Classic examples of such phenomena which have been studied include mercury-sensitized sodium, indium and thallium fluorescence. In these cases, the mercury atoms are excited by the resonance line at 253·7 nm.

Excitation transfer from metastable helium atoms to neon atoms is important in that it provides the basic mechanism for the helium–neon gaseous laser (Chapter 13, p. 235). The metastable helium atoms produced by electron impact excitation (Chapter 6, p. 107) have a high probability of transferring this excitation to neon atoms in collisions, so that an excess population of excited neon atoms is built up and maintained.

8.9. Ionization by Metastable Atom Impact

If the excitation energy $E_e(A)$ of an atom or molecule A is greater than the ionization energy $E_i(B)$ of an atom or molecule B then the following process may occur even at very low impact energies

$$A' + B \rightarrow A + B^+ + e. \qquad (8.31)$$

In such a process, the energy carried off by the electron resulting from the ionization of B is equal to $E_e(A) - E_i(B)$, so there is no change in the kinetic energy of relative motion of A and B. In practice, the most important cases are those in which the excited atom is in a metastable state, the process then being known as *Penning ionization*.

The most effective metastable atoms are those of helium which posses the highest excitation energies, 19·77 eV for 2^3S and 20·6 eV for 2^1S, and are capable of ionizing all atoms except neon. Laboratory measurements have been made by flowing afterglow (see Chapter 11, p. 207) and other techniques, of cross-sections for ionization of Ar, Kr, H_2, N_2 and CO by 2^3S metastable atoms of helium under gas-kinetic conditions. Values found range from 7×10^{-20} m^2 for argon to about $1·4 \times 10^{-20}$ m^2 for hydrogen.

Because of the considerable concentrations of metastable atoms which build up in electric discharges and in discharge afterglows, their reactions with other species which may be present are of considerable importance. In particular, in the early stages of an afterglow, the reaction

$$A' + A' \rightarrow A^+ + B^-, \qquad (8.32)$$

involving collisions between two metastable atoms, has been shown to be a significant source of free electrons which tends, in the early stages, to offset the decay of electron concentration due to recombination and diffusion (see Chapter 7, p. 115).

8.10. Production of Ion Pairs in Neutral–Neutral Collisions

The reaction

$$A + B \rightarrow A^+ + B^-, \tag{8.33}$$

in which two neutral systems are transformed on collision into a positive and negative ion pair may occur provided sufficient energy is available from kinetic energy of relative motion. If $E_i(A)$ is the ionization energy of A and $E_a(B)$ the electron affinity of B, the minimum energy required is $E_i(A) - E_a(B)$. Hence, if it is possible to measure the cross-section for a process such as (8.33), as a function of relative kinetic energy, the threshold for the reaction may be obtained and hence $E_i(A) - E_a(B)$. $E_i(A)$ will usually be known from other measurements, so this procedure offers the possibility of determining $E_a(B)$, which is less readily obtained otherwise.

In practice, the determination of the onset energy is very difficult if it is larger than a few eV, because the cross-section then rises very gradually from the threshold. Hence, if it is desired to determine an electron affinity with some precision, the system A must be chosen so that $E_i(A)$ is as small as possible. For this reason, many experiments have been carried out in which A is an alkali metal atom and particularly a caesium atom, which has the lowest ionization energy, 3·89 eV, of any element, (see Chapter 3, table 3.1), only a little greater than the highest atomic electron affinity (see Chapter 3, p. 57) 3·6 eV, that of chlorine.

For experiments of this kind, it is necessary to work with atomic or molecular beams of a few eV energy. It is difficult to produce such relatively low-energy beams by charge transfer (see p. 141), though this has been achieved for Cs by Helbing and Rothe. An alternative source is from sputtering. Figure 8.4 shows the arrangement using such a source in the experiments of Baede, Auerbach and Los.

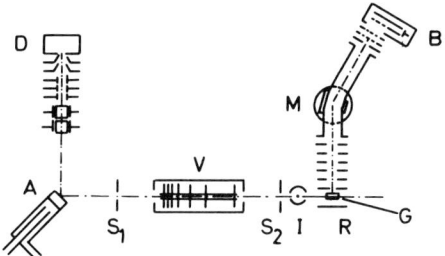

Figure 8.4. Arrangement of apparatus used by Baede, Auerbach and Los.

Argon ions produced in the discharge chamber D were accelerated by application of a few hundred volts and collimated by passage through an electrode system to impinge on an alkali metal surface A. Atoms sputtered from the surface by the impact of the argon ions were collimated by passage through the slits S_1 and S_2, between which a velocity selector V (see Chapter 10, p. 183) was placed. In this way alkali metal atoms of chosen velocity issued as a beam into the experimental chamber. Here they collided either with molecules in a bulk volume of gas or in a second beam G crossing the alkali atom beam at 90°. In either case, the positive and negative ions produced could be collected on plate electrodes held at voltages respectively negative and positive with respect to the alkali atom beam (cf. Chapter 12, p. 220 and figure 12.3). If the target systems are molecules, more than one negative ion may result, and in such cases the relative probability of production of the different ions may be obtained by sampling the ions passing through a small orifice in the negative ion collector plate and analysing them with a mass spectrograph (see Appendix 7). Thus in figure 8.4, R is a plate held at negative potential to repel negative ions, formed in the interaction region, through the extraction electrode system to the mass analyser M. Ions leaving M are accelerated by a further electrode system to enter the multiplier B in which they are recorded (Appendix 6). I is a surface ionization detector (Chapter 10, section 10.4) for measuring the flux of alkali metal atoms.

Figure 8.5 shows results obtained by Baede, Auerbach and Los for the

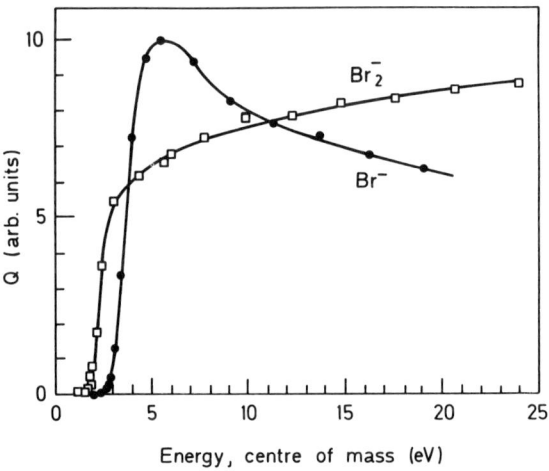

Figure 8.5. Relative cross-sections for production of Br^- and Br_2^- ions is collisions between K atoms and Br_2 molecules as a function of the kinetic energy in the centre of mass system observed by Baede, Auerbach and Los.

reactions

$$K + Br_2 \rightarrow K^+ + Br_2^- \qquad (8.34)$$

$$\rightarrow K^+ + Br^- + Br. \qquad (8.35)$$

For the determination of the threshold energies, allowance must be made for the velocity spread in the alkali atom beam and the thermal motion of the target molecules. Even then the cross-section may vary so gradually above the threshold that its location is uncertain. Nevertheless, useful results have been obtained by various experimenters using different types of source, different alkali metal atoms and both static gas and crossed beam targets. Electron affinities have been obtained for Cl_2, Br_2, I_2, O_2, NO and NO_2 which are consistent with those obtained by other methods (see in particular Chapter 13, p. 255).

CHAPTER 9
semi-classical collisions between atoms

We now turn to consider in some detail the elastic collisions between gas atoms with energies in the thermal or near-thermal range. In the general discussion on p. 139 of the previous chapter, it was pointed out that, because of the relatively large masses of gas atoms, these collisions are nearly classical. Thus the wavelength of relative motion is short compared with the range of interaction between the atoms, with the exception only of helium atoms at low temperatures. Apart from this exception, the viscosity and diffusion coefficients of gases may be calculated by classical mechanics, provided the atomic interactions are known. Nevertheless, it appears from a more detailed analysis that the wave aspect of the collisions shows up, not only in the scattering through very small angles but also in the occurrence of a number of interference effects which are revealed by experiments carried out with sufficiently good angular and energy resolution. Many, but not all, of these effects arise from the characteristic form of the interaction between atoms and their observation assists in its determination. Knowledge of atomic interactions is important not only for the theory of gases but also for the understanding of the properties of condensed matter.

Although some of the relevant theory was first worked out in the early 1930s, the wealth of interference effects to be expected was not realized until the work of Ford and Wheeler in 1959. Since that time, the experimental side of the subject, which began in the late 1920s, has taken advantage of many new techniques and now provides remarkably accurate data about the non-classical features. Some of these features are analogous to the light scattering by water droplets which gives rise to rainbow and glory effects in meteorology.

In this chapter, we shall give an account of the theory which shows how it is possible to determine to a good approximation by classical methods *both* the moduli and phases of the scattered amplitudes involved. The non-classical aspect emerges when these amplitudes are combined with allowance for the phase differences between them. Much of the theory, which is referred to as semi-classical, may then be developed on descriptive lines. In the following chapter, we shall describe something of the experimental methods which are now in use for studying these collisions, as well as the results obtained and their relation to the atomic interactions concerned.

Semi-classical Collisions between Atoms

Although we shall be dealing with collisions between two systems, of finite masses M_1 and M_2, both of which are in motion with relative velocity v, the problem is equivalent to that presented by the collision of a particle of mass $M, = M_1 M_2/(M_1 + M_2)$, moving with velocity v in the field of a fixed centre of force which is the same as that between the atoms (see Chapter 1, p. 6). The conversion from relative (centre of mass) to laboratory co-ordinates is then as described on p. 8.

9.1. *The Form of the Interaction Energy Between Gas Atoms*

The interactions between atoms have already been discussed in Chapter 4, section 4.1. It was pointed out there that, while in general two atoms may interact in more than one way, for rare gas atoms there is a unique interaction of the general form shown in figure 9.1 (cf. Chapter 4, figure 4.1(*b*)). For simplicity, we shall restrict our discussion in this chapter to such cases, although in Chapter 12 we consider some cases involving a neutral and ionized atom in which more than one interaction is possible.

At a short distance r, rare gas atoms repel each other strongly with a force which decreases rapidly as r increases so that $V(r)$ vanishes at $r = r_0$ say. As the separation increases, the force becomes weakly attractive so that $V(r)$ reaches a minimum value $-\varepsilon$ at $r = r_m$. It then tends to zero as r increases further, so that, at large r, it falls off as $-C/r^6$, where C is a positive constant, known as the Van der Waals constant.

The depth ε of the potential minimum is a few hundredths of an electron-volt at most which, as explained in Chapter 4, p. 60, is too weak to lead to permanent formation of diatomic molecules at ordinary temperatures.

It is not possible to express the interaction in a simple algebraical form but, as a useful approximation, for illustrative purposes it is often written as

$$V(r) = \frac{A}{r^n} - \frac{C}{r^6}, \tag{9.1}$$

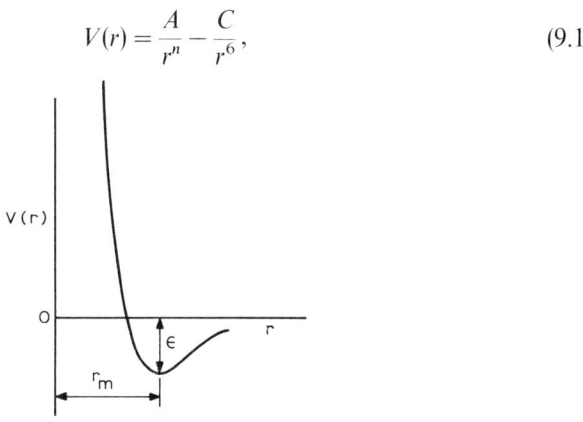

Figure 9.1. Typical form of the interaction energy $V(r)$ between rare gas atoms.

where n is > 6 and is usually taken at a first guess as 12. C may be estimated theoretically, while A and n are adjustable.

9.2. Classical Theory of Collisions between Gas Atoms—the Deflection Function

In Chapter 1, section 1.4, we showed how, in classical theory, the differential cross-section for scattering of particles of mass m by a centre of force could be expressed in terms of the relation between the impact parameter p and the angle of scattering θ. The impact parameter p (see figure 1.6, Chapter 1) is the closest distance to which the incident particle would approach the centre if it were undeviated from the initial direction of motion. If a unique value of p exists for a given θ, the differential cross-section per unit angle is given by

$$J(\theta) = 2\pi p \left|\frac{dp}{d\theta}\right|. \tag{9.2}$$

In general however, as we shall see, the same scattering angle will result from collisions with a number of different impact parameters. (9.2) is then generalized to

$$J(\theta) = 2\pi \left\{ p_1 \left|\frac{dp}{d\theta}\right|_{p=p_1} + p_2 \left|\frac{dp}{d\theta}\right|_{p=p_2} + \ldots \right\}, \tag{9.3}$$

where the sum is taken over the different values of p. It will be recalled also that the differential cross-section per unit solid angle $I(\theta)$ is given by

$$I(\theta) = J(\theta)/2\pi \sin \theta.$$

We now show how the relation between p and θ may be calculated according to the classical theory of orbits of a particle moving under the action of a centre of force which exerts upon it an interaction of potential energy $V(r)$ when it is at a distance r.

Basically the calculation depends on the use of two conservation laws—the total energy of the particle and its angular momentum (see Chapter 1, section 1.5) must both remain constant throughout the motion. We notice first that the motion must be confined to the plane containing the initial radius vector joining the centre to the particle and the initial direction of motion of the particle. To take advantage of the two conservation laws, it is then concenient to work in terms of polar co-ordinates (r, ϕ), in this plane with origin O at the scattering centre as shown in figure 9.2.

In terms of the usual cartesian co-ordinates (x, y) we have

$$x = r \cos \phi, \qquad y = r \sin \phi,$$

and the distance δs between two neighbouring points P, P' with co-ordinates (r, ϕ), $(r + \delta r, \phi + \delta \phi)$ respectively, is given by

$$\delta s^2 = \delta r^2 + r^2 \delta \phi^2, \tag{9.4}$$

Semi-classical Collisions between Atoms

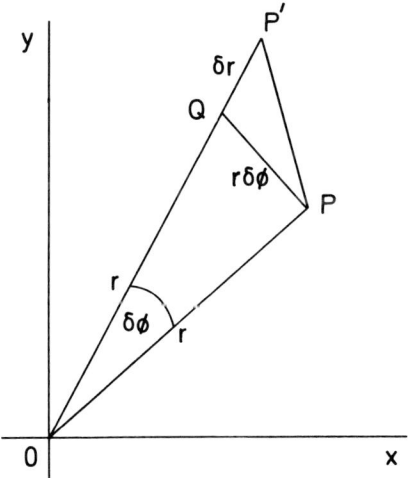

Figure 9.2. In the limit in which $\delta\phi \to 0$, PQ = $r\delta\phi$ and is perpendicular to OP.

as may be seen from figure 9.2. Thus δr and $r\delta\phi$ are the components of the displacement from (r, ϕ) to the neighbouring point respectively resolved along and perpendicular to the radius vector OP.

If the particle moves from P to P' in a time δt, its velocity u at P is $\delta s/\delta t$ and this may be resolved into components $\delta r/\delta t$ along OP and $r\delta\phi/\delta t$ perpendicular to OP. Denoting these components as u_r, u_ϕ respectively we have, in the limit when P' tends to P,

$$u_r = dr/dt, \quad u_\phi = r\, d\phi/dt, \tag{9.5}$$

$$u^2 = u_r^2 + u_\phi^2. \tag{9.6}$$

The conservation of energy requires that the sum of the kinetic and potential energies of the particle remains constant throughout. Hence

$$\tfrac{1}{2}mu^2 + V(r) = E = \tfrac{1}{2}mv^2, \tag{9.7}$$

where v is the velocity of the particle when it is initially at an infinite distance from the centre so $V(r)$ vanishes.

The conservation of angular momentum requires that

$$mru_\phi = L, \text{ a constant}. \tag{9.8}$$

Now

$$u_r = \frac{dr}{dt} = \frac{dr}{d\phi}\frac{d\phi}{dt}$$

$$= \frac{u_\phi}{r}\frac{dr}{d\phi}, \tag{9.9}$$

so that, from (9.8)

$$u_r = \frac{L}{mr^2}\frac{dr}{d\phi}.\tag{9.10}$$

Substituting in (9.6) for u_r from (9.10) and u_ϕ from (9.8) we have

$$\frac{d\phi}{dr} = \frac{L}{r^2}P(r),\tag{9.11}$$

where

$$P(r) = \pm\{2m(E-V) - L^2/r^2\}^{-1/2}.\tag{9.12}$$

The negative sign for the square root must be taken because ϕ decreases as r increases. Integrating (9.12) we have

$$\phi = -L\int^r \frac{P(r)}{r^2}dr,\tag{9.13}$$

where it is now understood that

$$P(r) = +\{2m(E-V) - L^2/r^2\}^{-1/2}.$$

To determine the limits in the integral, we note that the particle cannot penetrate closer to the centre than r_0, the largest value of r for which

$$E - V(r) - \frac{L^2}{2mr^2} = 0.\tag{9.14}$$

This is because it follows from (9.6) and (9.7) that

$$\tfrac{1}{2}mu_r^2 = E - V(r) - \tfrac{1}{2}mu_\phi^2,$$

$$= E - V(r) - \frac{L^2}{2mr^2},\tag{9.15}$$

and so, since u_r^2 can never be negative, the minimum value of the right-hand side of (9.15) is zero. Also, as the particle approaches the centre from infinity, the closest distance of approach r_0 will be the largest root of the equation (9.14) for r.

If now we choose the polar axis (that along which $\phi = 0$) as the line through the scattering centre parallel and in the opposite sense to the direction of incidence (see figure 9.3), we have $\phi = 0$ as $r \to \infty$, so (9.13) becomes

$$\phi = -L\int_\infty^r \frac{P(r)}{r^2}dr.\tag{9.16}$$

Referring to figure 9.3, we see that in the collision the particle has been deflected through an angle ϑ where

$$\vartheta = \pi - 2\alpha,\tag{9.17}$$

Semi-classical Collisions between Atoms

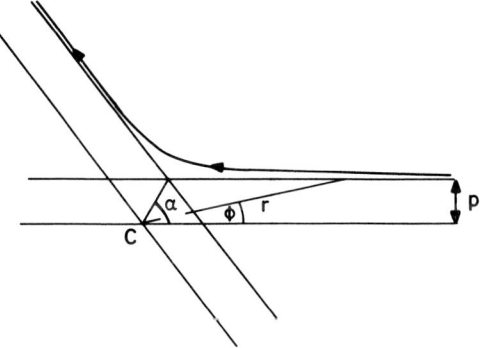

Figure 9.3. A typical orbit of a particle moving from infinity with impact parameter p past a repulsive centre of force C.

α being the value of ϕ when $r = r_0$. From (9.16) we have

$$\alpha = L \int_{r_0}^{\infty} \frac{P(r)}{r^2} dr, \qquad (9.18)$$

so that

$$\vartheta = \pi - 2L \int_{r_0}^{\infty} \frac{P(r)}{r^2} dr, \qquad (9.19)$$

or, in terms of the impact parameter p,

$$\vartheta = \pi - 2p \int_{r}^{\infty} \left(1 - \frac{p^2}{r^2} - \frac{V}{E}\right)^{-1/2} \frac{dr}{r^2}. \qquad (9.20)$$

When $V = 0$, $\vartheta = 0$, as it must be, because on making the substitution $r = p \sec \beta$ we have

$$\int_{p}^{\infty} \left(1 - \frac{p^2}{r^2}\right)^{-1/2} \frac{dr}{r^2} = \frac{1}{p} \int_{0}^{\pi/2} (1 - \cos^2 \beta)^{-1/2} \frac{\sin \beta}{\cos^2 \beta} \cos^2 \beta \, d\beta$$

$$= \frac{1}{p} \int_{0}^{\pi/2} d\beta$$

$$= \pi/2p. \qquad (9.21)$$

The quantity on the right hand side of (9.19), which is a function of the impact parameter p, is known as the *deflection function* and is central to the classical theory of scattering. However, the angle of deflection is not necessarily the same as the angle of scattering θ and we must determine the relation between them before our analysis is complete.

9.3. Relation Between Angle of Deflection and Angle of Scattering

In the experimental study of atomic collisions, it is only possible to measure

Atomic and Molecular Collisions

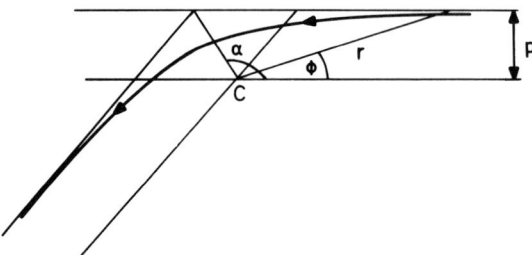

Figure 9.4. A typical orbit of a particle moving from infinity with impact parameter p past an attractive centre of force C, the angle of deflection being $\pi - 2\alpha$.

the angle θ between the initial and final directions of relative motion which will always be between 0 and π. It is easy to see that the angle of deflection as defined by (9.20) may be positive or negative and in either case its magnitude may be greater than π.

If V is never negative, as is the case when the centre exerts a *repulsive* force on the particle at all distances, it is true that ϑ lies between 0 and π and it may then be equated to the angle of scattering.

The conclusion does not follow in general when V may be negative over a certain range of r, i.e., if the centre is *attractive* over this range. Under these circumstances ϑ according to (9.17) may be < 0 while $|\vartheta|$ may exceed π. The reason for this may be understood by contrasting the orbits shown in figures 9.4 and 9.5 with that in figure 9.3.

In the repulsive case, the particle can never be closer to the centre than the impact parameter p, so that the orbit must always be as in figure 9.3. The radius vector to the particle from the centre describes an angle 2α, $< \pi$, as the particle pursues its trajectory from an infinite distance on one side of the centre to an infinite distance on the other side. However, when V may be negative over a range of values of r, these restrictions no longer apply and we may have such situations as are shown in figures 9.4 and 9.5 in which the point

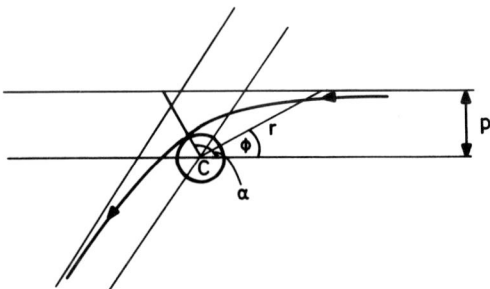

Figure 9.5. A typical orbit of a particle moving from infinity with impact parameter p past an attractive centre of force C. In this case the particle makes one complete revolution about the centre before moving off to infinity once more. The angle of deflection is $-\pi - 2\alpha$ where α is as shown.

of closest distance of approach lies between the centre and the asymptote. The angle turned through by the radius vector is $> \pi$ in figure 9.4 and $> 3\pi$ in figure 9.5. In these cases, the formula (9.20) gives the total angle turned through by the particle during the collision, including the possibility that the particle may make one or more revolutions around the scattering centre. It also takes account of the sense of rotation of the radius vector to the particle, being reckoned positive when it is clockwise and negative when anticlockwise. Thus in figure 9.4 ϑ, as defined by (9.20), is between 0 and $-\pi$ and in figure 9.5, between $-\pi$ and -3π, whereas in figure 9.3 it is positive and less than π.

In figure 9.4 the angle of scattering would be measured as $|\vartheta| - \pi$ and in figure 9.5 as $|\vartheta| - 2\pi$. Allowing for the fact that it is not possible in atomic collision experiments to distinguish between positive and negative values of ϑ, we may write in general

$$\theta = |\vartheta| - s\pi, \qquad (9.22)$$

where s is the largest integer for which (9.22) is > 0.

We now have the complete prescription for determining θ given p, E and $V(r)$. In the following section we apply it to interactions of the form (9.1) typical of that between gas atoms.

9.4. The Angles of Deflection and of Scattering for Interactions Between Rare Gas Atoms

We may readily sketch out the form of the relation between the deflection function ϑ and the impact parameter, for interactions of the form shown in figure 9.1, as follows.

In head-on collisions for which $p = 0$, the particle will penetrate to the closest distance of approach and then be reflected back by the strong repulsion,

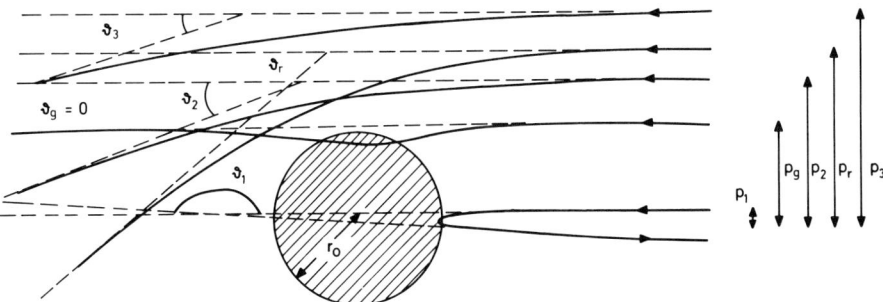

Figure 9.6. The orbits according to classical mechanics, pursued by a particle, moving under the action of a centre of force of the form (9.1). The shaded region is that in which $V(r) > 0$, i.e., the repulsive region. The impact parameters and the corresponding angles of deflection ϑ are shown. When $p = p_g$, $\vartheta_g = 0$ and so is a glory angle, while, for $p = p_r$, $|\vartheta|$ has the maximum value ϑ_r which is therefore a rainbow angle.

so that $\vartheta = \pi$. As p increases, the effect of the attractive region will become more and more important until, at $p = p_g$, it cancels out the effect of the repulsion. For this value of p, $\vartheta = 0$ (see figure 9.6). The attraction will continue to dominate the situation as p increases further, so that ϑ becomes negative and increases in magnitude (see figure 9.6). However, this increase will not continue indefinitely, because the attraction begins to decrease for $r > r_m$. $|\vartheta|$ therefore increases to a maximum at $p = p_r$ (see figure 9.6) and then decreases to approach zero asymptotically as $p \to \infty$. This behaviour of ϑ is illustrated in figure 9.7. It is clear that, even though $|\vartheta|$ may remain between 0 and π for

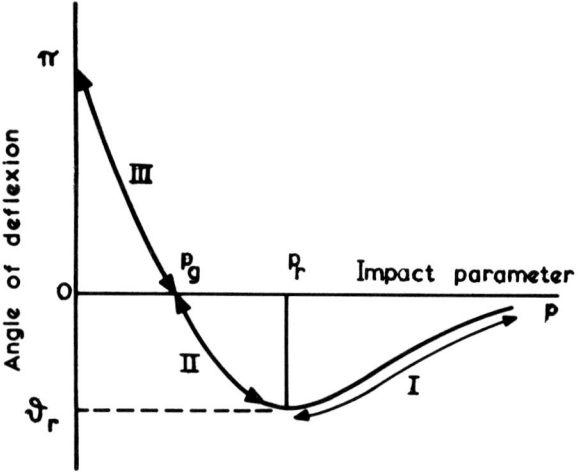

Figure 9.7. Form of the relation between impact parameter p and deflexion angle ϑ derived from (9.6).

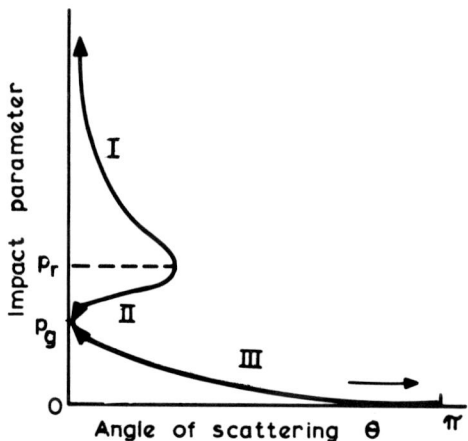

Figure 9.8. Form of the relation between impact parameter p and angle of scattering θ derived from (9.7).

all p, the relation between ϑ and the angle of scattering θ will not be a single-valued one.

To show this, we now plot in figure 9.8 the relation between p and θ derived from (9.22), for the case in which the maximum value of $|\vartheta|$ for $\vartheta < 0$ is $< \pi$ and equal to θ_r. For angles of scattering $\theta < \theta_r$, there are three possible values of p, but for $\theta > \theta_r$ only one. We note also that, at θ_r, $\dfrac{d\theta}{dp}$ vanishes so that $\dfrac{dp}{d\theta}$ tends to infinity. This means that the classical differential cross-section $I(\theta)$ also tends to infinity as $\theta \to \theta_r$. Because of the analogous situation which arises in the optical case of rainbow formation, which we shall next describe, θ_r is referred to as the *rainbow angle*.

9.5. The Optical Rainbow

Optical rainbows arise through the reflection and refraction of light by spherical water drops. Consider a beam of light incident on a droplet of radius a as shown in figure 9.9. Since the wavelength of the light will be small compared to the drop radius, it should be a good approximation to use geometrical optics and consider individual ray paths. These may be distinguished by an impact parameter as shown in figure 9.9, just as for scattering of particles by a centre of force. To determine the intensity of back scattering due to single internal reflection of rays within the drop, we may work in terms of the variation of the deflection angle ϑ of a ray with p.

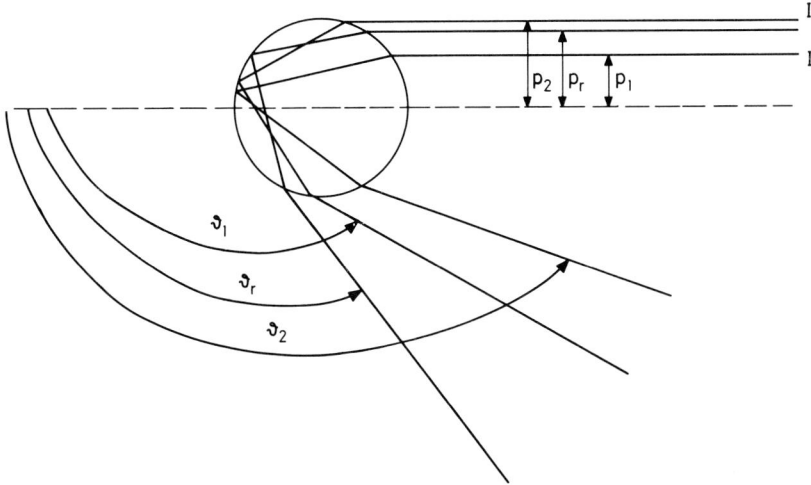

Figure 9.9. Paths of light rays of different impact parameter through a drop of refractive index $\mu > 1$. For the ray with $p = p_r$ the deflexion angle ϑ_r is a minimum so $\dfrac{d\vartheta}{dp} = 0$ and ϑ_r is the rainbow angle.

Within the range of validity of geometrical optics there is no difficulty in calculating ϑ. Let μ be the refractive index of water. Then, for a ray incident at an angle α on the drop, we have

$$\sin \alpha = p/a, \tag{9.23}$$

while, if β is the angle of refraction,

$$\mu \sin \beta = \sin \alpha. \tag{9.24}$$

The ray path will then pass through the drop to be reflected internally at B and pass out of the drop again at C. It is easy to see that the angle of deflection ϑ (see figure 9.9) is given by

$$\vartheta = \pi - 4\beta + 2\alpha, \tag{9.25}$$

which may be expresses as a function of p by means of the relations (9.23) and (9.24). ϑ will be a minimum considered as a function of p when

$$2 \frac{d\beta}{d\alpha} = 1. \tag{9.26}$$

Since from (9.24)

$$\mu \cos \beta \frac{d\beta}{d\alpha} = \cos \alpha, \tag{9.27}$$

(9.26) will be satisfied when

$$2 \cos \alpha = \mu \cos \beta. \tag{9.28}$$

Using (9.24) once again, this gives

$$\sin^2 \alpha = \tfrac{1}{3}(4 - \mu^2) = \sin^2 \alpha_m \text{ say}. \tag{9.29}$$

Thus from (9.23) ϑ will be a minimum when

$$p = p_m = a \left(\frac{4 - \mu^2}{3} \right)^{1/2}. \tag{9.30}$$

For water $\mu \simeq \tfrac{4}{3}$ so

$$\sin \alpha_m = \left(\frac{20}{27} \right)^{1/2}, \quad \sin \beta_m = \left(\frac{5}{12} \right)^{1/2} \tag{9.31}$$

from which $\alpha_m \simeq 59°$, $\beta_m \simeq 40°$ and $\vartheta_m \simeq 138°$.

Figure 9.10 shows the variation of ϑ with p for water drops, taking $\mu = \tfrac{4}{3}$. Just as for the atom–atom collisions, the intensity of back scattering by the water droplet will, according to geometrical optics, have a singularity when $\vartheta = 138°$. Although, as we shall see, the singularity is moderated when the wave aspect of the phenomenon is properly included, the intensity of back

Semi-classical Collisions between Atoms

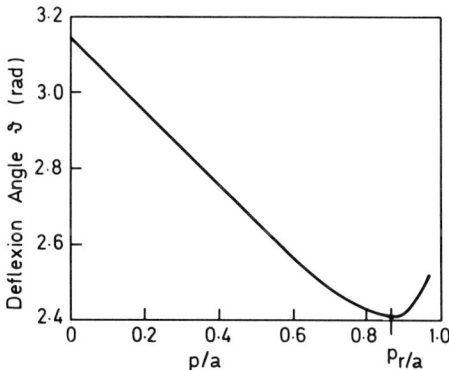

Figure 9.10. The relation between the impact parameter p and deflection angle ϑ for light rays scattered by a spherical drop of radius a after two internal reflections.

scattering will have a strong maximum for ϑ close to ϑ_m. In fact, because the refractive index varies somewhat with the wavelength λ, ϑ_m will also depend on μ so that back-scattered white light will be dispersed into a coloured rainbow with the red arc uppermost.

We shall refer again to the optical rainbow in section 9.9 when we are discussing the wave theory of the phenomenon in relation to the collision problem.

9.6. The Glory Singularity

Turning once more back to figure 9.8, we note that the angle of scattering θ vanishes when $p = p_g$. Since the differential cross-section $I(\theta)$ contains a factor $\csc\theta$ the classical cross-section will tend to infinity as $p \to p_g$. This is known as a *forward glory* singularity as distinct from a *backward glory* which arises when $\theta = \pi$. Again the singularities are moderated by wave effects, but nevertheless the intensity of scattering remains high in their vicinity.

Optical glory effects are of frequent occurrence. Thus backward glory scattering is responsible for the bright halos round shadows and is made use of, for example, in reflecting traffic signs in which the scattering takes place from small plastic beads.

9.7. The Classical Differential Cross-Section

To build up the classical differential cross-section $I(\theta)$, we consider the three branches I, II and III distinguished in figure 9.8, each of which can be regarded as making a separate additive contribution as shown in figure 9.11. For $\theta > \theta_r$, only Branch III gives a contribution, one which is finite everywhere except as $\theta \to \pi$. Branches I and II, which only contribute for $\theta \leq \theta_r$, give contributions which are singular also at the rainbow angle. Adding the three gives the resultant cross-section shown in figure 9.11.

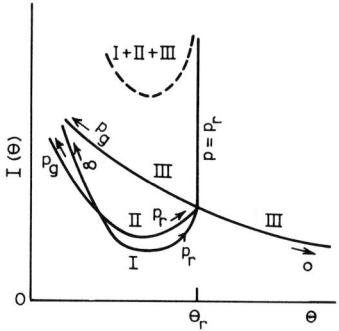

Figure 9.11. Classical differential cross-section $I(\theta)$ per unit solid angle derived from the relation shown in figure 9.8. Branches I, II and III are derived from the curves I, II and III of that figure. For $\theta < \theta_r$ the cross-section is the sum I + II + III of the contributions from each branch while for $\theta > \theta_r$ it is given by III alone.

9.8. Orbiting

Before going on to discuss the modification of the classical cross-sections by wave phenomena, we must note that, while in figure 9.7 the magnitude of the rainbow angle ϑ_r is $< \pi$, this need not be so in general. Indeed, under certain important circumstances, it may tend to infinity, giving rise to what is known as *orbiting*.

To see physically how and when this arises, we first note that the radial velocity u_r of the particle of mass m, total energy E and angular momentum L, in the collision is given by

$$\tfrac{1}{2}mu_r^2 = E - V(r) - L^2/2mr^2, \qquad (9.32)$$

or, since

$$L^2 = m^2v^2p^2$$
$$= 2mEp^2, \qquad (9.33)$$

by

$$\tfrac{1}{2}mu_r^2 = E - V_{\text{eff}}(r) \qquad (9.34)$$

where

$$V_{\text{eff}} = V(r) + p^2E/r^2. \qquad (9.35)$$

V_{eff} is the effective potential energy for the radial motion, p^2E/r^2 representing the contribution from the centrifugal force. Thus

$$-\frac{\mathrm{d}}{\mathrm{d}r}(p^2E/r^2) = 2p^2E/r^3$$
$$= mu_\phi^2/r,$$

Semi-classical Collisions between Atoms

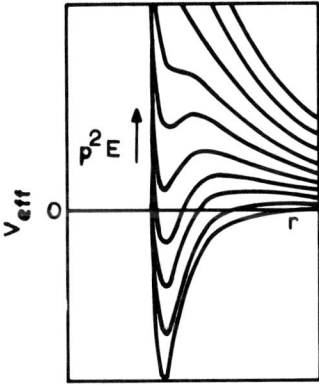

Figure 9.12. V_{eff} given by (9.35) as a function of r for different values of p^2E which increase in proceeding from the lowermost to the uppermost curve.

u_ϕ being the transverse velocity L/mr (see (9.8)).

In figure 9.12, V_{eff} is illustrated as a function of r for different values of p^2E, V being of the form shown in figure 9.1. For large p, V_{eff} falls off monotonically with r but, as p decreases, a maximum and minimum appear. If the radial velocity u_r, given by (9.34), is small over an appreciable range of r, the angular motion will carry the particle several times round the centre before the effect of the radial velocity in changing r becomes significant. Ultimately, for a particular value of p, the radial velocity may vanish at a maximum of V_{eff}, in which case the number of revolutions performed will tend to infinity. This is the phenomenon of *orbiting* and corresponds in a plot, such as that in figure 9.7, to the magnitude of the rainbow angle ϑ_r tending to infinity.

The critical impact parameter p_c for orbiting is important because, in collisions with $p \leq p_c$ the particle will spend a long time close to the centre, so that energetically possible reactions are likely to occur. This aspect is discussed further in Chapter 11.

The form taken by the classical scattered intensity when orbiting occurs is considerably more complex than for a rainbow and will not be considered further here.

9.9. The Quantum Scattering Formula near the Classical Limit—The 'Semi-Geometrical' Wave Theory of the Optical Rainbow

To see how the classical differential cross-section shown in figure 9.11 is modified when account is taken of the wave nature of the phenomenon, it is instructive to begin by considering the relatively simple case of the optical rainbow.

Essentially the main modification is introduced when allowance is made for the phase differences between rays which, while emerging with the same deflection angle, have pursued paths of different length within the droplet. Thus referring to figure 9.9, the rays I and II both emerge with nearly the same deflection angle ϑ, but the latter has traversed a smaller distance within the drop. If l_1 and l_2 are these distances, there will be a phase difference $2\pi(l_1 - l_2)/\lambda'$ between them on emergence, λ' being the wavelength λ/μ within the drop, in the notation of section 9.5.

The total amplitude scattered at the deflection angle ϑ can now be written

$$|f_1|\exp(i\eta_1) + |f_2|\exp(i\eta_2), \qquad (9.36)$$

where

$$\eta_1 = 2\pi l_1/\lambda', \eta_2 = 2\pi l_2/\lambda', \qquad (9.37)$$

and the scattered intensity becomes

$$|f_1|^2 + |f_2|^2 + 2|f_1||f_2|\cos(\eta_1 - \eta_2). \qquad (9.38)$$

Except close to the rainbow angle. $|f_1|^2 + |f_2|^2$ is just the intensity predicted by classical theory. However, because there are two values of p for each value of ϑ, the phase difference between the two emergent rays introduces an oscillating interference term which leads to maxima and minima in the angular distribution. Under nearly classical conditions, the phase difference may be calculated very simply, without resort to any wave theory.

Near ϑ_r the scattered amplitude must be calculated using wave theory. At $\vartheta_r, \dfrac{d\vartheta}{dp} \to 0$ but, when $\dfrac{d^2\vartheta}{dp^2}$ is taken into account, it may be shown that the scattered intensity, while large near ϑ_r, nevertheless remains finite, in contrast

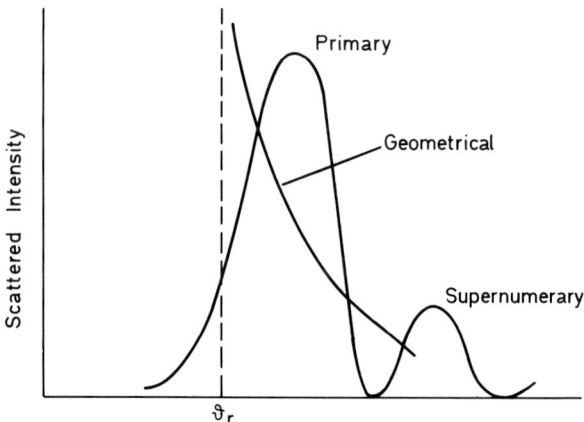

Figure 9.13. Typical variation with angle of deflection of the intensity of radiation of a particular wavelength scattered from a water droplet when account is taken of the wave nature of the radiation.

Semi-classical Collisions between Atoms

to the purely classical theory. Taking into account these modifications, the intensity scattered by the drop varies with angle as shown in figure 9.13. Thus there is now, not only a primary rainbow maximum, as indicated, which occurs close to ϑ_r, but a number of supernumerary maxima arising from the interference term in (9.38).

We refer to this theory as 'semi-geometrical' because the wave effects are included in terms of interference between rays whose relative phases are calculable in simple geometrical terms. It is valid only when the wavelength of the radiation is small compared with the radius of the drop.

9.10. *The Semi-Classical Wave Theory of the Scattering of Particles*

Corresponding to the semi-geometrical theory of the optical rainbow, we would expect to have a semi-classical wave theory of the scattering of particles. This is indeed the case, although the problem is a little more complicated because the scattering potential varies with distance and has no sharp boundary.

Instead of the constant wavelength within the droplet, we use the concept of the local wavelength (see Chapter 2, p. 19) which varies from point to point along the path of the particle which is still considered to be well defined. Thus, consider the scattering of particles of mass m and velocity v by a fixed centre of force which interacts with it with a potential energy $V(r)$. At infinity the wavelength is simply $\lambda(\infty) = h/mv$. This is replaced at a point distant r from the centre by

$$\lambda(r) = h/mu(r), \qquad (9.39)$$

where $u(r)$ is the velocity at r given by

$$\tfrac{1}{2}mu^2 = E - V(r).$$

This concept will be useful provided the change of $V(r)$ in a distance $\lambda(r)$ is small compared with $V(r)$, i.e., that

$$\frac{\lambda(r)}{V(r)} \frac{dV}{dr} \ll 1, \qquad (9.40)$$

a condition which is automatically satisfied within the uniform water drop.

We are now in a position to calculate, to the accuracy of this approximation, the phase change produced in passage along a trajectory of impact parameter p with respect to the centre of force. Thus if $N_1(p)$ is the number of wavelengths contained within the actual trajectory and $N_0(p)$ the corresponding number for the same impact parameter if $V(r)$ were zero, then the phase change $C(p)$ is given by

$$2\pi\{N_1(p) - N_0(p)\}. \qquad (9.41)$$

It remains to calculate N_1 and N_0.

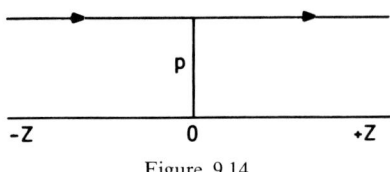

Figure 9.14.

Suppose that, when $r > R$, the effect of the centre of force on the trajectory can be neglected. For N_0 we have simply, in the notation of figure 9.14,

$$N_0 = \frac{1}{\lambda(\infty)} \int_{-z}^{z} dz \qquad (9.42)$$

where $z^2 = R^2 - p^2$. Since $z^2 + p^2 = r^2$ we have

$$z\, dz = r\, dr, \qquad (9.43)$$

so

$$N_0 = \frac{2}{\lambda(\infty)} \int_p^R \frac{r\, dr}{(r^2 - p^2)^{1/2}}$$

$$= \frac{2}{h} \int_p^\infty \frac{2mE\, dr}{\{2mE - L^2/r^2\}^{1/2}}. \qquad (9.44)$$

For N_1 we must integrate along the actual trajectory and use the local wavelength, so

$$N_1 = \int \frac{ds}{\lambda(r)}, \qquad (9.45)$$

where ds is now an element of length along the trajectory given (see (9.4)) by

$$ds^2 = dr^2 + r^2\, d\phi^2,$$

$$= dr^2 \left\{ 1 + r^2 \left(\frac{d\phi}{dr}\right)^2 \right\}. \qquad (9.46)$$

Also from (9.11)

$$\frac{d\phi}{dr} = \frac{L}{r^2} P(r),$$

$$L = mvp,$$

$$P(r) = \{2m(E - V) - L^2/r^2\}^{-1/2}.$$

Hence

$$1 + r^2 \left(\frac{d\phi}{dr}\right)^2 = \left\{ \frac{2m(E - V)}{2m(E - V) - L^2/r^2} \right\} \qquad (9.47)$$

Semi-classical Collisions between Atoms

and

$$ds = dr \left\{ \frac{2m(E-V)}{2m(E-V) - L^2/r^2} \right\}^{1/2}. \tag{9.48}$$

On substitution in (9.45) from (9.39) and (9.48) we have

$$N_1 = \frac{2}{h} \int_{r_0}^{R} \frac{2m(E-V)\,dr}{\{2m(E-V) - L^2/r^2\}^{1/2}}, \tag{9.49}$$

where r_0 is the outermost zero of the integrand.

In the limit $R \to \infty$, both N_0 and N_1 tend to infinity but $N_0 - N_1$ remains finite, so that we may write for the phase difference

$$c(p) = \frac{8\pi m}{h} \left[\int_{r_0}^{\infty} \frac{(E-V)\,dr}{\{2m(E-V) - L^2/r^2\}^{1/2}} - \int_{p}^{\infty} \frac{E\,dr}{\{2mE - L^2/r^2\}^{1/2}} \right]. \tag{9.50}$$

Within the semi-classical approximation, the contribution to the scattered amplitude arising from the trajectory considered will now be

$$|f(\theta)| \exp(iC(p)), \tag{9.51}$$

where $|f(\theta)|^2$ is the classical value of the scattered intensity $I(\theta)$ as in (1.31) of Chapter 1, i.e.,

$$|f(\theta)| = \left\{ \frac{1}{\sin\theta} p \left| \frac{dp}{d\theta} \right| \right\}^{1/2}. \tag{9.52}$$

It may be shown that, under the conditions assumed, the full wave formula for the scattered amplitude

$$f(\theta) = -\frac{i\lambda(\infty)}{4\pi} \sum_{l} i^l (2l+1)[(\exp 2i\eta_l) - 1] P_l(\cos\theta) \tag{9.53}$$

discussed in Chapter 2, does indeed reduce to (9.51) with $|f(\theta)|$ and $C(p)$ given by (9.52) and (9.50) respectively.

Just as for the optical rainbow, the analysis must be extended to deal with scattering at and close to the rainbow angle θ_r. As for the optical case, the classical singularity at θ_r no longer appears but the scattering nevertheless has a strong maximum close to θ_r.

Having derived the form (9.51) for the contribution of the amplitude scattered at an angle θ from a trajectory of impact parameter p, the total amplitude is obtained, as for the optical rainbow, by adding the amplitudes from each impact parameter p_1, p_2 which leads to the same scattering angle θ on classical theory. The only additional feature, as seen from figure 9.8, is that there is now a range of values of $\theta(< \theta_r)$ in which contributions come from not just two but three values of p. Because of this, the rainbow pattern shown in figure 9.15 is modulated by rapid oscillations due to interference with the scattered

Atomic and Molecular Collisions

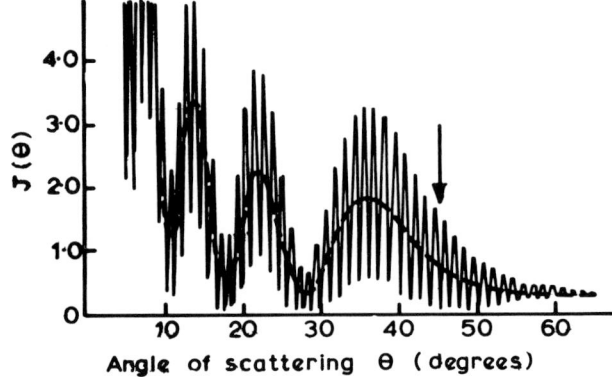

Figure 9.15. Typical form of the variation of the scattered intensity $J(\theta)$ per unit angle in the neighbourhood of a rainbow, for collisions in which the relation between impact parameter p and angle of scattering θ is of the form shown in figure 9.8. The arrow indicates the rainbow angle θ_r. The rapid oscillations are due to interference of the rainbow scattering with the scattering from trajectories in branch III of figure 9.8. The broken curve results when this interference is averaged out. It is of the same general form as for the optical rainbow (see figure 9.13) when θ is replaced by $\pi - \theta$.

amplitude arising from the trajectory in the branch III of figure 9.8. The scattered intensity then has the general form shown in figure 9.15, with the positions of the rainbow maxima as indicated. It may be compared with the observed angular distributions shown in Chapter 10, figures 10.11 and 10.12.

9.11. Glory Undulations

The incorporation of the glory effect into the wave theory of scattering leads to an interference phenomenon which is most clearly manifest in the variation of the total cross-section with impact velocity.

To see how this arises, it is best to consider a slightly simplified picture, of scattering by a potential $V(r)$ which vanishes for $r > a$. The measurement of the total cross-section essentially requires the determination of the radius of the cylindrical shadow region behind the scatterer when it is bombarded by the beam of particles. Classically this involves measuring the minimum separation of two beams on diametrically opposite sides of the cylinder which are undeviated. However, a trajectory for which the impact parameter is p_g in figure 9.8, a glory trajectory, is also undeviated but will be distinguished from a 'truly' unaffected trajectory by a phase change due to passage through the region in which $V(r)$ is finite. This will, in general, decrease as the impact energy increases. Interference will take place between the glory trajectory and the unaffected trajectories which will lead to fluctuations in apparent size of the scattering region, the scattering cross-section, as the impact energy increases. The amplitude associated with the glory trajectory may be written in

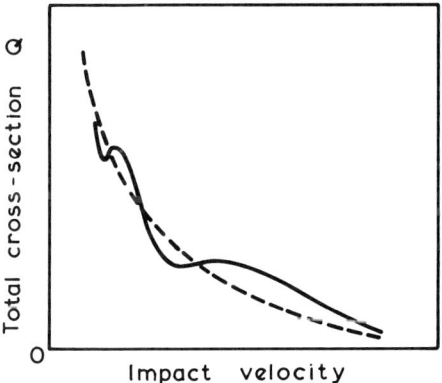

Figure 9.16. Effect of glory scattering on the variation of the total cross-section with impact velocity.
———, variation when a glory exists;
- - - - -, variation if the effect of the glory is neglected.

the form

$$|f_g(0)| \exp(iC_g) \tag{9.54}$$

where $|f_g(0)|$ is finite, in contrast to the classical theory. It can then be shown that the total collision cross-section can be written as

$$Q = \bar{Q} + \Delta Q \tag{9.55}$$

where $\bar{Q}$ is a mean cross-section varying gradually with impact velocity while ΔQ varies as cos C_g. The general form of the variation of Q with impact velocity is thus as shown in figure 9.16. Observed examples are shown in figures 10.7 and 10.10 of Chapter 10.

9.12. Collisions Between Similar Atoms—Symmetry Interference Effects

The interference effects discussed above occur whether or not the colliding atoms are of the same kind. If the colliding atoms are indistinguishable then additional effects arise.

In a scattering experiment in which, say, a beam of atoms A collides with target atoms of the same kind, initially at rest, suppose one of the beam atoms A is scattered in a direction making an angle Θ with the incident direction. This means that, at the same time, a target atom A is projected in a direction making an angle $\frac{\pi}{2} - \Theta$ with that of incidence (see Chapter 1, p. 8). In terms of relative coordinates $\theta_r = 2\Theta$, when an atom is scattered through an angle θ there must also be an atom projected through $\pi - \theta$. If it is not possible to distinguish the scattered and projected atoms, we would expect that the intensity observed per unit solid angle in the direction θ will be obtained simply by adding the contributions from scattering and projection. Thus, if $I(\theta)$ is the differential cross-section for scattering through an angle θ, what we would expect to observe would be

$$I(\theta) + I(\pi - \theta) = |f(\theta)|^2 + |f(\pi - \theta)|^2. \tag{9.56}$$

According to wave mechanics, however, this is not so. As might be expected, we must combine not scattered intensities but scattered amplitudes. It is a little more surprising that it is not always just a matter of *adding* amplitudes. For certain atoms, including the helium atoms of mass number 4, which possess no net angular momentum in their normal states, it is correct to add the amplitudes. In such cases, the differential cross-section becomes

$$|f(\theta) + f(\pi - \theta)|^2. \tag{9.57}$$

New interference effects will arise because of the phase difference between $f(\theta)$ and $f(\pi - \theta)$. These occur both in the observed angular distribution of scattered atoms and in the total cross-section as a function of impact velocity.

In particular, they are not confined to scattering angles less than the rainbow angle so that oscillations in the angular distribution at large angles of scattering (though less than $\pi/2$) can be identified as arising from symmetry interference. Observed examples are illustrated in figure 10.18 of the following chapter.

If the atom possesses a net angular momentum with quantum number $\frac{1}{2}$, as for example the helium isotope of mass 3, (9.57) must be replaced by

$$\tfrac{1}{4}|f(\theta) + f(\pi - \theta)|^2 + \tfrac{3}{4}|f(\theta) - f(\pi - \theta)|^2. \tag{9.58}$$

The contrast between the observed interference effects observed in ^{4}He–^{4}He and ^{3}He–^{3}He collisions is illustrated in figure 10.16 of Chapter 10.

It is important to realize that these interference effects only arise when the colliding systems are indistinguishable. No effects of this kind occur in ^{4}He–^{3}He collisions, for example.

The existence of interference effects arising from indistinguishability of incident and target species was first predicted by Mott in 1928 in relation to the scattering of alpha particles by helium nuclei. This prediction was verified experimentally by Chadwick in the following year. Direct observation of the effect in near-thermal collisions between neutral atoms, hampered by severe requirements of energy and angular resolution, was not achieved until 1969, by Dondi, Scales, Torello and Pauly using apparatus described on p. 194 of Chapter 10. Symmetry interference effects arise also in collisions of positive ions with atoms of the same species and are discussed in Chapters 11 and 12.

9.13. *Application to Determination of Atomic Interactions*

Observations of total and differential cross-sections as functions of impact energy provide an important method for determining the interaction energy between pairs of atoms.

In terms of the form of interaction shown in figure 9.1, the magnitude of the rainbow angle is determined mainly by the ratio ε/E of the well depth of the potential to the impact energy. Thus for an interaction of the form (9.1) with $n = 12$

$$\vartheta_r \simeq -2\cdot05\varepsilon/E. \qquad (9.59)$$

Hence merely from the location of the primary rainbow maximum, a good approximation to ε may be obtained. The separation between the primary and supernumerary rainbows is determined mainly by r_m/λ where λ is the wavelength of relative motion.

Because the glory trajectory penetrates to smaller values of r than does the rainbow trajectory (see figure 9.2), it follows that the glory undulations provide different information about the scattering potential. The glory phase C_g is determined mainly by the product εr_m and for an interaction of the form (9.1) with $n = 12$ is given by

$$C_g = 0\cdot536\pi(2\pi\varepsilon r_m/hv)\{1 - 0\cdot25(\varepsilon/E)^{1/2}\}. \qquad (9.60)$$

While analysis of rainbow scattering and the associated interference effects, as well as of the glory undulations, provides a useful approximation to the scattering potential, observed data now available are so precise that the analysis may be carried further, to determine the deflection function directly. Alternatively, elaborate analytical forms for the interaction, involving a number of adjustable parameters are assumed and the parameters adjusted to give the best fit with the observed scattering data. In such cases, the values of ε and r_m, derived quickly from the rainbow and glory effects, may be used to give initial approximations to certain of the parameters.

CHAPTER 10
the experimental study of atom—atom collisions at thermal energies

In the preceding chapter it was shown from theoretical arguments based on semi-classical considerations that a great number of interesting results could be obtained from experimental measurement of total and differential cross-sections for collisions between atoms moving with energies characteristic of gas atoms at ordinary temperatures and below. Not only do such measurements offer the most direct means available for determining the interaction energy $V(r)$ between two atoms as a function of separation r between their nuclei, but they also provide confirmation of the predicted rich variety of interference effects arising from the wave aspect of the collisions.

The first experiments of this kind were carried out in the 1930s, particularly by Stern and his collaborators at Hamburg, but it is only in recent years that the technique has developed to such a stage that a thorough study of the interference effects, even for the collisions of quite heavy atoms, has become possible in detail. At the same time, it is becoming possible to determine atomic interactions to quite good accuracy.

In this chapter we first describe something of the techniques involved in these experiments and then give an account of some of the results obtained and their relation to the theoretical considerations of Chapter 9.

10.1. Experimental Methods—General Remarks

Just as for other scattering experiments designed to measure total and differential cross-sections as functions of impact energy and angle of scattering, it is necessary to have means available for the production of beams of the atoms concerned which are well defined geometrically and in which the kinetic energies of the atoms are confined within a narrow range of values which, in this case, are not far from those corresponding to thermal energies ($0.03T$ eV where T is the kelvin temperature). Such beams are said to be velocity-selected. It is further necessary to have means for measuring not only the particle flux in the beams but also the very much smaller fluxes of scattered atoms. For the latter purpose, the main problem is that of distinguishing the wanted signals from usually much larger background signals. To reduce the latter as much as possible, it is necessary to use high pumping speeds in the vacuum system.

Experimental Study of Atom–Atom Collisions

Even provided with these facilities, it is important that the angular resolution available should be high in order that the true total cross-section should be measured with sufficient accuracy. As a rough criterion, if θ_0 is the minimum angle of scattering measured in the laboratory system which is detectable, then, if the error in the measured cross-section should not exceed 10%,

$$\theta_0 \simeq 300/a(MT)^{1/2} \tag{10.1}$$

where M is the atomic mass of the incident atoms, T their equivalent temperature (in K) and a the sum of the gas-kinetic radii of the colliding atoms derived from viscosity or other properties of the gas, measured in nanometres (nm). Thus at room temperature for caesium atoms scattered by argon atoms θ_0 is about 0·14°.

Measurements of total cross-sections may be carried out either by observing the absorption of an atomic beam in a cell containing the target gas or by passage through an intersecting beam of the target atoms. Differential cross-sections can be measured effectively only in crossed-beam experiments so that the interacting region is well defined.

While a complete experimental study of the elastic collisions between gas atoms would involve the absolute measurement of the differential cross-section as a function of angle of scattering and relative impact velocity, many experiments have been directed towards the measurement of the total cross-section Q as a function of relative velocity. Even if in such measurements the absolute value of Q is not measured, sufficiently accurate observations of relative values at different velocities can lead to results of much interest.

Thus, referring to Chapter 9, p. 174, if the respective slowly-varying and oscillatory components $\bar{Q}$ and ΔQ of Q can be separated out, the variation of $\bar{Q}$ with v will check whether the interaction effective in the scattering is that of Van der Waals varying as r^{-6} for large r. The existence of the oscillatory term establishes the glory effect and provides information about the interaction parameter εr_m (see figure 9.1). Absolute measurements of $\bar{Q}$ provide, in addition, the magnitude of the Van der Waals interaction, i.e., the parameter C in (9.1) of Chapter 9.

If l is the path of the beam through the attenuating gas and n the concentration of the target atoms, the fractional loss F of intensity in the beam after passing through the gas will be $\exp\{-nQl\}$ (see Chapter 5, p. 74). It follows that, if F_1 and F_2 are the fractional losses under two different conditions of beam velocity the ratio Q_1/Q_2 of the cross-section will be given by

$$Q_1/Q_2 = \ln F_1/\ln F_2. \tag{10.2}$$

Thus, whereas measurement of the absolute value of Q requires knowledge of n and l, the relative values depend only on the relative values of $\ln F$. This is important, because in practice n, and l especially, may be difficult to measure accurately. These difficulties are particularly acute if the crossed beam method is used.

The problem of working under conditions in which the relative impact velocity is well defined is not solved even when the atoms in the incident beam have well defined kinetic energies. This is because, in contrast say to experiments on electron scattering by atoms, the kinetic energies of the target atoms are comparable with those of the atoms in the beam. In principle, this could be overcome in a crossed-beam experiment by ensuring that the atoms in the target beam also have well-defined energies. However, velocity selection can be carried out only at the expense of beam flux. The scattered intensity is already low in a crossed-beam experiment in which the energies of the target atom are not selected but have a thermal spread, so to work with both beams with selected energies is normally impracticable. Nevertheless, much may be done by cooling the target atoms by liquid nitrogen or even by liquid hydrogen or helium so their root mean square velocities are low compared with the selected velocity of the incident atoms. Even if this is not done, it is possible to allow for the thermal distribution of velocities of the target atoms. From effective cross-sections obtained using (10.2), cross-sections for a definite relative impact velocity may be derived by a suitable analysis.

We now consider in a little more detail how velocity-selected atomic beams are prepared and how they are detected and their intensity measured.

10.2. The Production of Atomic Beams

For the collision experiments with which we are concerned here, the atomic beams must be unidirectional and the atoms composing them must not collide with each other. The conditions are therefore of effusive flow of gas (Appendix 1). The first experimental production of an atom beam of this kind was carried out by Dunoyer in 1911, but the main development of the technique was due to Stern and his collaborators over the decade from 1923 to 1933. In recent years the most outstanding new feature has been the introduction of the supersonic nozzle source suggested by Kantrowitz and Gray in 1950.

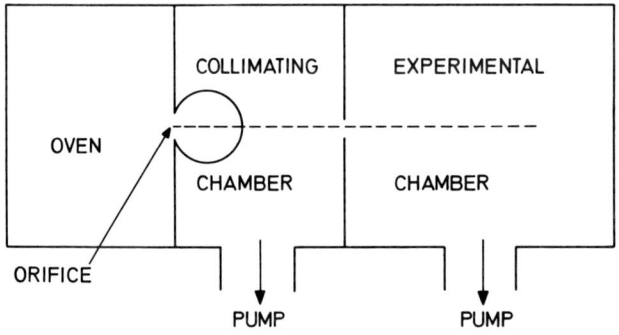

Figure 10.1. An effusive source of atomic beams.

Experimental Study of Atom–Atom Collisions

Until that time the general principles employed in the production of an atomic beam were as illustrated in figure 10.1. The source consisted of a small chamber containing a gas or vapour at a pressure of a few torr. The atoms effused from the chamber through a slit or orifice, known as the source slit, usually about 0·02 mm wide and 10 mm or more high. At the chamber pressure the mean free path of the atoms is longer than the slit width and this is the condition that the flow should be effusive and not hydrodynamic (see Appendix 1).

Thus the mean free path l is given by $1/nQ$, where n is the number of atoms per unit volume and Q is the effective mean collision cross-section. At 1 torr pressure, $n = 3 \cdot 8 \times 10^{22}$ m^{-3} so that, for Q equal to say 10^{-18} m^2, l would be $2 \cdot 7 \times 10^{-5}$ m or 0·03 mm. The number δJ of molecules which emerge per second from the orifice in directions within the solid angle $\delta\omega$ about an angle θ normal to the plane of the slit is given by

$$\delta J = \frac{\delta\omega}{4\pi} n\bar{v} \cos\theta A_s, \tag{10.3}$$

where n is the number of atoms per unit volume, $\bar{v}$ is the mean molecular velocity inside the source and A_s the area of the source slit. The total number emerging per second, obtained by integrating (10.3) over all solid angles, is then

$$J = \tfrac{1}{4} n\bar{v} A_s. \tag{10.4}$$

A fraction of the emergent atoms is then selected by a second slit of area A_c from which emerges an approximately ribbon-shaped beam parallel to the line joining the centres of the slits. The number of atoms which pass through this collimating slit per second is now given by

$$I = \frac{1}{4\pi} \frac{A_c}{D^2} n\bar{v} A_s, \tag{10.5}$$

where D is the distance between the slits.

In practice, it is found that before serious departure from effusive flow conditions sets in, the concentration n of atoms in the source chamber can be increased until the mean free path in the chamber is approximately equal to the source slit width. At higher concentrations, atoms are lost from the beam through scattering in a cloud which forms just outside the slit.

For atoms of elements normally gaseous and monatomic at the temperatures involved, such as those of the rare gases, the source chamber may simply be a capillary through which the gas flows. If the atoms belong to elements which have a low vapour pressure at the working temperatures but the vapours are largely monatomic, as for the alkali metals, a heated oven containing a sample of the solid element may be used as source.

An oven source is also appropriate for the production of an atomic beam from a gaseous element such as hydrogen (H_2) which is normally diatomic.

Under suitably low pressure and high temperature conditions, a useful fraction of molecules may be dissociated to give at least a mixed atomic and molecular beam. Alternatively, dissociation may be produced by application of an electric discharge to the gas in the source chamber. This will be necessary for such gases as nitrogen, the molecules (N_2) of which are very strongly bound. However, it has the disadvantage of introducing metastable atomic and molecular species into the beam which lead to spurious effects in scattering experiments. On the other hand, it is possible to prepare a mixed beam of normal and metastable helium atoms by applying a discharge to helium gas in a source chamber. As it is possible to measure the metastable atom flux separately, this provides a means of studying collisions of metastable excited helium atoms with other atoms or molecules.

The important advance made by replacing effusive flow by supersonic flow through a nozzle, as suggested by Kantrowitz and Gray, is that a large part of the energy of random motion is converted into that of directed motion along the beam—the supersonic velocity U being large compared with the root mean square random velocity. This leads, for two reasons, to a gain in intensity for a given concentration of atoms in the source. Thus the supersonic flow is directed towards the source slit so that a larger fraction passes through while, of that fraction, for the same reason, a larger fraction passes through the collimating slit. The first of these leads to a gain by a factor of 5 or so and the second a further factor of 10 or so, giving an overall gain by a factor of 50 or more.

In addition to this advantage, the fact that the supersonic flow speed is high compared with the random velocities means that the velocity distribution in the beam is sharper than for pure effusion. The problem of velocity selection is therefore less difficult for the nozzle beam and in some cases the sharpening of the velocity distribution is sufficient in itself (see figure 10.3). It increases with the Mach number of the supersonic flow, i.e., with the ratio of the flow velocity to the velocity of sound in the gas.

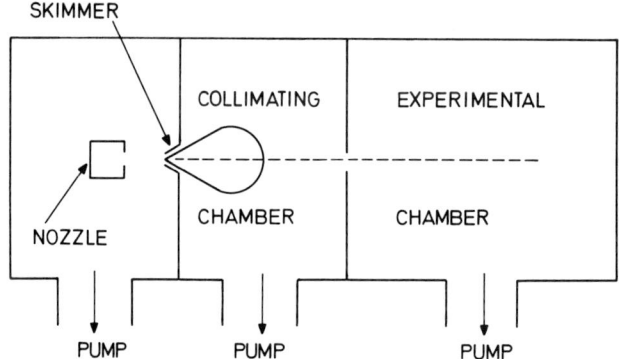

Figure 10.2. A supersonic nozzle source of atomic beams.

A brief account of supersonic flow through a nozzle is given in Appendix 3. Figure 10.2 shows schematically how a nozzle source is arranged, it being seen that the source slit is replaced by a skimmer, the angle of which must be suitably chosen.

In most recent experiments, nozzle sources are in use, though a great deal of valuable results were obtained in earlier work using pure effusion.

10.3. Velocity Selection

The beam issuing from an effusion source possesses a thermal distribution of velocities. While a nozzle beam has a narrow velocity distribution, it is still necessary, in many cases, to introduce some means of selecting from the beam only those atoms whose velocities deviate much less from the mean. Figure 10.3 illustrates these distributions for air at 300 K as source gas, both in the case of effusive flow and for a nozzle beam with Mach number equal to 4.

Velocity selection is usually done by the use of mechanical selectors which operate on the same principle as that used by Fizeau in his slotted disc method of measuring the velocity of light. Essentially, two toothed wheels rotate about a common axle. Atoms which pass through a slot in the first wheel, in the direction of the axle, will also pass through a slot in the second wheel if the velocity of the atom falls between narrow limits determined by the geometry of the wheels, the distance between them and their speed of rotation.

Thus in figure 10.4 the surfaces of the slotted wheels are shown unrolled on a plane. Each wheel is of radius r and thickness d and they are at a distance L apart. Each slot is of width l_1, and each tooth of width l_2. The selector is spun with the wheels rotated with respect to each other through an angle ϕ.

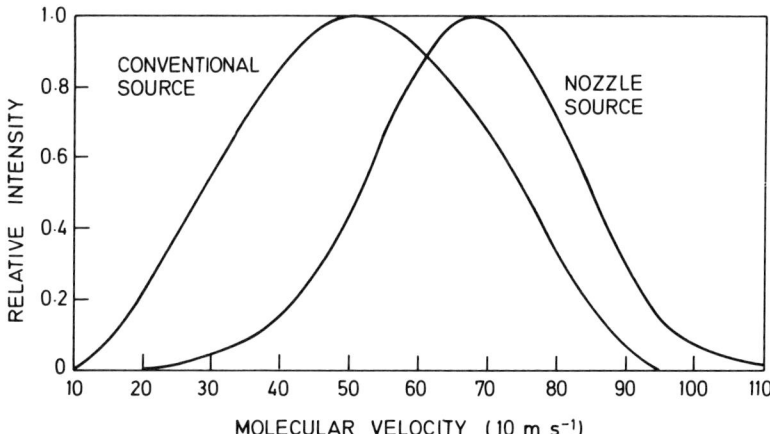

Figure 10.3. Typical velocity distributions of molecules from a conventional effusive source and from a nozzle source.

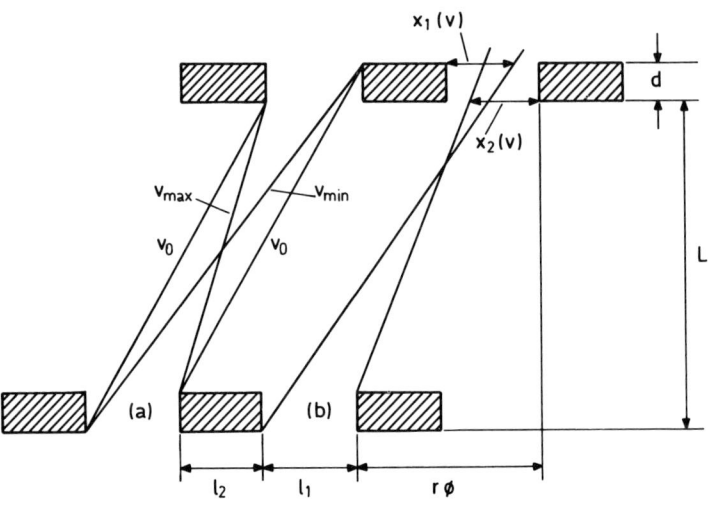

Figure 10.4. The geometrical arrangement of a slotted disc velocity selector.

If a velocity v_0 is to be selected then the wheels must be spun at an angular velocity ω so that, in the time taken for a wheel to rotate through ϕ an atom with this velocity will just traverse the distance L, i.e., we must have

$$v_0/L = \omega/\phi. \qquad (10.6)$$

Because of the finite wheel thickness and slot length, velocities between v_{max} and v_{min} will be transmitted, where

$$v_{max}/(L-d) = \omega/(\phi - l_1/r), \qquad (10.7)$$

$$v_{min}/(L+d) = \omega/(\phi + l_1/r). \qquad (10.8)$$

The velocity resolution R may be defined by

$$R = (v_{max} - v_{min})/2v_0, \qquad (10.9)$$

With the two-wheel system, an atom with velocity v_s say will also be transmitted if

$$v_s/L = n\omega/\phi \qquad (10.10)$$

where n is an integer. To eliminate transmission of atoms with velocities v_0/n where n is an integer > 1, the number of rotating slotted wheels is increased to six or more. One of these must be placed at a separation from the first of L/n_1, where n_1 is half integral.

In a typical velocity selector used in the experiments of Buck and Pauly described below (p. 188), seven discs were used with the following dimensions

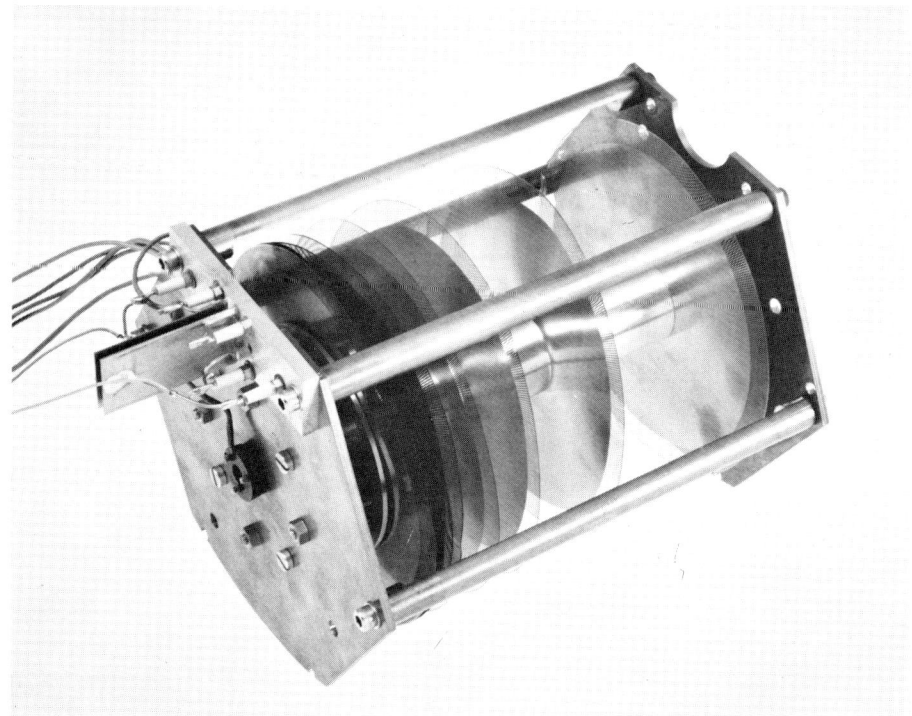

Figure 10.5. Velocity selector designed by Trujillo for selection of high velocity atomic beams.

in mm

$$l_1 = 0{\cdot}32, l_2 = 0{\cdot}38, d = 1{\cdot}6, r = 71{\cdot}3, L + d = 85{\cdot}5. \tag{10.11}$$

The angle $r\phi/L$ was 0·1228 radians. With this selector, a frequency of revolution of 1 Hz selected a velocity of 3·75 m s^{-1}. Rotation rates could be chosen in the range 0 to 500 Hz. The velocity resolution defined by (10.9) was 1·3% while the fraction of incident atoms transmitted under the working conditions was 18·5%.

The maximum velocity which can be selected by mechanical systems of this kind is limited by the angular velocity which can be used. This must be below the limit for fracture of the disc material by the centrifugal force. As an indication of where the limit lies, a selector designed and constructed by Trujillo (see figure 10.5) can be rotated as fast as 1320 Hz and selects velocities up to 2×10^4 m s^{-1}, the highest available at the time of writing.

10.4. Detection of Atomic Beams

Because they do not possess an electric charge, the problem of measuring the flux of atoms in a beam at thermal velocities is less readily solved than for an electron or ion beam.

There is little difficulty, however, for beams of the heavier alkali metal atoms, K, Rb and Cs, because such atoms are ionized on collision with the surface of a filament heated to a temperature of 1000–2000 K. Thus a suitable detector is a fine tungsten wire of, say, 0·03 mm diameter normal to the beam direction, which may be moved parallel to itself in the plane perpendicular to the beam. By extracting, with a suitable negative potential, the positive ion current, due to collision with atoms of the beam, a scan may be made of the intensity across the atom beam. It is possible by suitable treatment of the wire to extend the method to sodium by working with an oxide-coated filament, such as a hot platinum wire, bathed in oxygen. Because of the convenience of this method, many experiments have been carried out in the first instance using alkali metal atoms as one of the colliding partners.

To extend quantitative methods of detection to other atoms and molecules, it is necessary to resort to what is known as a universal ionization detector. In such a device, a fraction of the atoms or molecules in the beam is ionized by an electron beam and the ions are admitted into a mass analyser which measures the current of each ionic species produced. Because the probability that an atom or molecule will be ionized by electron impact is low, the current leaving the analyser for the detector is very small. It is usual therefore to observe this current with an electron multiplier. The ions concerned produce secondary electrons by impact with the first electrode and these are amplified in successive stages as described in Appendix 6. A universal detector is adequate when relative intensity measurements are sufficient but is very difficult to calibrate for absolute measurements.

In crossed-beam scattering experiments in which either one or both of the beams are not readily condensed on cooled surfaces, a major problem is the strength of the signals which arise from scattering by background gas and from other sources. These noise signals normally exceed the wanted signals by some orders of magnitude. It is therefore essential to use phase-sensitive detection (see Chapter 13, p. 246), for which purpose the intensity of one or both beams in modulated at a frequency of 50–500 Hz by a chopper wheel and the standard procedure may then be applied.

10.5. *Some Typical Experiments and Results—Scattering of Alkali Metal Atoms*

A considerable number of measurements have now been carried out, but we will confine ourselves to choosing for illustration some of the more recent in which the energy and angular resolution is high. We begin by describing some of the work carried out on the scattering of alkali metal atoms for which surface ionization detectors may be used.

Most recent experiments, whether directed towards the measurement of

Experimental Study of Atom–Atom Collisions

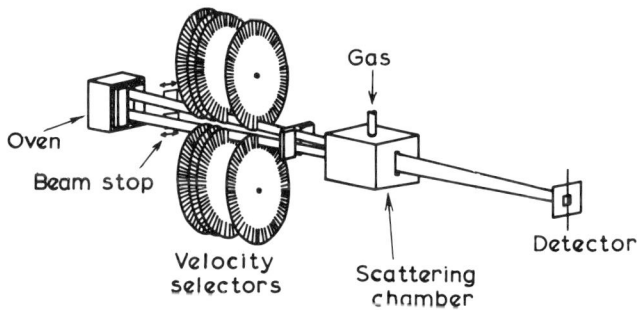

Figure 10.6. The apparatus used by von Busch, Strunck and Schlier for measuring relative effective collision cross-sections for K and Cs atoms with rare gas atoms.

total or of differential cross-sections, have used crossed-beam techniques. However, figure 10.6 illustrates the general arrangement of the apparatus used by von Busch, Strunck and Schlier for measuring with good resolution the variation with relative velocity of the total collision cross-sections for K and Cs atoms with rare gas atoms. In these experiments, the attenuations of two velocity-selected beams of metal atoms passing through the scattering chamber containing the rare gas were compared. Two ribbon-shaped beams issued from the same oven and, after passage through separate velocity selectors of slotted disc type and the scattering chamber filled with the rare gas at an appropriate pressure, were monitored by the same hot wire detector.

After passage through the scattering chamber the intensity of a beam of atoms of velocity v_1 will be

$$I_n(v_1) = I_0(v_1) \exp\{-n Q_{\text{eff}}(v_1) l\} \tag{10.12}$$

where $Q_{\text{eff}}(v_1)$ is the effective total cross-section in the rare gas for metal atoms of velocity v_1, l is the path length through the gas and $I_0(v_1)$ is the intensity before passing through the rare gas. $Q_{\text{eff}}(v_1)$ is an average over the velocity distribution of the rare gas atoms. It is not difficult to show that if measurements are made at two gas concentrations n and n' for beams of selected velocities v_1 and v_2, then writing $I_n(v_1)/I_0(v_1) = F_n(v_1)$,

$$Q_{\text{eff}}(v_1)/Q_{\text{eff}}(v_2) = 1 + \ln\left\{\frac{F_n(v_1)F_{n'}(v_1)}{F_n(v_2)F_{n'}(v_2)}\right\} \bigg/ \ln\{F_n(v_2)/F_{n'}(v_2)\}. \tag{10.13}$$

The oven and collimator slits were 50 μm wide and the detector slit 100 μm, the detector wire was 100 μm thick and the measured ion currents were between 10^{-11} and 10^{-12} A.

Although in these experiments the measured cross-section $Q_{\text{eff}}(v)$ are averaged over the velocity distribution of the rare gas atoms, the glory undulations (see Chapter 9, p. 174) are not completely smoothed out. This may be seen from figure 10.7, which shows, for K–Kr collisions, a plot of $\Delta Q_{\text{eff}}(v)/Q_{\text{eff}}(v)$ where ΔQ_{eff} is the fluctuating part (cf. (9.55) of Chapter 9) due to the glory effects.

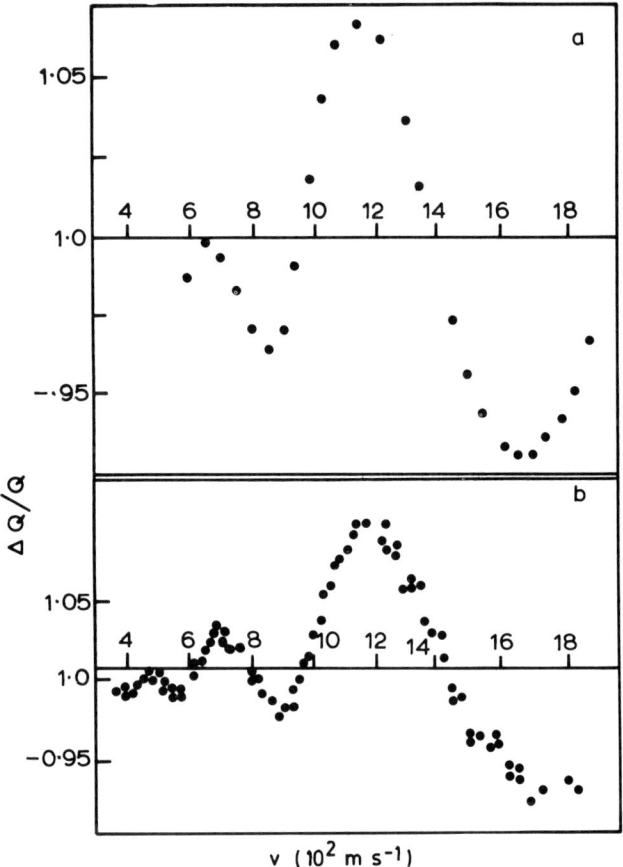

Figure 10.7. Fluctuating part ΔQ of the total cross-section for collisions between K and Kr atoms as a function of relative velocity.
(a) Observed by von Busch, Strunck and Schlier.
(b) Observed by Beck and Loesch.

In the same figure, a corresponding plot is shown from the observations of Beck and Loesch, who used the crossed-beam technique. Quite good agreement is obtained. The maxima and minima occur at nearly the same relative velocities, are of comparable magnitude and are of the form expected for glory undulations.

We now describe the crossed-beam experiments of Buck and Pauly on the collisions between sodium and mercury atoms. In these experiments, both the variation of the total cross-section with relative velocity and of the differential cross-sections over a wide angular range were measured. Figure 10.8 shows the general arrangement of the apparatus for the differential cross-section measurements.

Experimental Study of Atom–Atom Collisions

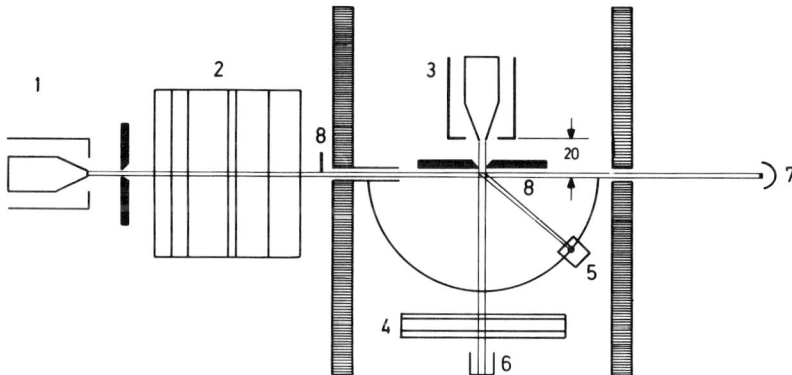

Figure 10.8. Arrangement used by Buck and Pauly for the measurement of differential cross-sections for collisions between sodium and mercury atoms.
1, Sodium vapour oven; 2, Primary velocity selector; 3, Mercury vapour oven; 4, Secondary velocity selector; 5, Scattered sodium detector; 6, Mercury beam monitor; 7, Monitor for main sodium beam.

Nozzle sources were used for both beams. However, as the sodium beam was velocity-selected, the main advantage of the nozzle source for it was the high intensity. On the other hand, the mercury beam was not velocity-selected and the nozzle source was operated at a Mach number as high as 19·5 in order that the velocity distribution should be narrowed. This distribution was measured by a second velocity selector after passage through the scattering chamber and figure 10.9 shows the results obtained. Taken together with the large mass ratio of mercury to sodium, the contribution of the spread in this distribution to that of the relative velocity was so small that the half width of the latter was only about 2% of the most probable value, not much greater than that arising from the finite resolution of the sodium beam velocity selector.

The scattered sodium beam was measured by surface ionization on a platinum ribbon 0·5 mm wide and 1·5 mm high which was continuously sprayed with oxygen.

To reduce the background scattering to insignificant proportions, exposed surfaces were cooled to liquid nitrogen temperatures.

Figure 10.10 shows the observed variation of the effective total cross-section with relative impact velocity, uncorrected for finite angular resolution. After including this correction and analysis into the form (see (9.55) of Chapter 9)

$$Q = \bar{Q} + \Delta Q, \qquad (10.14)$$

the variation of $\Delta Q/Q$ with relative velocity shown in figure 10.10 is obtained. No less than seven glory maxima are exhibited.

Measured angular distributions at five different relative impact energies are shown in figure 10.11. The intensity per unit angle $J(\theta)$ proportional to $I(\theta) \sin \theta$, is shown, $I(\theta)$ being the differential cross-section per unit solid angle

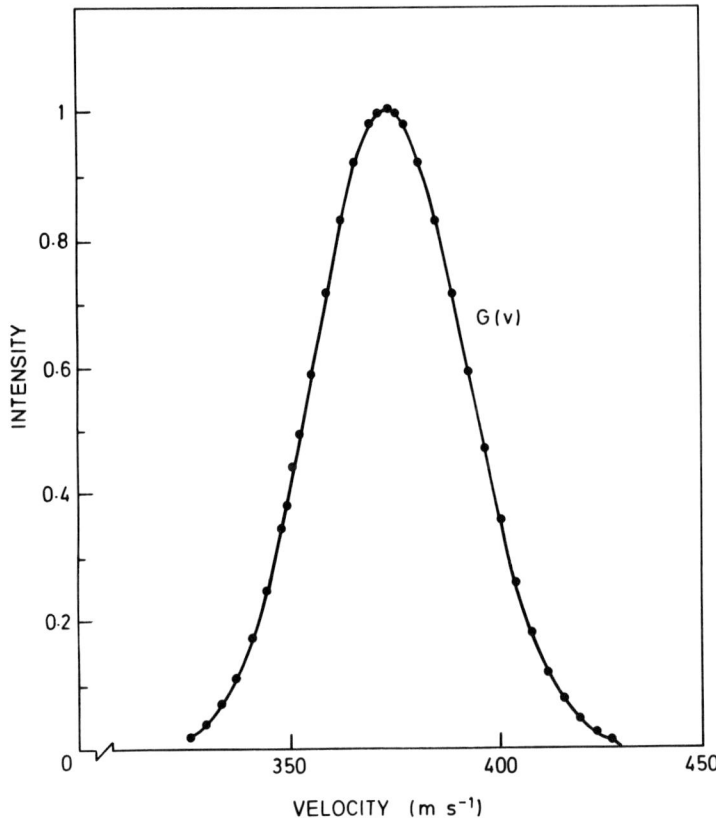

Figure 10.9. Velocity distribution of the mercury atom beam in the experiments of Buck and Pauly as measured with the secondary velocity selector (see figure 10.8).

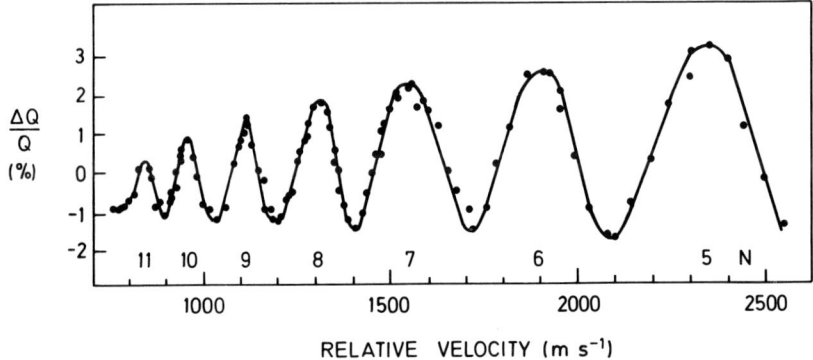

Figure 10.10. Fluctuating part ΔQ of the total cross-section for collisions between Na and Hg atoms as a function of relative velocity, observed by Buck and Pauly using the apparatus shown in figure 10.8.

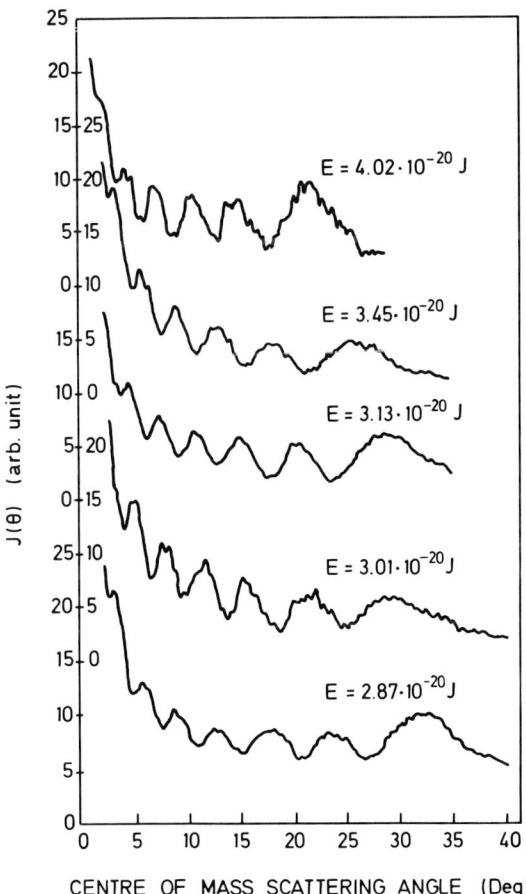

Figure 10.11. Distributions per unit angle (in the CM system) of scattered intensity in collisions between sodium and mercury atoms at different impact energies as indicated. Observed by Buck and Pauly (see figure 10.8).

in the centre of mass system. It is convenient to introduce the factor $\sin \theta$ because it moderates the otherwise very rapid rise of scattered intensity at small angles.

The main maxima exhibited are rainbow maxima. In each case the maximum at the largest value of θ is the primary rainbow, the other being supernumerary rainbows, of which up to six are visible. Superposed on the broad rainbow maxima are much more rapid fluctuations which arise, as explained in Chapter 9 (see figure 9.15), from interference between rainbow scattering and scattering at a much smaller impact parameter (i.e., in figure 9.16 between the trajectories with $p = p_1$ and those with $p = p_2$ and p_3).

Figure 10.12 shows some measurements made in earlier experiments by Hundhausen and Pauly in which the angular and energy resolution was in-

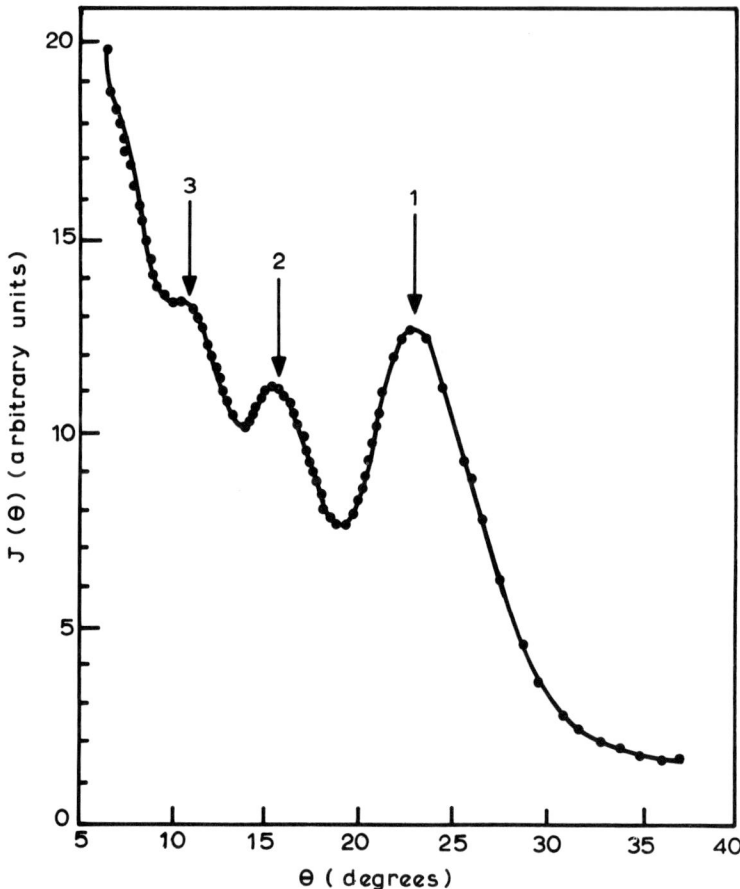

Figure 10.12. Distributions per unit angle (in the CM system) of scattered intensity in collisions between sodium and mercury atoms at an impact energy of 8.4×10^{-20}/J observed by von Hundhausen and Pauly using an apparatus of lower energy resolution than that of Buck and Pauly so that the fine structure oscillations are averaged out and the primary rainbow (marked 1) and supernumerary rainbows (2 and 3) are clearly seen (cf. figure 9.15)

adequate to resolve the rapid oscillations. The primary and two supernumerary rainbows are clearly visible.

Using their data, Buck and Pauly derived the interaction $V(r)$ between Na and Hg atoms which is shown in figure 10.13. The parameters ε and r_m were found to be

$$r_m = (4.72 \pm 0.5\%) \times 10^{-10} \text{ m}, \qquad (10.15)$$

$$\varepsilon = (8.79 \pm 3\%) \times 10^{-21} \text{ J},$$

$$= 5.49 \times 10^{-2} \text{ eV},$$

Experimental Study of Atom–Atom Collisions

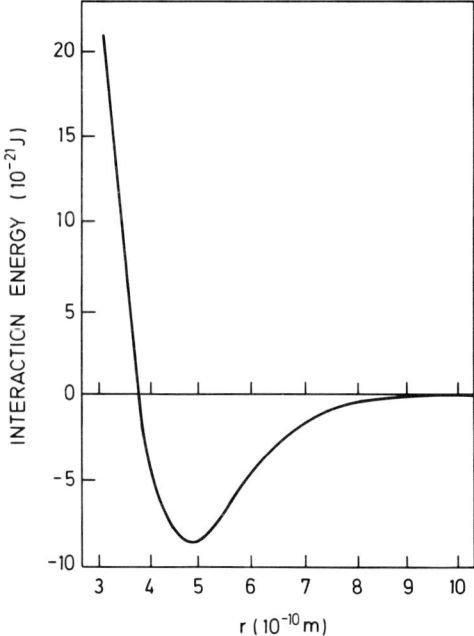

Figure 10.13. Interaction energy between sodium and mercury atoms derived from the experiments of Buck and Pauly (see figures 10.9–10.11).

but the shape of the interaction cannot be represented adequately by a simple form such as (9.1) of Chapter 9.

10.6. Collisions Between Helium Atoms—Symmetry Oscillations

Although it had been realized as long ago as 1930 that symmetry interference effects would arise in the collisions between indistinguishable atoms, and in 1932 Massey and Mohr had calculated many of the effects to be expected in the case of helium atoms, it was not until 1969 that they were first observed by Dohmann and a little later by Dondi, Scoles, Torrello and Pauly, in the variation of the total cross-section with relative velocity. Observation of symmetry interference effects in the differential cross-section followed in 1970, and both types of experiment have been used to determine with some accuracy the interaction between helium atoms.

Figure 10.14 shows the arrangement used by Dondi, Scoles, Torrello and Pauly to measure the variation of the total cross-section with relative velocity. In these experiments the absorption of a velocity-selected nozzle beam of helium atoms in helium gas contained in a scattering chamber cooled to 2 K in a liquid helium bath was observed. The nozzle beam was operated at a Mach number around 10, while the temperature of the source could be held constant at any

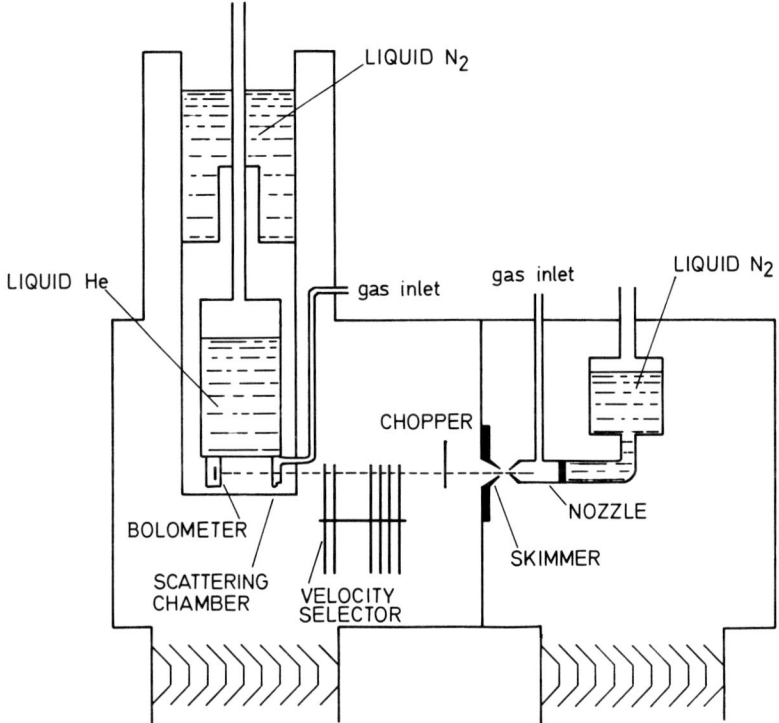

Figure 10.14. Arrangement used by Dondi, Scoles, Torrello and Pauly to measure the variation of the total cross-section with relative velocity for collisions between helium atoms at low temperatures.

point in the range 75–200 K. The nozzle temperature was measured with platinum-resistance thermometers. Before entering the nozzle chamber, helium was passed through silica gel cooled by liquid nitrogen. This reduced the impurity content, as measured by a mass analyser, to less than 0·02%. The temperature of the scattering chamber walls was measured by a thermocouple with the reference junction maintained at the temperature of the liquid helium bath.

The pressure of the gas in the scattering gas was maintained constant by expanding the helium from a 6-litre volume into the scattering chamber through a capillary leak while the pressure in the reservoir was monitored with a quartz manometer.

Background effects were reduced by phase-sensitive detection (see p. 246) using a chopper wheel to modulate the primary beam at 30 Hz.

The intensity of the beam after passage through the scattering chamber was measured by a bolometer of the type described in Appendix 6.

With this arrangement, the contribution to the relative impact velocity

Experimental Study of Atom–Atom Collisions

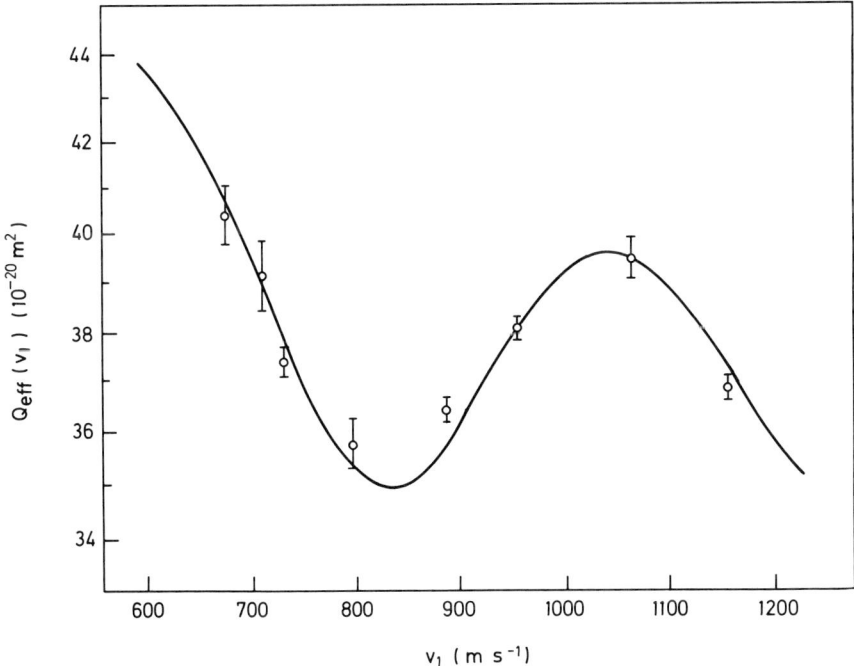

Figure 10.15. Variation with relative velocity of the total cross-section for ^{4}He–^{4}He collisions showing for the first time the undulations due to symmetry effects. Observations by Dondi, Scoles, Torrello and Pauly (see figure 10.14).
●, observed points, with probable error; ——, calculated using an interaction of type (9.1) with $n = 12$

from the atoms in the scattering chamber was very small, so the width of the relative velocity distribution was determined by the resolution of the velocity selector, about 5%.

Figure 10.15 shows the observed effective cross-section as a function of velocity. The symmetry oscillations are very apparent.

Convincing evidence that the effects observed are due to symmetry interference has been provided from the experiments of Bennewitz, Busse, Dohmann and Schröder who compared cross-sections measured for collisions between ^{3}He atoms with those for ^{4}He. As explained in Chapter 9, p. 176, for the ^{3}He isotope of helium for formula for the scattered intensity differs from that for the more abundant ^{4}He isotope. Figure 10.16 shows the contrast between the cross-section for ^{4}He–^{4}He collisions, which is that normally observed because ^{4}He is by far the most abundant isotope in normal helium, and that for ^{3}He–^{3}He. The marked difference in behaviour is in agreement with theoretical expectation (see figure 10.16) based on the He–He interaction shown in figure 10.20, which is consistent with the observed data on normal He–He total and differential cross-sections, as explained below.

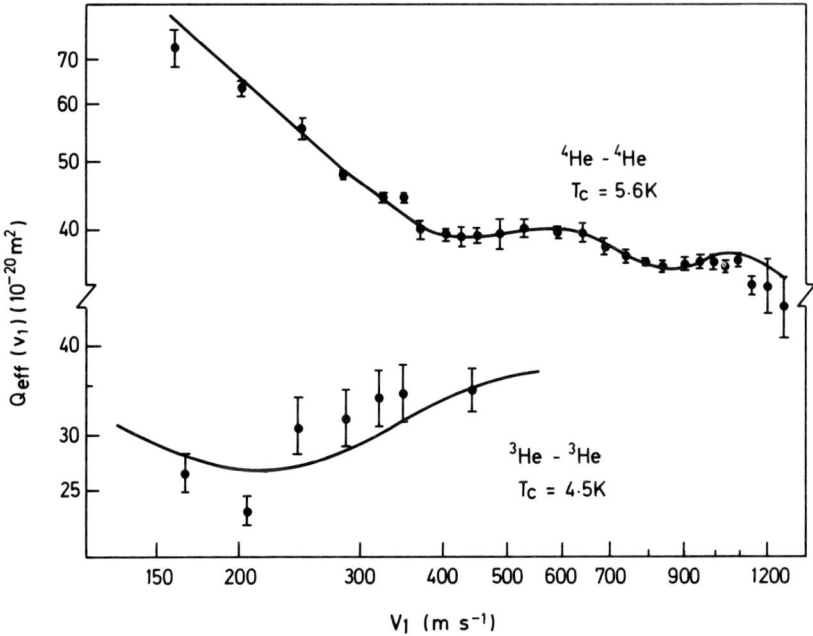

Figure 10.16. Variation with relative velocity v_1 of the total cross-sections for ^{4}He–^{4}He and for ^{3}He–^{3}He collisions at target gas temperatures T_c.
●, observed points, with probable error (Bennewitz, Busse, Dohmann and Schröder)———, calculated using the He–He interaction shown in figure 10.19.

High resolution measurements of the differential cross-sections for ^{4}He–^{4}He collisions were made by Siska, Parson, Schafer and Lee using the arrangement shown in figure 10.17. Two nozzle beams were used, crossing at right angles. These were operated at Mach numbers between 15 and 25 so that the velocity spread in each beam was about 6–8%. No velocity selection was used. The angular spread of each beam is indicated in figure 10.17.

Helium atoms scattered through an angle Θ were detected by a universal ionization detector which was rotatable so that the variation of scattered intensity with Θ could be measured.

While phase-sensitive detection was used, with chopper modulation of the primary beam at 30 Hz, it was found that the modulated background could be considerably reduced by catching the primary beam in a beam trap.

The collision energy was controlled through the nozzle temperatures, which in turn could be adjusted either by passing water or cold dry nitrogen through cooling tubes, and measured by thermocouples.

In later experiments Farrar and Lee worked down to much lower relative impact velocities by using helium atoms produced by passing helium at a pressure of 400 torr through copper tubing immersed in a reservoir of liquid hydrogen. The nozzle temperature was 23 K.

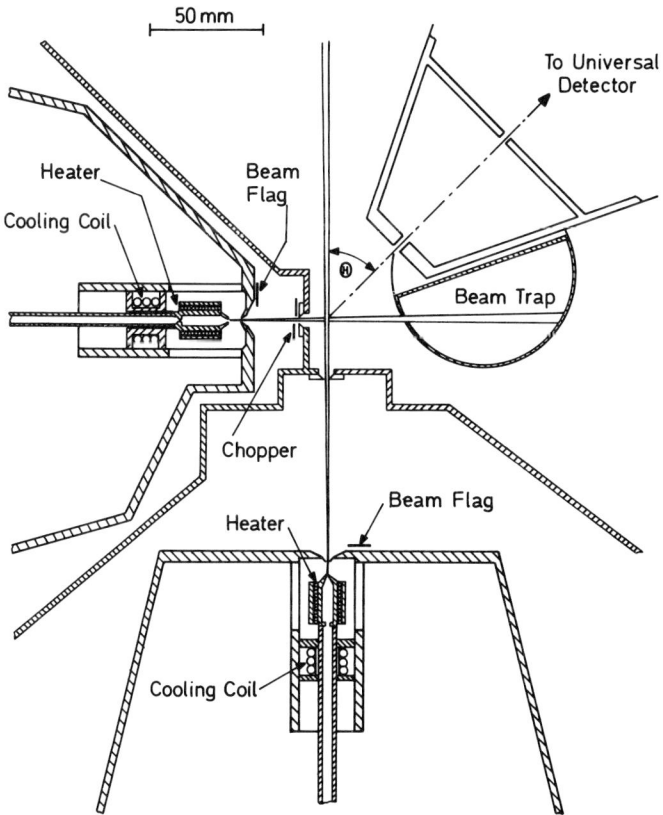

Figure 10.17. Arrangement used by Siska, Parson, Schafer and Lee to measure differential cross-sections for ^{4}He–^{4}He collisions, with high energy and angular resolution.

Figure 10.18 shows angular distributions in the laboratory system measured at two mean impact energies. The symmetry oscillations, which characteristically persist out to large scattering angles, are very obvious in the curve obtained by Siska, Parson, Schafer and Lee at an energy of $10 \cdot 1 \times 10^{-21}$ J. At the much smaller impact energy observed by Farrar and Lee, the wavelength of the relative motion is so large that the semi-classical approximation is no longer satisfactory and the full theory, involving the formula (2.50) of Chapter 2, must be used in attempting theoretical analysis of the data.

Using the measurements both of total and differential cross-sections, it is possible to derive the interaction energy $V(r)$ between helium atoms at separations r larger than about 0·23 nm. Figure 10.19 shows the derived form of $V(r)$ for which the parameters ε and r_m are given by

$$\varepsilon = 0 \cdot 152 \pm 0 \cdot 003 \times 10^{-21} \text{ J}, \quad (10.17)$$

$$r_m = 2 \cdot 963 \times 10^{-10} \text{ m}. \quad (10.18)$$

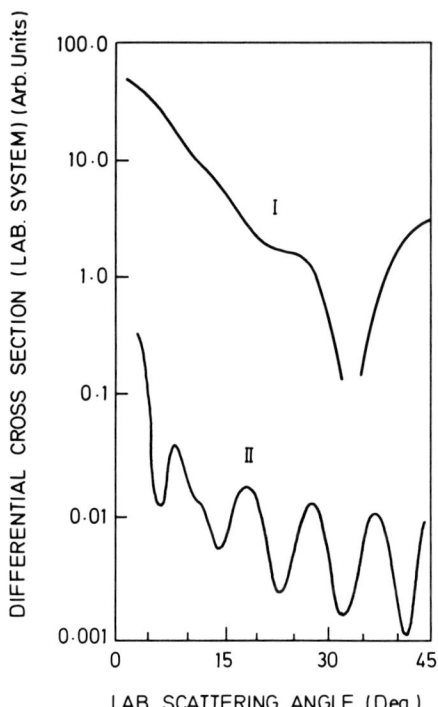

Figure 10.18. Differential cross-sections in the laboratory system for collisions between helium atoms,
I, for an impact energy of 0.799×10^{-21} J, observed by Farrar and Lee;
II, for an impact energy of 10.1×10^{-21} J observed by Siska, Parson, Schafer and Lee.

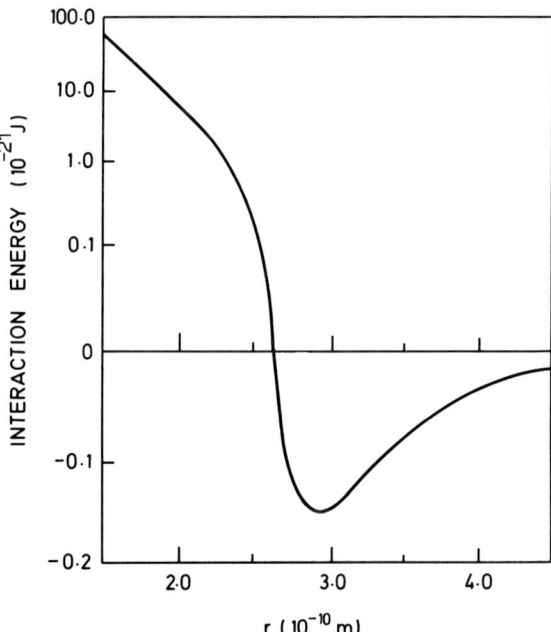

Figure 10.19. Interaction between helium atoms derived from collision experiments and from the equation of state of helium gas. Note the change of energy scale from linear to logarithmic above 0.1×10^{-21} J.

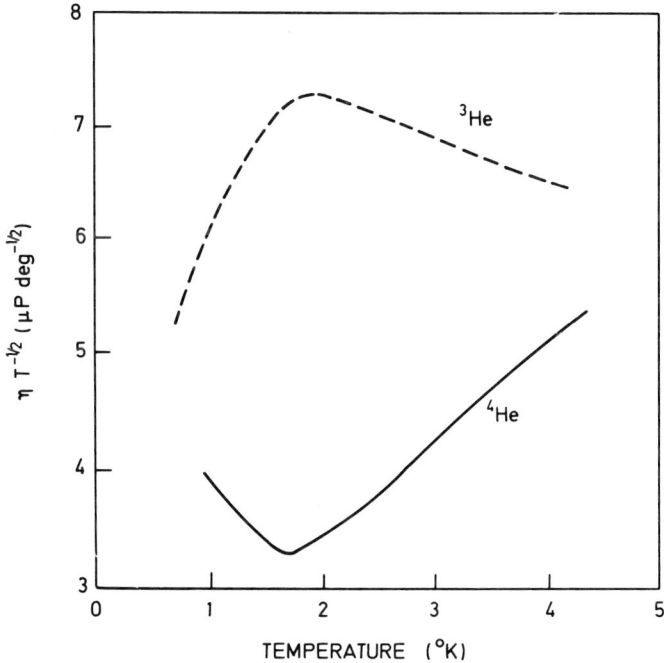

Figure 10.20. Calculated variation with absolute temperature T at low temperatures of $\eta T^{-1/2}$ where η is the viscosity (in micropoise, $\mu\mathrm{P} = 10^{-7}$ kg m^{-1} s^{-1}) for ^{4}He and ^{3}He.

For $r < 2 \cdot 3 \times 10^{-10}$ m, $V(r)$ has been determined from the equation of state of helium gas. The agreement obtained with the observed data for the cross-sections using $V(r)$ in conjunction with wave theory is shown in figure 10.16.

At temperatures below 20 K, the relative velocities of the atoms in helium gas are so low that the wave aspect of their relative motion cannot be ignored in the calculation of the viscosity, thermal conductivity and other transport properties of the gas. Because of the different symmetry properties, the viscosity and diffusion cross-sections, (Chapter 8 (8.2) and (8.3) respectively) are different at these low impact energies for ^{4}He and for ^{3}He. These differences, as shown in figure 10.20, have been observed, and they are in agreement with theoretical expectation.

CHAPTER 11
thermal collisions between ionized and neutral atomic and molecular systems

Knowledge of the cross-sections for different kinds of collisions between ionized and neutral atoms and molecules is of key importance for the interpretation and understanding of many natural physical phenomena such as electrical discharges, the formation and behaviour of the ionized regions in the outer atmospheres of the Earth and planets and the formation of molecules in interstellar space. Reactions involving ions are also important in the design of certain lasers. They are central to the design of certain devices for the generation of power from controlled nuclear fusion and for the acceleration of charged particles for studies of nuclear structure. Apart from this wide variety of applications, the study of ionic collisions is of interest for its own sake in providing a deeper understanding of atomic and molecular physics.

Much of the information required refers to collisions at thermal or near-thermal energies. For the experimental study of these collisions, swarm techniques must be used as for the corresponding electron collisions (see Chapter 5, p. 69). In this chapter we confine our discussion to collisions at these low energies. A richer variety of possible types of collision occurs than for electrons and a number of special techniques have been developed to deal with them.

Applications to controlled nuclear fusion and to particle acceleration involve collisions at much higher kinetic energies. These may be carried out with beam techniques, some illustrations of which are discussed in the following chapter.

11.1. *The Long-range Polarization Force*

One special feature introduced when one of the colliding systems is an ion is the long range polarization force which is always present.

Just as discussed in Chapter 3 for electron–atom interactions (see also Chapter 5, p. 86) polarization of the target by the ionic charge introduces an attractive interaction energy which at large values of r is given by $-\frac{1}{2}\alpha e^2/r^4$, where α is the polarizability of the target. This will be valid for collisions in which the relative velocity of impact is small compared with the orbital velocities of the electrons in the target. Otherwise they are unable to adjust sufficiently rapidly to the varying interaction with the incident ion to produce the full polarization of the target. Taking for the electron orbital velocity a typical value of 10^6 m s^{-1}, the kinetic energy of the lightest ions H$^+$ or H$^-$ with this velocity is as high as 5 000 eV. Hence we assume that the polarization force assumed will be valid up to impact energies at least as high as 1 000 eV and indeed much higher for most ions.

As a crude representation of the interaction between an ion and an atom between which no strong chemical attraction exists, we may take

$$V(r) = -\tfrac{1}{2}\alpha e^2/r^4 \quad , r > a,$$
$$\to \infty \quad , r < a, \qquad (11.1)$$

so that the two systems do not penetrate closer than $r = a$. Just as for collisions between neutral atomic systems, the elastic scattering can be taken as given by classical theory except at very small angles of scattering. According to classical theory, the hard-core repulsion for $r < a$ will not be important in determining the dynamics of an impact when the impact energy is much less than the polarization attraction at $r = a$, i.e., if

$$\tfrac{1}{2}Mv^2 \ll \tfrac{1}{2}\alpha e^2/a^4, \qquad (11.2)$$

where M is the reduced mass of the colliding systems and v their relative velocity.

11.2. The Condition for Orbiting

In Chapter 9, p. 168, we described the phenomenon of orbiting in which the relative position vector of the colliding system revolves many times about the scattering centre in the collision. The condition for orbiting to occur was obtained in terms of the effective radial interaction energy, $V_{\text{eff}}(r)$, including that due to centrifugal force, when the impact parameter is p, i.e.,

$$V_{\text{eff}}(r) = V(r) + p^2 E/r^2, \qquad (11.3)$$

where E is the kinetic energy of relative motion. For orbiting to occur, the radial velocity $[2\{E - V_{\text{eff}}(r)\}/M]^{1/2}$ must vanish when $V_{\text{eff}}(r)$ is at an accessible maximum.

If we put $V(r) = -\tfrac{1}{2}\alpha e^2 r^4$, orbiting will occur when $p = p_c$, where

$$-\tfrac{1}{2}\alpha e^2 r^{-4} + p_c^2 E r^{-2} = E, \qquad (11.4)$$

$$2\alpha e^2 r^{-5} - 2p_c^2 E r^{-3} = 0. \qquad (11.5)$$

From these equations we find

$$p_c = (2\alpha e^2/E)^{1/4}. \qquad (11.6)$$

This result is of importance because, when orbiting occurs, the systems spend a relatively long time close to each other, so that there is opportunity for any reactions between them to occur with considerable probability. We may

Atomic and Molecular Collisions

define an 'orbiting cross-section'

$$Q_{or} = \pi p_c^2 \tag{11.7}$$

which should be in many cases an upper limit to the cross-section for any reaction. More will be said about this on p. 206.

11.3. The Mobility of Ions in Gases

Just as for collisions of electrons with atoms or molecules, very low-energy impacts must be studied by swarm methods. Thus the drift motion of a swarm of positive or negative ions in a gas has been an important subject for study since the beginning of the present century.

As compared with the drift of an electron swarm, the problem of analysing the motion of ions of molecular mass in a gas under the action of a uniform electric field is complicated by the fact that the masses of the ions and gas molecules are comparable. However, in the steady state, the ions acquire a constant drift velocity u which will be a function of the ratio F/p at constant gas temperature T, F being the electric field strength and p the gas pressure. Results of measurements for positive and negative ions are usually expressed in terms of the *mobility* at a fixed gas pressure which is the drift velocity for unit electric field strength. Thus if μ is the mobility u/F, μp will be a function of F/p at a fixed gas temperature. μ is usually given in m s^{-1} per V m^{-1} (i.e., m^2 s^{-1} V^{-1}) at 1 atmosphere pressure and a temperature of 0°C. Measurements at other temperatures are usually reduced so as to refer to the pressure at the standard temperature. If the collision frequency v_c is defined by

$$v_c = nQ_d v, \tag{11.8}$$

where n is the number of gas atoms or molecules per unit volume, Q_d the diffusion (momentum transfer) cross-section defined by (8.3) of Chapter 8 for impacts of relative velocity v, then the value of μ in the limit of vanishing applied field is given by

$$\mu = e(M_1 + M_2)/v_c M_1 M_2, \tag{11.9}$$

where M_1 and M_2 are the masses of the ions and target atoms or molecules respectively. If v_c is not a constant, it must be replaced by a suitable mean value of (11.8). Thus, as for electrons, the mobility or drift velocity depends on the diffusion, rather than the total, collision cross-section.

For the interaction (11.1), the mobility for zero field was calculated as long ago as 1905 by Langevin, who found

$$\mu = (1 + M_1/M_2)^{1/2} \{\rho(K-1)\}^{-1/2} g(\lambda), \tag{11.10}$$

where

$$\lambda = \{8\pi nkTa^4/(K-1)e^2\}^{1/2}. \tag{11.11}$$

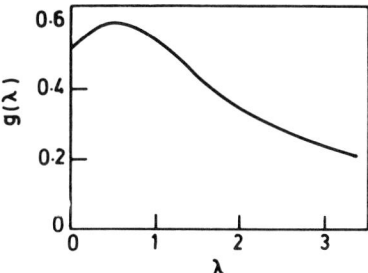

Figure 11.1. The function $g(\lambda)$.

Here T is the gas temperature, ρ the gas density and n the number of gas molecules per unit volume. K is the dielectric constant of the gas at standard temperature and pressure, related to the polarizability α by

$$\alpha = (K-1)/4\pi n. \qquad (11.12)$$

$g(\lambda)$ is a function illustrated graphically in figure 11.1.

For small λ, $g(\lambda)$ varies very slowly with λ so that μ is nearly proportional to $(1 + M_1/M_2)^{1/2}$, independent of the hard core radius a and the temperature T. In the limit $\lambda \to 0$, $g(\lambda) \to 0 \cdot 510$ and with $n = 2 \cdot 69 \times 10^{25}$ m^{-3}, the number of gas atoms per m³ at s.t.p.,

$$\mu \to 3 \cdot 59 \times 10^{-3}/(\alpha M)^{1/2} \text{ m}^2 \text{ V}^{-1} \text{ s}^{-1}, \qquad (11.13)$$

where α is the polarizability of the gas atom in units a_0^3, a_0 being the radius of the first Bohr orbit in hydrogen (Chapter 3, p. 48). M is the reduced mass $M_1 M_2/(M_1 + M_2)$ in units of the proton mass.

Although measurements of mobilities of positive ions in gases have been carried out since the beginning of the century, many results have been vitiated due to failure to exclude impurities, including particularly water vapour, molecules of which cluster to the ions and hence reduce their mobility. Furthermore, for the above theory to be applicable, the ions and atoms must be of different species, A$^+$ and B say, and the ionization energy of A must be less than that of B so that charge exchange effects (see p. 204) cannot occur. Even when these conditions are satisfied ionic reactions may take place between the ions and gas atoms, particularly at high gas pressure (compare the formation of polyatomic positive ions in discharge afterglows, Chapter 7, p. 123). Because of these complications, there is not a great deal of reliable experimental material with which to check the conclusions outlined above.

The best evidence is provided from the experiments on the mobilities of alkali metal ions in rare gases carried out in 1938 by Tyndall and Powell using a shutter method operating on the same principle as for measurements of electron mobility described in Chapter 5, p. 71. Table 11.1 gives values of

	Ion/Gas	He	Ne	Ar	Kr	Xe
Observed values	Li	38.6	25.2	11.4	9.4	7.3
	Na	41.9	26.8	11.5	9.3	7.5
	K	41.0	27.4	11.7	9.6	7.4
	Rb	39.3	27.2	11.7	9.5	7.4
	Cs	37.0	25.5	11.5	9.5	7.4
Calculated values		30.5	21.9	10.8	8.9	6.9

Table 11.1. Comparison of observed and calculated values of $M^{1/2}\mu$ (in 10^{-4} m² V⁻¹ s⁻¹) for alkali metal ions in rare gases at 291 K.

$M^{1/2}\mu$ derived from these observations for $T = 291$ K for different ion–atom pairs, compared with the calculated limiting values $35.9 \times 10^{-4}\alpha^{-1/2}$.

It will be seen that, as the mass of the gas atom increases, the limiting behaviour is more and more nearly approached, although even for Xe the calculated limit is about 1.08 times smaller than the observed. This may be due to an error in the measurement of the gas pressure.

Further evidence which illustrates the importance of the polarization term is provided by the observed variation of $M^{1/2}\mu$ with temperature. In the theoretical limit, no such variation should be observed and over the range 477–195 K for Cs^+–Xe and Rb^+–Ar none is observed.

11.4. Ionic Reactions

It has been remarked above that the interpretation of mobility experiments is complicated in many cases by the occurrence of ionic reactions. As mentioned earlier, such reactions are of importance in many physical phenomena. A number of possible types of reaction occur and it is convenient to begin by providing a general description of these different types under four main headings.

(a) Charge transfer

In collision with a neutral molecule B, a positive ion A^+ may transfer its charge to produce B^+

$$A^+ + B \rightarrow A + B^+. \tag{11.14}$$

This transfer is brought about by transfer of an electron from B to A^+.

If the atom A and ion B^+ are in their ground states the energy released is equal to $E_i(A) - E_i(B)$, where $E_i(A)$, $E_i(B)$ are the ionization energies of A and B respectively. When $E_i(A) < E_i(B)$, there will be a threshold energy E_t for the reaction, where

$$E_t = E_i(B) - E_i(A), \tag{11.15}$$

but when $E_i(A) > E_i(B)$ the reaction can occur at very low impact energies. If

A and B^+ are formed in excited states with excitation energies $E_e(A)$, $E_e(B)$ respectively, then the threshold energy will become

$$E_t = E_i(B) - E_i(A) + E_e(A) + E_e(B). \tag{11.16}$$

Of special interest is the case in which the neutral molecule is of the same species as the positive, so that we have

$$A^+ + A \to A + A^+. \tag{11.17}$$

In such a case, if the products are in their ground states, no internal energy change occurs and it is only possible to distinguish the ions produced in this way from those directly scattered,

$$A^+ + A \to A^+ + A, \tag{11.18}$$

through their angular distribution and then only under certain circumstances. This aspect of the subject will be discussed particularly in the following chapter (p. 215) in relation to beam experiments on charge exchange.

Similar possibilities arise with negative ions, the reaction

$$A^- + B \to A + B^-, \tag{11.19}$$

being possible provided B forms a stable negative ion and the energetic relations are satisfied. Thus if $E_a(A)$, $E_a(B)$ are the electron affinities of A and B respectively and $E_a(A) > E_a(B)$, the threshold energy for the reaction is

$$E_t = E_a(A) - E_a(B). \tag{11.20}$$

Beam experiments have been carried out with negative ions to measure differences in electron affinities from observation of the threshold energy for charge transfer.

(b) *Rearrangement reactions*
A variety of collisions of the following kind are possible:

$$A^+ + BC \to AB^+ + C, \tag{11.21}$$

$$\to AC^+ + B. \tag{11.22}$$

These are referred to as rearrangement reactions because they involve rearrangement of the atoms between the colliding systems. It is obvious that many further possibilities arise when the colliding systems are polyatomic. Similar possibilities arise with negative ions.

(c) *Associative detachment*
With negative ions detachment of an electron may occur on impact through a reaction such as

$$A^- + B \to AB + e, \tag{11.23}$$

the electron being ejected with an energy $D(AB) - E_a(A)$, $D(AB)$ being the dissociation energy of AB. Processes of this kind are the inverse of the dissociative attachment processes described in Chapter 7, p. 112. However, whereas attachment occurs to molecules in vibrational states which may be excited at the temperature of the experiments, the molecules AB may be formed, in a reaction such as (11.23) in other vibrational states.

(d) *Three-body association reactions*
In the presence of a third body, C say, reactions of the kind

$$A^+ + B + C \rightarrow AB^+ + C \qquad (11.24)$$

are possible.

Without the third body present, a complex AB^+ formed by association of A^+ and B will be unstable and will, if left alone, break up again into its constituents. However, a third body can take away sufficient energy from the system to leave stabilized AB^+, that is to say AB^+ with total energy less than the sum of the energies of the separated systems A^+, B or A, B^+.

The rate at which an ion may undergo a three-body collision will be proportional to the concentrations of both B and C unless the concentration of C is so large that a system C is always to be found within the interaction region when A^+ and B collide. In such a case, there will be no dependence on the concentration of C.

Similar considerations apply to three-body collisions involving negative ions.

Three-body associative reactions are important in the initiation of ion clustering which leads to the formation of complex ions (Chapter 14, p. 273). Because of their large permanent dipole moments, water molecules may readily be associated with positive or negative ions to form clustered ions of the type $A^+(H_2O)_n$ or $A^-(H_2O)_n$, where n may be as large as 6.

11.5. *Orbiting and Ionic Reaction Rates*

We have already referred to the importance of orbiting for ionic reactions. We can expect that the cross-section for an ionic reaction with a neutral atom of polarizability α will be comparable with the orbiting cross-section πp_c^2 where

$$\pi p_c^2 = \pi(2\alpha e^2/E)^{1/2}, \qquad (11.25)$$

E being the kinetic energy of relative motion of the interacting systems. If ζ is the probability that, when orbiting occurs, a particular reaction will take place, then the cross-section for that reaction will be

$$\pi\zeta(2\alpha e^2/E)^{1/2}. \qquad (11.26)$$

Since the collision velocity $v = (2E/M)^{1/2}$, where M is the reduced mass, the reaction rate will be

$$2\pi\zeta(\alpha e^2/M)^{1/2}. \qquad (11.27)$$

Hence if ζ varies little with E the rate will be independent of the impact energy and proportional to $(\alpha/M)^{1/2}$. Quite a number of ionic reactions do behave more of less in this way.

11.6. Measurement of Ionic Reaction Rates under Thermal or Near-Thermal Conditions

By suitable design of mobility experiments, it is possible to derive information about ionic reaction rates when the mean ion energies range from thermal to a few eV. A number of other methods have also been derived but, for measurements at thermal energies, the technique of the flowing afterglow is probably the most versatile and reliable. We shall therefore select it for a detailed description.

The principle of operation may be illustrated by considering first the following experiment. Helium is pumped at high speed V through a cylindrical tube. At some section of the tube, He^+ ions are produced by causing a discharge to occur. These ions travel down the tube in the afterglow which flows with the main helium gas. At some later section, gas which reacts with the He^+ ions may be injected into the stream at a known, adjustable, rate. The ions reaching the end section of the tube and sampled through a fine hole are mass-analysed while the main stream is pumped out at the side (see figure 11.2). Let $(He^+)_0$ be the concentration of He^+ ions just before the injection nozzle. Then in the absence of any injected molecules, the He^+ concentration will decay with some time constant τ_0, so that at the end of the tube, the concentration will be

$$[He^+]_1 = [He^+]_0 \exp(-T/\tau_0), \qquad (11.28)$$

where T is the flow time from the nozzle to the end of the tube.

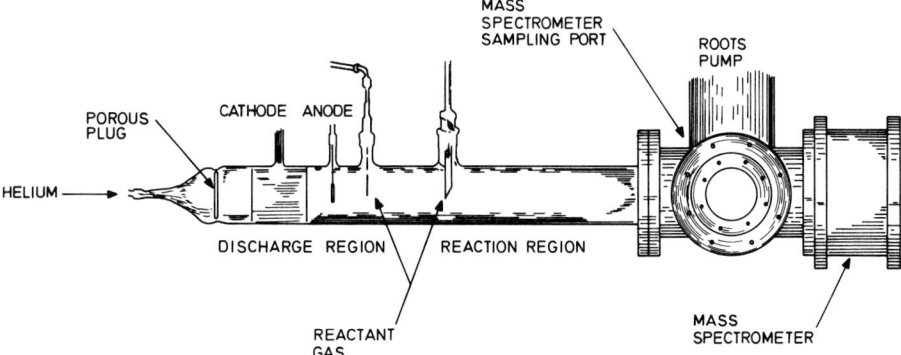

Figure 11.2. Arrangement of a flowing afterglow experiment to measure ionic reaction rates.

If, now, a gas of molecules M which react with He^+ is injected at the nozzle, there is an additional loss of He^+ at a rate $\lambda[M][He^+]$, where $[M]$ is the concentration of M. Then the rate of loss of He^+ concentration is given by

$$\frac{d(He^+)}{dt} = \left\{\frac{-1}{\tau_0} - \lambda[M]\right\}[He^+], \qquad (11.29)$$

$$= \frac{-1}{\tau}[He^+] \text{ say.}$$

The concentration of He^+ at the end of the tube after this additional loss is now given by

$$[He^+]_2 = [He^+]_0 \exp(-T/\tau), \qquad (11.30)$$

so

$$\ln(He^+)_2 = \ln(He^+)_0 - \frac{T}{\tau_0} - \lambda[M]T. \qquad (11.31)$$

Since the He^+ current, $i(He^+)$, in the spectrograph is proportional to $[He^+]_2$, a plot of $\ln i(He^+)$ against $[M]$ will give a straight line of slope λT. Since T is known from the flow speed V, the rate coefficient λ for the reaction of He^+ with M may be obtained.

With flow rates of around 10^2 m s^{-1}, which are attainable with fast pumps, and tube lengths of some tens of centimetres, T is of the order of milliseconds, which is generally very suitable when the gas pressure is a few tenths of a torr.

It is easy to extend the technique to reactions with other ions, using two injectors. A suitable gas is introduced through the first to convert He^+ ions to ions A^+ whose reactions with some molecule B are to be studied. Molecules B are then injected at adjustable known rates through the second injector. For example, suppose it is desired to investigate reactions between atomic nitrogen ions N^+ and those of some other molecules M. Injection of nitrogen gas through the first nozzle converts He^+ to N^+ by the reaction

$$He^+ + N_2 \rightarrow He + N^+ + N. \qquad (11.32)$$

If the molecules M are now injected from a second nozzle sufficiently far downstream for all the He^+ ions to have reacted, it is now possible to determine the rate coefficient for reactions of M with N^+ by observing the current of N^+ ions received in the spectrometer as a function of M.

An alternative procedure is to add the parent gas to the helium stream before it passes through the discharge.

A number of precautions have to be taken to avoid spurious effects. Thus, if the reactions of ions A^+ with molecules M are being studied by injection of neutral molecules A into the helium flow, it is essential that *any* reactions which produce A^+ at an appreciable rate should have ceased to be important by the

Thermal Collisions of Atoms and Molecules

time the section at which the molecules M are injected is reached. With helium as buffer gas, account must be taken of the fact that metastable helium atoms can produce ions in collision with neutral molecules by Penning ionization (see Chapter 8, p. 152). Again, the resonance radiation (Chapter 13, p. 150) from excited helium atoms is energetic enough to produce A^+ downstream of the final nozzle, by photoionization. To eliminate these effects the discharge producing the afterglow is operated on a pulsed instead of a continuous basis. The resonance photons travel much faster down the tube than do the ions and neutral molecules, so their effects are not detected if observations are only made after a suitable time delay since cessation of the discharge.

For many applications, the rates of reactions of ions with atomized species such as atomic oxygen or nitrogen are required. Thus, in the interpretation of phenomena occurring in the Earth's atmosphere, the rates of the reactions

$$O^- + O \to O + e, \tag{11.33}$$

and

$$N_2^+ + O \to NO^+ + N, \tag{11.34}$$

are of major importance (see Chapter 14, p. 273). It is a great advantage of the flowing afterglow technique that it may be used to measure these rates. A considerable concentration of atomic nitrogen may be built up in the normal molecular gas by subjecting it to an electric discharge generated by microwave power. The mixture of atomic and molecular species may then be introduced to the flow tube through a side arm, as usual. If the actual inflow rate of the atoms alone can be measured, the reaction rates of the ions under study with the atoms can be determined in the usual way.

The technique used for these measurements is one of titration. Thus nitric oxide (NO) is introduced into the mixed stream at an adjustable and measurable flow rate. This produces the reaction

$$N + NO \to N_2 + O. \tag{11.35}$$

The flow rate of NO which just leads to complete recombination of the N atoms can be determined visually as follows.

If the N atoms are in excess, they react with the O atoms formed from (11.35) to form NO molecules in three-body collisions

$$N + O + M \to NO + M, \tag{11.36}$$

M being a third molecule. NO molecules are formed in excited states and blue light is emitted from these molecules. On the other hand, if the NO molecules are in excess they react with O atoms to produce NO_2

$$NO + O + M \to NO_2 + M. \tag{11.37}$$

Again some of the NO_2 molecules are formed in excited states and emit greenish blue light.

Atomic and Molecular Collisions

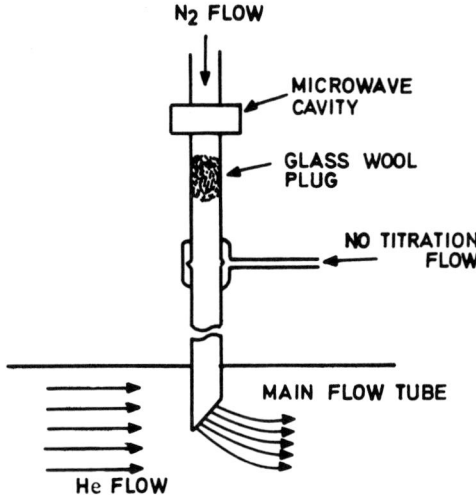

Figure 11.3. The arrangement used to inject N atoms in known concentration into the main flow tube for measurement of reaction rates by the flowing afterglow technique.

To take advantage of these photochemical effects, the NO flow rate is slowly increased until the blue light, from (11.36), suddenly fades out, in which case the NO and N concentrations are equal. This may be checked by verifying that further increases of the NO flow rate lead to emission of the greenish blue light arising from (11.37).

The reaction (11.35) also provides a source of atomic oxygen with a known flow rate. Figure 11.3 illustrates the general arrangement of an inlet system for N or O atoms.

In many cases, a particular ion can react with a particular neutral molecule in more than one way. Evidence about the relative importance of different reactions can be obtained by comparison of the rate of decrease of concentration of the reacting ion with the rates of increase in the concentrations of various other ions as sampled by the mass spectrometer.

11.7. Results of Measurements of Ionic Reaction Rates

The rates of a great number of ionic reactions have now been measured by the flowing afterglow and other techniques. We cannot hope here to do more than mention a very small selection of these. Many of the afterglow experiments were directed towards the measurement of the rates of reactions likely to play some role in the behaviour of the Earth's ionosphere, involving positive and negative ions of atmospheric atoms and molecules (see Chapter 14, p. 274). We shall choose for discussion reactions of this type under four headings—charge transfer, associative detachment, rearrangement and cluster formation.

Thermal Collisions of Atoms and Molecules

Reaction	Energy released/eV	Rate constant/10^{-15} m^3 s^{-1}	
		Measured	Orbiting limit
(a) $He^+ + N_2 \rightarrow N_2^+ + He$	9·0	1·5	1·7
(b) $O^+ + O_2 \rightarrow O_2^+ + O$	1·3	0·04	0·9
(c) $N_2^+ + N \rightarrow N^+ + N_2$	1·05$_5$	< 0·01	0·82
(d) $O_2^+ + NO \rightarrow NO^+ + O_2$	2·8	0·80	0·75
(e) $NO^- + O_2 \rightarrow O_2^- + NO$	0·4	0·9	0·76
(f) $O^- + O_3 \rightarrow O_3^- + O$	0·6$_5$	0·53	—

Table 11.2. Measured rate constants for selected charge transfer reactions at room temperature, compared with the orbiting limit.

11.8. Charge Transfer Reactions

Table 11.2 gives rate constants measured for a number of charge transfer reactions all of which are exothermic—that is to say, energy is released by the reaction, so that it would be expected to proceed at a reasonably rapid rate under thermal conditions.

For all but the last reaction (f), the orbiting limit given by (11.27) with $\zeta = 1$ can be calculated, the polarizability of the neutral target being known. It appears that, for the reactions (a), (d) and (e), the observed results are not inconsistent with a value of ζ close to 1, but for (b) and (c) ζ is very much smaller. The reason for this is not clear.

11.9. Associative Detachment

Rates for four associative detachment reactions are given in table 11.3. In all of these cases, the measured rates agree with those obtained from (11.27) with ζ being between 0·25 and 0·5, an entirely reasonable result.

11.10. Rearrangement Collisions

The rates of some rearrangement collisions of ionospheric interest are given in table 11.4. Again the reaction rates for (a) and (c) as well as that of the cluster

Reaction	Energy released/eV	Rate constant/10^{-16} m^3 s^{-1}	
		Measured	Orbiting limit
$O^- + O \rightarrow O_2 + e$	3·6	1·9	7
$O^- + NO \rightarrow NO_2 + e$	1·6	5	10
$O^- + CO \rightarrow CO_2 + e$	4·0	5	10
$O_2^- + O \rightarrow O_3 + e$	0·6	3·3	6
$O_2^- + N \rightarrow NO_2 + e$	4·1	5	8

Table 11.3. Measured rate constants for selected associative detachment collisions at room temperature, compared with the orbiting limit.

Reaction	Measured	Orbiting limit
(a) $N_2^+ + O \to NO^+ + N$	0·14	0·65
(b) $O^+ + N_2 \to NO^+ + N$	0·001	1·0
(c) $O_4^- + NO \to NO_3^- + O_2$	0·25	0·70
(d) $O_4^- + CO_2 \to CO_4^- + O_2$	0·43	–
(e) $O_2^- \cdot H_2O + NO \to O_2^- \cdot NO + H_2O$	0·31	0·71

Table 11.4. Rates of some rearrangement collisions of ionospheric interest.

exchange reaction (e), are what would be expected from the orbiting model with ζ between 0·2 and 0·4. (b) is anomalous, requiring ζ to be of the order 10^{-3}. As for (b) and (c) of table 11.2, the reason for this is unknown.

11.11. Cluster Formation

In the course of the discussion in Chapter 7, p. 123, of the study of electron loss in discharge afterglows, attention was drawn to the fact that the ions present under such conditions are often more complex than would be expected at first sight. Similar conditions exist in the lower ionosphere at altitudes below 70 km and in general in gases at pressures above 0·1 torr. A considerable amount of attention has actually been paid to the nature and rates of the reactions which lead to formation of complex ion clusters from simple ions. Some account of this work is given in Chapter 14, p. 273, in relation to the interpretation of ionospheric data.

CHAPTER 12

the collisions of energetic ions in gases—charge transfer

In the preceding chapter we discussed collisions of ions in gases at thermal or near-thermal energies, for the experimental study of which it is necessary to use swarm techniques. We now consider collisions at much higher energies (of the order of some keV) so that the scattering of ion beams of well defined energy and geometry may be studied in the laboratory.

A great amount of experimental and theoretical work has been devoted to the study of these collisions, so we must confine our attention here to one or two selected aspects. We have already discussed the rates of charge transfer reactions at near-thermal energies (Chapter 11, p. 204 and p. 211). It is appropriate to choose this type of reaction as the one to concentrate on in this chapter, partly because of the importance of these reactions at keV energies for controlled fusion experiments and in such phenomena as the polar aurora and partly because of their intrinsic interest—a number of new concepts are introduced.

In particular, we will deal largely but not exclusively with symmetrical charge transfer

$$A^+ + A \to A + A^+,$$
$$A^- + A \to A + A^-,$$

in which an electron is transferred without excitation of either colliding system. Because of the symmetry, the sum of the internal energies of the two systems remains unchanged.

We shall begin by giving a brief theoretical account of the phenomena to be expected in these symmetrical charge transfer collisions. This draws heavily on the semi-classical theory of atom–atom collisions but requires extension of the theory in some directions.

12.1. *Symmetrical Charge Transfer—Theoretical Discussion—H^+–H Collisions*

To introduce the subject in the simplest way we consider collisions between a proton and a hydrogen atom in which only one electron is involved. We shall further assume at first that the incident and target protons are distinguishable, so it is possible to observe whether the electron emerges bound to the former or latter proton.

As a close approximation to this situation, we can replace the target proton by a deuteron, which has the same positive charge but twice the mass. The binding energy of an electron in deuterium therefore differs by very little (3.6×10^{-3} eV, see Chapter 3, p. 51) from that in hydrogen. The reaction

$$H^+ + D \rightarrow D + H^+,$$

can then be regarded for present purposes as one of symmetrical charge transfer. Modifications introduced if the target atom is a hydrogen atom in which the proton is experimentally indistinguishable from the incident proton are discussed in section 12.4 below (cf. also pp. 218 and 226, where collisions of ^{4}He ions with ^{4}He atoms are discussed comparatively).

12.2. The H^+–H Interactions

In the discussion of atom–atom collisions in Chapter 9, we specifically

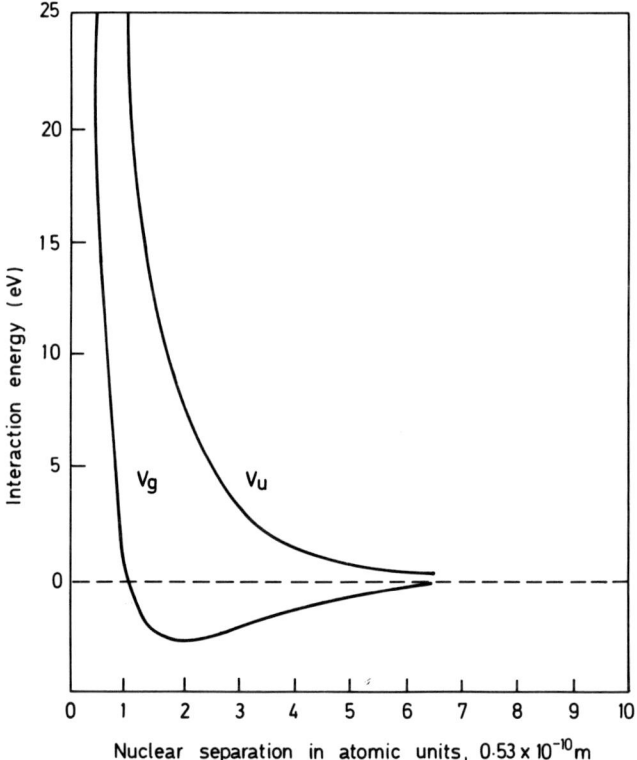

Figure 12.1. The interactions between a proton and a hydrogen atom.

excluded cases in which two atoms in their ground states may interact in more than one way. We must immediately drop this restriction in dealing with collisions between ions and atoms of the same species, because in all such cases an ion and atom in their ground states can interact in more than one way. Thus in the H^+–H case there are two possible forms of interaction as is also the case for He^+–He (Chapter 4, p. 64). For Ne^+–Ne as well as heavier rare gas ions and atoms there are four.

Figure 12.1 shows the two interactions for H^+–H which are distinguished by the suffixes g, u respectively. These stand for the German words *gerade* (even) and *ungerade* (odd) and refer to the fact that the curve with the deep minimum corresponds to an electronic state in which the wave function remains unchanged if the nuclei are exchanged, whereas the corresponding wave function for the u state changes sign under this exchange.

The g curve has a depth 2·69 eV at a separation of $2a_0$. This is a much deeper attraction than any rising from polarization and leads to formation of stable H_2^+ molecules. At large separations, both interactions tend to the form (11.1) of Chapter 11, where α is the polarizability of atomic hydrogen. Apart from this weak attraction at large separations, the u curve is repulsive. There is an equal chance that an H^+ ion and an H atom will interact according to either the g or the u curve, i.e., as $V_g(R)$ or as $V_u(R)$, where R is the nuclear separation.

12.3. Elastic Scattering and Charge Transfer

Consider a collision between an H^+ ion and a normal H atom in which the relative velocity is v. We may calculate the scattered amplitude, given the interaction $V(R)$ between them, exactly as described in Chapters 2 and 9, but how are we to take account of the fact that there is an even chance that the interaction will follow $V_g(R)$ or $V_u(R)$?

The reduced mass M of the system is closely $\tfrac{1}{2}M_p$ where M_p is the proton mass. Suppose that $f_g(\theta)$ and $f_u(\theta)$ are the amplitudes for scattering of a particle of mass M and velocity v by a centre of force exerting a potential energy V_g, V_u respectively. The differential cross-section per unit solid angle in the CM system for all collisions between a proton and a hydrogen atom which do not involve any transfer of energy to internal motion is then given by

$$I_t = \tfrac{1}{2}|f_g(\theta)|^2 + \tfrac{1}{2}|f_u(\theta)|^2 \tag{12.1}$$

i.e., it is the mean of the cross-sections for collisions under the g and u interactions respectively.

This result is not surprising and also not very informative. In particular, because an average of intensities and not of amplitudes is involved, no interference effects will arise between the waves scattered by the different interactions. However, we must remember that contributions to the collisions considered in (12.1) come not only from direct elastic scattering in which no electron transfer occurs but also from symmetrical charge-transfer collisions. When this is analysed, it is found that the differential cross-sections I_d and I_{tr} for these respective collisions are given by

$$I_d = \tfrac{1}{4}|f_g(\theta) + f_u(\theta)|^2, \text{ for direct elastic scattering}, \tag{12.2}$$

$$I_{tr} = \tfrac{1}{4}|f_g(\theta) - f_u(\theta)|^2, \text{ for charge transfer}. \tag{12.3}$$

The existence of the two interactions is thus associated with the possibility of symmetrical charge transfer.

As the expressions for I_d and I_{tr}, both involve direct combination of amplitudes, which will in general be of different phase, we expect interference effects to occur in both. It is to be noted that, as it must,

$$I_t = I_d + I_{tr}. \tag{12.4}$$

The probability P_{tr} of charge transfer in collisions in which the scattering angle in the CM system is θ is now given by

$$P_{tr} = I_{tr}/I_t = I_{tr}/(I_{tr} + I_d). \tag{12.5}$$

Just as for atom–atom collisions, the conditions encountered in any experiment will usually be semi-classical. In the simplest case in which there is an unique relation between the impact parameter p and the scattering angle θ for both interactions, we may write (Chapter 9, p. 173).

$$f_g(\theta) \simeq \{I_g^{cl}(\theta)\}^{1/2} \exp\{iC_g(\theta)\}, \tag{12.6}$$

and similarly for $f_u(\theta)$. $I_g^{cl}(\theta)$ is the classical scattered intensity for the interaction V_g, given as in Chapter 2, p. 11, by

$$I_g^{cl}(\theta) = \operatorname{cosec} \theta \left| p \frac{dp}{d\theta} \right|_g, \tag{12.7}$$

where the suffix g indicates that p and $\dfrac{dp}{d\theta}$ are calculated for the impact parameter p which leads to scattering through an angle θ by the interaction V_g. $C_g(\theta)$ is the phase angle given by the formula (9.50) of Chapter 9 with $V(r)$ replaced by V_g.

At the relatively high impact energies with which we are now concerned, both $f_g(\theta)$ and $f_u(\theta)$ fall off rapidly with increasing θ at large θ, a result of some importance when we consider the effect of the indistinguishability of ions produced by charge transfer from those directly scattered.

12.4. Effect of the Identity of the Nuclei

There is no difficulty in principle in carrying out a laboratory experiment to measure the cross-section for symmetrical charge transfer as a function of the velocity of the incident ions when the ions and atoms, while being of the same species, are of different isotopes. In such a case, say in a collision between protons and deuterium atoms, we can always distinguish a scattered proton from a deuteron produced by charge transfer because their masses are different. If, however, we are studying proton collisions with hydrogen atoms, how can we know whether a proton observed in a collision as scattered through an angle θ in the centre of mass system is really an elastically scattered proton and not the original atom scattered through an angle $\pi - \theta$ after suffering charge transfer?

If the nuclei are identical then, according to wave mechanics, the overall wave function for the system, including both electronic and nuclear motion, must possess special symmetry properties. For protons in particular, it must be antisymmetrical for interchange of any pair of protons. The result is that interference will occur between the amplitudes for the two indistinguishable events which can result in ions being observed in the scattering direction θ. The differential cross-section for such collisions between H^+ and H is now given by

$$\tfrac{1}{4}|f_g(\theta) + f_g(\pi - \theta) + f_u(\theta) - f_u(\pi - \theta)|^2$$
$$+ \tfrac{3}{4}|f_g(\theta) - f_g(\pi - \theta) - f_u(\theta) + f_u(\pi - \theta)|^2, \tag{12.8}$$

which is a fairly obvious extension of the corresponding formula (9.58) of Chapter 9. In practice, since the functions f_g and f_u fall off very rapidly with increasing θ, the small angle scattering is largely given by

$$\tfrac{1}{4}|f_g(\theta) + f_u(\theta)|^2 \tag{12.9}$$

and the large angle by

$$\tfrac{1}{4}|f_g(\pi - \theta) - f_u(\pi - \theta)|^2. \tag{12.10}$$

These are just the cross-sections for direct elastic scattering and for charge exchange in the case when the nuclei, while possessing the same force fields, are distinguishable. Because of this, we may effectively observe charge transfer even when the nuclei are identical by measuring the intensity of ions scattered through large angles in the relative system, that is to say at angles close to 90° in the laboratory system. At the same time, it must be realized that, in fact, interference effects will arise and are detectable at intermediate angles for which the direct and charge exchange scattering are of comparable intensity. No such effects will arise if the nuclei are of different isotopes.

12.5. Symmetrical Charge Transfer in General

Similar considerations apply to collisions between other ions and atoms of the same species, though the final formulae allowing for the identity of the nuclei will differ from case to case.

Atomic and Molecular Collisions

For He$^+$–He collisions, there are again two interactions V_g, V_u although in this case V_g is the repulsive one. For collisions involving only the isotope of mass number 4, (12.8) is replaced by

$$|f_g(\theta) - f_g(\pi - \theta) - f_u(\theta) + f_u(\pi - \theta)|^2. \tag{12.11}$$

On the other hand for ^{3}He$^+$–^{3}He$^+$ collisions (12.8) remains valid.

12.6. Experimental Methods and Results—Ion Sources

Experiments may be carried out to measure total cross-sections for charge transfer, differential cross-sections for elastic scattering of ions by gas atoms and the probability of charge transfer as a function of angle of scattering and incident ion energy. After first describing briefly a typical ion source for such experiments, we then describe a typical experimental arrangement used for each of the three types of measurement and discuss the results which have been obtained in relation to the theoretical considerations discussed above.

A considerable variety of ion sources are used in practice for studying the collisions of ions with gas atoms and molecules. It would be out of place here to attempt a comprehensive description of these sources. Instead, we follow our usual procedure in selecting a typical arrangement to illustrate the principles involved.

In general, positive ions are produced at near-thermal energies in an electric discharge in the gas whose ions are to be studied. These ions are extracted from the discharge plasma by a negative electric potential. The emergent stream will usually contain ions of different mass and sometimes different charge and it will not be sufficiently well defined in energy, or geometrically. Ions of the

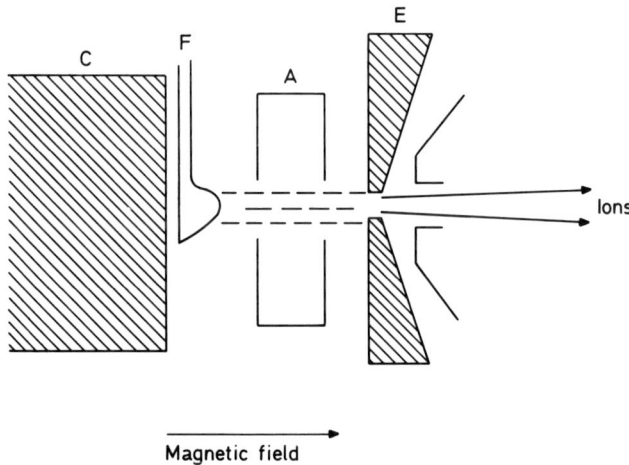

Figure 12.2. Typical arrangement of a Finkelstein ion source.

Collisions of Energetic Ions in Gases

desired species are selected by a mass analyser of some type (Appendix 7), are collimated through a suitable electrostatic lens system (Appendix 5) and rendered homogeneous in energy by passage through a suitable analyser such as described in Chapter 6, Section 6.3.

Figure 12.2 shows schematically a typical low current discharge source first used by Finkelstein in 1940. Gas at a pressure between 10^{-2} and 10^{-4} torr in the anode chamber A is ionized by electrons accelerated from the hot filament F by a suitable potential difference. Ions produced by electron impact pass out through an orifice in the electrode E, which is maintained at a negative potential with respect to A. By means of a magnetic field directed along the axis of the system, electrons are constrained to move parallel to the axis. These electrons are reflected back by the negative potential at E and thence again at F so that they oscillate back and forth many times. This increases the ionization yield per electron.

In all of these ion sources, it is essential to have very fast pumping to maintain the pressure difference between the chamber in which the gas is ionized and the region into which the ion stream is extracted. In the latter region, it is most important that the pressure be so low that the ions have little chance of suffering collisions and so being lost to the beam.

12.7. The Measurement of Total Cross-Sections for Charge Transfer

For these measurements, a beam of ions A^+ of well defined energy is fired through the gas of atoms A. Ions which can be taken to have resulted from charge transfer will be those which are observed moving in directions nearly normal to the beam. In classical terms, these are target atoms which have suffered glancing collisions with the incident ions so that they acquire a small velocity nearly normal to that of the beam, and which at the same time have lost an electron to the colliding ion.

In principle, then, it is only necessary to measure the current i_s of slow secondary ions produced in this way by the beam, of current i_0, in passing a small distance δl through the gas at concentration n. Q_{tr} is given by

$$nQ_{tr}\delta l = i_s/i_0, \qquad (12.12)$$

provided both are small so that only single collisions contribute.

In practice, it is necessary to distinguish the ions resulting from charge exchange from those produced through ionization of the atoms by ion impact, and to allow for the fact that secondary electrons may be produced when ions and electrons are collected at the receiving electrodes.

It is important to remember also that the charge-transfer cross-section derived from measurements of this kind will include contributions from reactions in which the ion and/or the atom are left in excited states. However, provided the ion energy is not greater than a few keV, these reactions will be

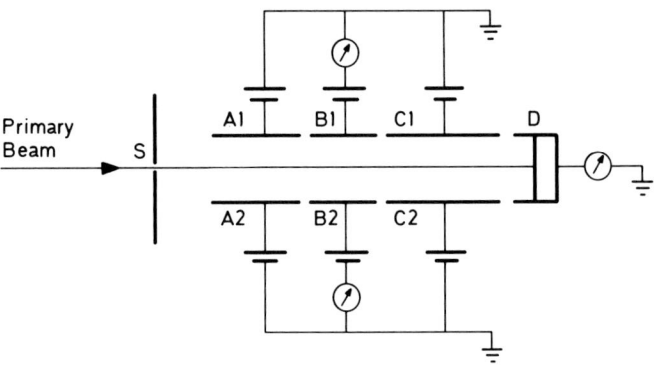

Figure 12.3. Arrangement of apparatus for measurement of total cross-sections for charge transfer collisions of ions in permanent gases.

unimportant and the observed Q_{tr} for ions and atoms of the same species will be closely that for the symmetrical charge transfer process with which we are concerned.

Figure 12.3 illustrates the arrangement commonly adopted for measurement of Q_{tr} for collisions in permanent gases. The primary ion beam, produced by a discharge-type ion source, enters the collision chamber containing the gas under investigation, through a slit S. It then passes between three pairs of parallel plate electrodes A1, 2, B1, 2 and C1, 2 before entering a collector D, often in the form of a Faraday cup (Appendix 6). The current recorded by D is the beam current i_0. An electric field is applied between the parallel plate electrodes, strong enough to collect on B1 and B2 all slow ions and electrons formed by the beam in passing between these plates. To ensure that the electric field between B1 and B2 is uniform and that the plates do receive all of the slow charged particles formed between them, and none formed elsewhere on the beam path, A1, 2 and C1, 2 act as guard electrodes, A1 and C1 being held at the same potential as B1, and A2 and C2 as B2.

We suppose that the only reaction which occurs in addition to symmetrical charge transfer is the production of single atomic ions,

$$A^+ + A \rightarrow A^+ + A^+ + e. \tag{12.13}$$

If i_c, i_i are the slow positive ion currents due to charge transfer and to ionization respectively, the current received by the negative electrode B1 will be

$$i_1 = i_i + i_c + i_s, \tag{12.14}$$

where i_s is the current due to secondary electron emission from B1. Again, if i_e is the current of slow electrons produced through (12.13), the current received by B2 will be given by

$$i_2 = i_e + i_s. \tag{12.15}$$

Collisions of Energetic Ions in Gases

But $i_e = i_i$ so that the required current

$$i_c = i_1 - i_2. \qquad (12.16)$$

This method is adequate provided the ion energies are not so high (many keV) that double ionization occurs, i.e., that two electrons are removed from an atom in a single impact with an ion. Precautions must be taken to ensure that the gas pressure encountered by the beam is never large enough to alter the composition of the beam through charge transfer nor to produce any other effect due to multiple collisions.

A crossed-beam method may be applied to measure Q_{tr} for collisions in which the atoms concerned are chemically unstable or do not form permanent gases. Such methods have been used for measurement of Q_{tr} for H^+–H and for collisions of alkali metal ions with alkali metal atoms.

In the arrangement used by Perel, Vernon and Daley, a beam of alkali metal ions intersected a beam of alkali metal atoms issuing from an oven at a known temperature (see Chapter 10, p. 181). Slow ions formed by charge transfer collisions were swept by a weak electric field on to a collector plate. The intensity of the ion beam was measured by a Faraday cup and of the atom beam by a surface ionization detector (cf. Chapter 10, p. 186). The collision volume was accurately defined by beam-collimating slits, and the uniformity of the ion beam across its width was checked by measurement of the current passing through a small slit which could be moved across the front of the Faraday cup. As usual, to reduce the serious background effects which are always associated with crossed-beam experiments, the atom beam was modulated so that selective amplification and phase-sensitive detection could be used (cf. Chapter 13, p. 246).

12.8. Results of Total Charge Transfer Cross-section Measurements

Figure 12.4 illustrates the observed cross-sections[†] for slow H^+–H collisions compared with those calculated. The corresponding comparison for H^-–H is given in figure 12.5. In both cases the agreement between theory and experiment is not unsatisfactory.

There is no difficulty in measuring cross-sections for unsymmetrical charge transfer using the same apparatus as that described above. Figure 12.6 shows measured cross-sections for protons in helium, neon and argon. The form of the cross-section–velocity curve is typical of asymmetrical cases in which an amount of energy ΔE of the order some electron-volts must be transferred to or from the kinetic energy of relative motion for the charge transfer reaction to occur. For protons in helium, ΔE is as high as 11 eV, the difference between the ionization energies of H and He. The corresponding values for protons in Ne and Ar are 8·0 and 2·2 eV. For the asymmetrical reactions, the cross-section is

---

[†] In figure 12.4 the square root of the cross-section rather than the cross-section itself has been plotted so as to make the diagram more compact.

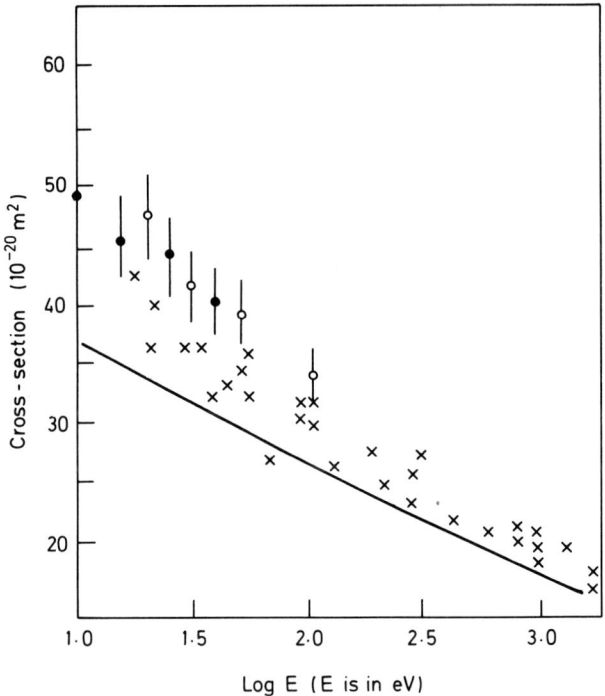

Figure 12.4. Comparison of observed and calculated cross-sections for the charge transfer reaction between protons and hydrogen atoms. ×, ●, ○, experimental observations; ——, calculated.

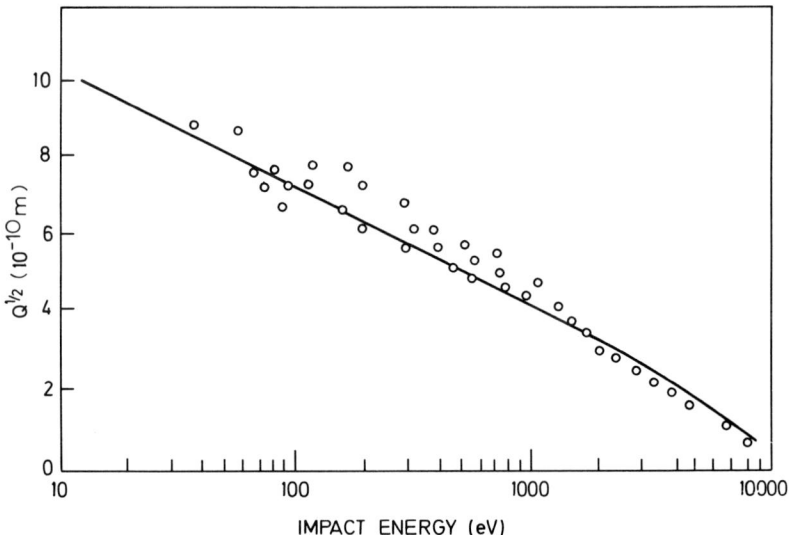

Figure 12.5. Comparison of observed and calculated cross-sections for the charge transfer reaction between H$^-$ ions and hydrogen atoms. ○, observed values; ——, calculated.

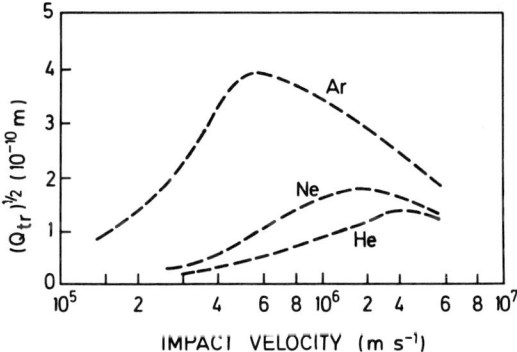

Figure 12.6. Observed cross-sections (Q_{tr}) for charge transfer reactions of protons in helium (He) neon (Ne) and argon (Ar) as indicated. Note that $(Q_{tr})^{1/2}$ is plotted instead of Q_{tr}.

very small at low ion velocities and reaches a maximum at a velocity corresponding to an energy of some keV, whereas for the symmetrical case it falls gradually as the velocity decreases. In general, the larger the value of ΔE, the larger the velocity at which the transfer cross-section is a maximum. This may be seen by comparing the forms of the observed cross-sections for protons in He, Ne and Ar shown in figure 12.6.

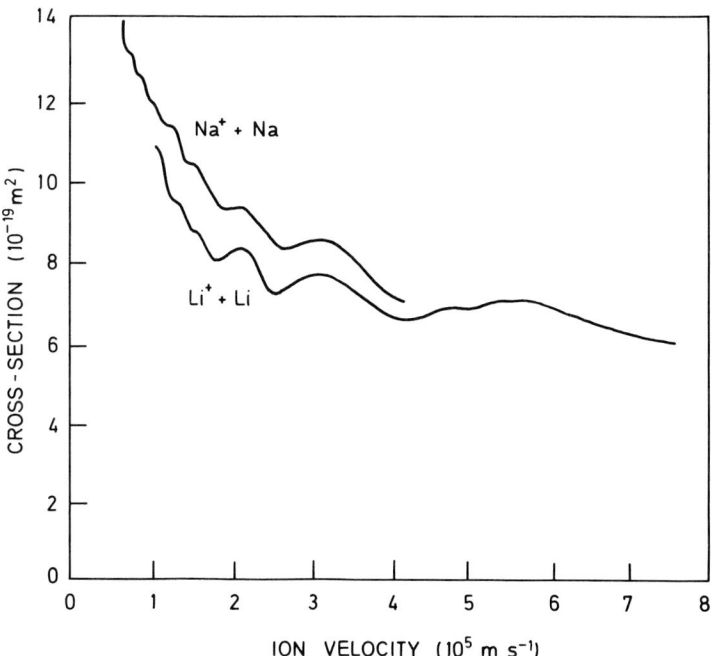

Figure 12.7. Observed cross-sections for charge transfer reactions of Li^+ in Li and of Na^+ in Na, showing 'glory' type oscillations

Atomic and Molecular Collisions

In all these cases, the cross-section varies smoothly with ion velocity, but figure 12.7 shows some observations by Perel and his collaborators which exhibit oscillations closely related to the 'glory' oscillations observed in total elastic cross-sections for collisions between gas atoms (cf. figure 10.10 of Chapter 10).

12.9. Measurement of Charge Transfer Probabilities and of Differential Scattering Cross-sections

The measurement of the probability P_{tr} of charge transfer for a fixed angle of scattering and ion beam energy is relatively simple.

Figure 12.8 illustrates the arrangement used by Everhart and his collaborators. The ion beam enters the collision chamber containing the target gas atoms through the hole a. The intensity of the beam may be measured by a removable Faraday cup f shown dotted in figure 12.9. Particles scattered into a certain narrow angular range about θ, defined by the exit hole c, pass through the hole d into an electrostatic analyser which separates them into different beams according to their charges. By means of a movable slit and photomultiplier detector (Appendix 6) the intensities of these beams may be separately measured. The flexible bellows which bridges the collision chamber makes it possible to observe at a number of chosen scattering angles θ. For measurements with H atom

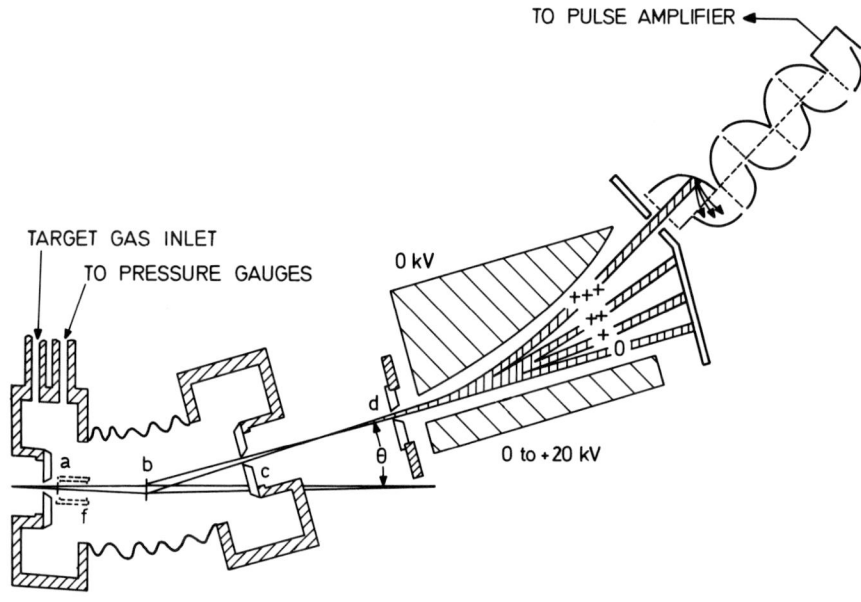

Figure 12.8. Arrangement of apparatus used by Everhart and his collaborators for measuring probabilities of charge transfer for ions scattered through a well defined angle by gas atoms.

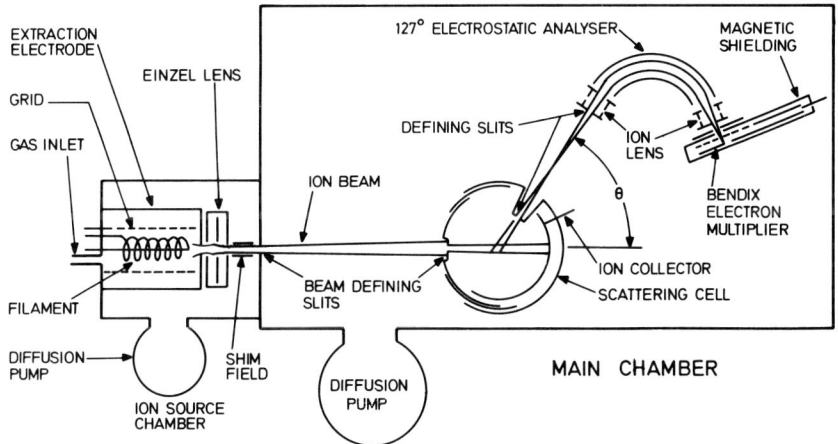

Figure 12.9. Arrangement of apparatus used by Aberth and Lorents for measuring differential cross-sections for scattering of He$^+$ ions in helium gas.

targets, the collision chamber was enclosed in a hot oven which produced dissociation of the enclosed H_2.

The measurement of differential cross-sections for ion–atom collisions is difficult because of the very rapid increase in intensity with angles of scattering. It was only in 1963 that the first successful results were obtained by Aberth and Lorents for He$^+$ ions with energy in the range 20–600 eV scattered through angles ranging from 1° to 36° in the laboratory system. Figure 12.9 illustrates the general arrangement of their apparatus.

The scattering cell was constructed of three concentric cylinders, the innermost being fixed, while the outermost, which contained an exit slit of dimensions 0·5 × 5 mm, could be rotated about the common axis. The fit between the cylinders was so close that the pressure in the main chamber (see figure 12.9) could be maintained at less than 1% of that in the scattering cell which was usually around 5×10^{-4} torr.

The intensity of the main beam was monitored by a collector plate inside the scattering cell, mounted on the outermost cylinder.

The scattered beam was analysed by a 127° electrostatic analyser (see Chapter 6, p. 93) with resolving power sufficient to distinguish the elastically scattered from the inelastically scattered ions. After passage through the analyser, the ions were detected by an electron multiplier (Appendix 6) coupled with a vibrating reed electrometer (Appendix 6) so that a current as small as 10^3 ions per second could be measured. The analyser and detector were mounted on a platform so they could be rotated with the outermost cylinder of the scattering cell.

The differential cross-section per unit solid angle in the laboratory system is given by

$$I(\theta) = i(\theta)/ni_0(\Delta\omega)_{\text{eff}}$$

where i_0 is the incident ion current, $i(\theta)$ the measured current when the detector is set to measure scattering through an angle θ in the laboratory system, and n the concentration of scattering atoms along the beam. The effective solid angle $(\Delta\omega)_{\text{eff}}$ must be calculated from the geometry of the system.

12.10. Results for He^+–He Collisions and Their Interpretation

Figure 12.10 illustrates typical experimental data points for scattering of $^4\text{He}^+$ ions of 100 and 600 eV energy by He atoms. For ions at 100 eV, regular oscillations are observed over almost the whole angular range and it is natural to regard these as arising from the interference between the f_g and f_u amplitudes as given by (12.9). It is to be noted that, between 10° and 20°, the amplitude of the oscillations is somewhat larger than would be expected but otherwise no marked departure from regularity is detectable.

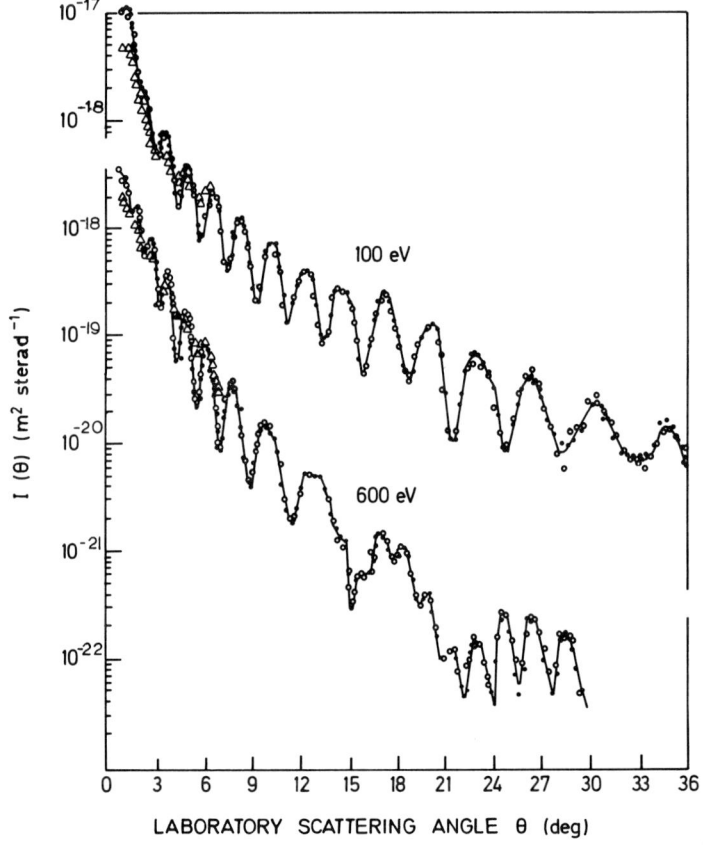

Figure 12.10. Observed differential cross-sections $I(\theta)$ per unit solid angle for elastic scattering of He^+ ions in He. Results are given for ions of energy 100 eV and 600 eV as indicated.

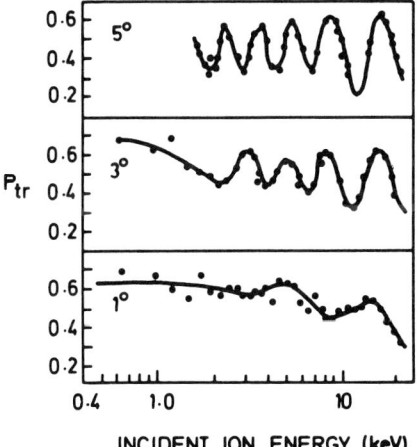

Figure 12.11. Probability of charge transfer P_{tr} as a function of scattering angle for He$^+$ ions in He.

For ions at 600 eV, regular oscillations are observed out to about 12° but at larger angles a more complicated pattern appears, indicating that some further effect is involved.

Evidence about the nature of these effects may be gained from further experiments. First of all, it should be possible to relate the maxima and minima of the charge transfer oscillations to those observed in the variation of the charge transfer probability P_{tr} with angle of scattering (see figure 12.11). According to (12.5) and (12.9), the fluctuations in $I_d(\theta)$ and $P_{tr}(\theta)$ should be in antiphase (i.e., differ in phase by 180°) and this is found to be so.

Assuming that the regular oscillations do arise in this way, further evidence about the nature of the additional effects can be obtained by comparing differential cross-sections observed for ^{4}He$^+$–^{4}He collisions with those for ^{4}He$^+$–^{3}He in which the atoms of the main isotope are replaced by those of the light isotope ^{3}He. Such a comparison is shown for ions of 600 eV energy in figure 12.12. It will be seen that, when the ions and atoms are of different isotopic species the oscillations in $I_d(\theta)$ remain regular throughout the entire angular range of the observations. This suggests strongly that the complex features present in $I(\theta)$ for ^{4}He$^+$–^{4}He collisions arise from the indistinguishability of the nuclei through the formula (12.11).

On the other hand, the less obvious irregularity observed between 10° and 20° for ions of 100 eV kinetic energy does not disappear when ^{4}He$^+$ ions are replaced by ^{3}He$^+$. The explanation of this feature requires consideration of inelastic as well as elastic scattering and is beyond the scope of this book.

12.11. *Other Reactions of Energetic Ions in Gases*

We have only barely touched on the wide range of collisions which may

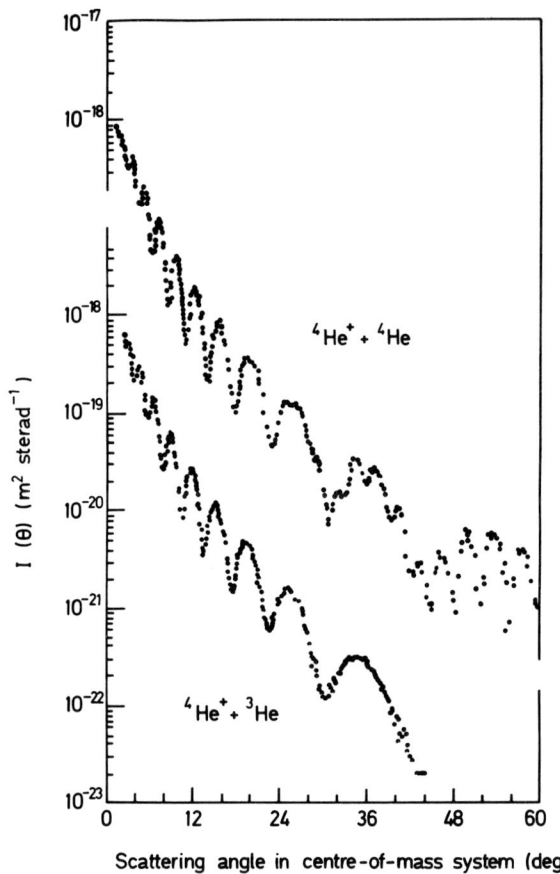

Figure 12.12. Comparison of observed differential cross-sections $I(\theta)$ per unit solid angle for elastic scattering of ^{4}He$^+$ ions of 600 eV energy in ^{4}He and in ^{3}He.

occur between energetic ions and atoms. The number of possibilities is much larger than for electron–atom collisions because the ions may differ in mass and charge. Thus we may have a collision between two atomic species A and B, one of which, A, is k-fold ionized which may be written

$$A^{k+} + B \rightarrow A^{m+} + B^{n+} + (m+n-k)e, \qquad (12.17)$$

the collision products being an m-fold ionized A atom, an n-fold ionized B atom and $(m+n-k)$ free electrons, k, m and n being all zero or positive integers. The cross-section for a process such as (12.17) may be written

$$_{k0}Q_{mn}^{AB}. \qquad (12.18)$$

If A and/or B form stable or metastable negative ions, we may generalize (12.18) to include the possibility of a collision between A$^-$ and B by taking

$k = -1$ and the possibility that A^- or B^- is one of the product ions by taking m or $n = -1$, respectively. It is only necessary that $(m + n - k)$ should not be less than 0.

A great deal of experimental work has been devoted to the measurement of cross-sections of this kind for impact energies of a few keV upwards. Data of this kind are required, for example, to determine the effect of impurities in hot plasmas which could be important for the design of devices for economic production of power from nuclear fusion. Although lack of space has limited us to consideration of only a very small sample of ion collisions, the techniques described are fairly representative of a wide range of experimental research.

CHAPTER 13
photon collisions with atoms and molecules—photoionization and photodetachment

13.1. Introduction

In preceding chapters we have discussed the collisions of electrons, of neutral atoms and of positive and negative ions with gas atoms, and in some cases, molecules. It remains to consider the effects which arise when the colliding particles are photons; in other words, effects due to interaction of atoms with electromagnetic radiation. These are of great importance for many applications, for example in atmospheric and solar physics, as well as in the design of lasers.

The best known and most probable collisions which occur are distinguished from those we have discussed earlier in that they involve actual absorption of the incident photon. If the frequency v of the radiation is less than E_i/h, where E_i is the ionization potential of the atom, absorption can take place by the atom in the ground state only if the frequency v is closely equal to that associated with a transition from the ground state of energy E_0 to an excited state with energy E_n, i.e.

$$hv = E_n - E_0. \tag{13.1}$$

This is known as *line absorption*. On the other hand, if $v > E_i/h$, absorption will produce ionization of the atom, the energy E_e of the ejected electron being such that

$$E_e = hv - E_i. \tag{13.2}$$

This is the process of *photoionization*, the probability of which will vary continuously with v. The threshold frequency E_i/h for a number of atoms is given in table 13.1. Even the lowest frequency, for Cs, falls in the ultraviolet, while for the rare gas atoms it is in the so-called vacuum ultraviolet. This means that the atmospheric absorption of the radiation is so strong that the optical system used in studying it must be contained within an evacuated chamber.

Equally well we may consider collisions between photons and negative ions A^- say. In general there will be no line absorption because of the paucity of bound excited states of negative ions but, if $v > E_a/h$, where E_a is the electron affinity, absorption of the photon may occur, leading to detachment of an electron from the negative ion, with kinetic energy E_e given by

$$E_e = hv - E_a. \tag{13.3}$$

Photon Collisions with Atoms and Molecules

Because the electron affinities of all atoms are less than 3·6 eV, and for many are less than 2 eV, the threshold frequencies for photodetachment are much lower than for photoionization. Thus for an electron affinity of 2 eV, the threshold frequency is $4·8 \times 10^{14}$ s^{-1} and wavelength 619 nm, in the visible (orange) part of the spectrum.

It is possible, if the frequency of the radiation is high enough, that the positive ion resulting from photoionization, or the neutral atom resulting from photodetachment, be left in an excited state. If the excitation energy of the state is E_{ex} the minimum frequency for such a process to occur will be given by

$$h\nu = E_i + E_{ex}, \text{ for photoionization}, \tag{13.4}$$

$$= E_a + E_{ex}, \text{ for photodetachment}. \tag{13.5}$$

Thus for a hydrogen atom, $E_{ex} = 10·1$ eV for the first excited state. The lowest frequency of radiation which will eject an electron from H$^-$ leaving the atom in this state will be given by (13.5) with $E_a = 0·75$ eV, i.e., a frequency of $2·62 \times 10^{15}$ s^{-1}, corresponding to a wavelength of 114 nm. This is to be compared with the threshold frequency for photodetachment leaving the atom in its ground state, the primary threshold, i.e. $0·181 \times 10^{15}$ s^{-1}, corresponding to a wavelength of 1650 nm. On the other hand, for many atoms such as oxygen, the corresponding positive ion O$^+$ has low-lying excited states so that photoionization of O leaving the ion O$^+$ in one of these states must be taken into account at frequencies not so far above the primary threshold. Thus for O$^+$ the lowest excited states have $E_{ex} = 3·3$ and $5·0$ eV respectively, so that the minimum frequencies for photoionization leaving the O$^+$ in these states are $4·05 \times 10^{15}$ and $4·6 \times 10^{15}$ corresponding to wavelengths 73·2 and 66·4 nm, not so much shorter than for the primary threshold, 92·2 nm (see table 13.1). These considerations are of importance in considering photoionization of atmospheric atoms and molecules by ultraviolet radiation from the Sun (see Chapter 14).

Atom	Ionization energy/ eV	Threshold frequency/ 10^{15} s^{-1}	Threshold wavelength/ nm
H	13·5	3·27	91·8
He	24·5	5·92	50·4
Ne	21·6	5·22	57·4
Ar	15·8	3·82	78·4
Kr	14·0	3·38	88·5
Xe	12·1	2·93	102
Li	5·4	1·30	230
Na	5·1	1·23	244
K	4·3	1·04	287
Rb	4·2	1·01	294
Cs	3·9	0·94	317
O	13·6	3·34	92·2
N	14·5	3·51	84·9

Table 13.1. Threshold frequencies and wavelengths for photoionization of various atoms.

Atomic and Molecular Collisions

Everything we have been discussing above has been concerned with single photon collisions. Until the development of lasers, it seemed beyond the reach of practical possibility to observe simultaneous absorption of two, let alone more, photons as, for example, photoionization in a three-body collision

$$A + hv + hv \to A^+ + e. \tag{13.6}$$

However, it is now possible to provide such high light intensities in focused laser beams that not only two-photon but multi-photon absorption has been observed.

For the two-photon process (13.6), the primary threshold frequency is now given by

$$hv = \tfrac{1}{2} E_i$$

and proportionally less for a multiphoton process. It has been possible in this way to ionize xenon with radiation from a ruby laser at a frequency of 0.43×10^{15} s^{-1}, which is only 1/7 of the primary threshold for single photon ionization given in table 13.1. More extreme examples have also been reported.

Finally, we have omitted from consideration other possibilities such as Raman scattering in which the incident photon is absorbed but a second photon is emitted with a different frequency leaving the atom in an excited state. Thus if v is the frequency of the incident radiation, v' of the scattered,

$$hv' = hv - (E_n - E_0), \tag{13.7}$$

where E_n is the energy of the excited state of the atom and E_0 of the ground state. Under these circumstances, v' is always less than v but if the radiation is scattered by excited atoms the atom may be de-excited by the incident radiation and the energy released carried off by an emitted photon. In that case the frequency of the photon will be given by

$$hv' = hv + (E_n - E_0) \tag{13.8}$$

and so will be greater than v.

The intensity of Raman scattering is very weak but has important applications in physical chemistry and elsewhere. It is not possible in this work to do more than draw attention to the existence of the phenomenon.

13.2. Line Absorption and Stimulated Emission

An atom in an excited state of energy E_n, if left alone, will eventually drop to a state of lower energy E_m emitting a photon of frequency $v_{nm} = (E_n - E_m)/h$. The probability of this happening in any given second is called the *spontaneous emission coefficient* A_{nm} (see Chapter 3, p. 46). Thus an assembly of a large number N_n of atoms in the excited state of energy E_n will radiate $N_n A_{nm}$ quanta of frequency v_{nm} per second.

For many transitions, A_{nm} is of the order 10^7–10^8 s^{-1}. These are called allowed transitions, in contrast to forbidden transitions for which A_{nm} may be 10^4 s^{-1} or less. We concern ourselves here only with allowed transitions.

Equally well we may consider the inverse process of absorption. Atoms in the state of energy E_m may absorb photons of frequency v_{nm}, in which case they will make a transition to the state of energy E_n. To specify the rate at which absorption takes place let $\rho(v)dv$ be the radiant energy per unit volume with frequency between v and $v + dv$. Thus, for example, for black body radiation at an absolute temperature T, $\rho(v)$ is given by the Planck formula

$$\rho(v) = \frac{8\pi h v^3}{c^3} \frac{1}{\{\exp(hv/kT)\} - 1}, \tag{13.9}$$

where h is Planck's constant, k Boltzmann's constant and c the velocity of light. If there are N_m atoms present in the state m, the number of atoms excited to state n per second by absorption of radiation of frequency v_{nm} is given by

$$N_m B_{nm} \rho(v_{nm}), \tag{13.10}$$

where B_{nm} is a second coefficient, related to the spontaneous emission coefficient A_{nm}, as we shall see. Special importance attaches to cases in which the state m is the ground state and B_{nm} is then related to the line absorption coefficient.

The absorption process we have been considering is the analogue in quantum theory of resonance absorption of electromagnetic waves by an oscillatory electron. If the electron is acted on by forces which cause it to oscillate with a natural frequency v, it will be strongly affected by electromagnetic radiation of the same frequency. In some circumstances energy will be absorbed from the electromagnetic waves. However, depending on the phase relations between the motion of the oscillating electron and the electromagnetic waves, it is also possible for the electron to *lose* energy to the waves. What does this correspond to in quantum theory?

Loss of energy by the electron means, in quantum theory, a transition to a lower state, while gain by the electromagnetic field means emission of energy in the form of photons. The quantum process is called *stimulated emission*. An atom in the state n exposed to radiation of frequency v_{nm} and energy density $\rho(v_{nm})$ may be stimulated to make a transition to the state m, emitting a further quantum of frequency v_{nm}. The rate at which this will occur for an assembly of N_n atoms is given by

$$N_n B_{mn} \rho(v_{nm}). \tag{13.11}$$

Einstein, in 1917, showed that the three coefficients A_{nm}, B_{nm} and B_{mn} are closely related. Thus, in thermal equilibrium at absolute temperature T, it may be shown from statistical arguments that

$$N_n/N_m = \exp\{-(E_n - E_m)/kT\}$$
$$= \exp\{-hv_{nm}/kT\}. \tag{13.12}$$

This equilibrium ratio must be maintained by a balance between absorption and emission processes involving the black body radiation. Thus, under these conditions, the total rate at which $n \to m$ transitions occur must be equal to that of the inverse $m \to n$ transitions. This requires

$$\rho(v_{nm})B_{nm}N_m = A_{nm}N_n + \rho(v_{nm})B_{mn}N_n, \qquad (13.13)$$

the second product on the right-hand side arising from stimulated emission. We now have

$$\rho(v_{nm}) = \frac{A_{nm}}{B_{nm}\exp(hv_{nm}/kT) - B_{mn}}. \qquad (13.14)$$

But this must agree with (13.9) so that

$$B_{nm} = B_{mn} = B, \qquad (13.15)$$

and

$$A_{nm}/B_{nm} = 8\pi h v_{nm}^3/c^3. \qquad (13.16)$$

It is to be noted that, if stimulated emission were not included in (13.13), the Planck formula (13.9) for $\rho(v)$ could not be obtained.

More detailed analysis shows that when stimulated emission occurs, the photon is emitted with the *same phase and direction of motion* as the stimulating photon, a result of great importance which makes possible the operation of lasers.

13.3. Stimulated Emission and Laser Operation

Under normal conditions the fraction of atoms in a gas or solid which are in excited states is very small so that stimulated emission is quite unimportant. Thus we have, for photons of frequency v_{nm},

$$\frac{\text{Rate of absorption}}{\text{Rate of stimulated emission}} = \frac{N_m}{N_n}. \qquad (13.17)$$

If, in some way, conditions could be arranged in which $N_n > N_m$ (a so-called *inverted population*), the gain in photons due to stimulated emission would exceed the loss due to absorption. We would in effect, have achieved *amplification* of the intensity of radiation of frequency v_{nm}. Advantage could also be taken of the fact that the stimulated photon is emitted with the same phase as that of the stimulating photon—in other words that they cohere.

Suppose, for example, we have a cylindrical cavity with semi-transparent, accurately parallel, reflecting end plates (see figure 13.1) containing a medium in which, in some way, there is maintained a number N_2 of atoms in an excited state of energy E_2, greater than that N_1 in a lower state of energy E_1. Photons of frequency $v_{12} = (E_1 - E_2)/h$ will be reflected back and forth between the end

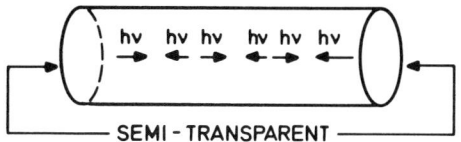

Figure 13.1.

mirrors to build up to a high intensity provided the following two conditions are satisfied.

(a) The distance d between the two mirrors is equal to an integral number of wavelengths of the radiation, i.e.,

$$d = n\lambda = nc/v_{12}. \qquad (13.18)$$

This condition ensures that the phase relations between the quanta which are being produced are just right to provide constructive instead of destructive interference.

(b) The photons must be moving accurately parallel to the axis of the cylinder. Otherwise, they will soon be lost through the surface of the medium and there can be no build-up.

If these conditions are satisfied and one of the end mirrors is slightly transparent, an intense beam of radiation of frequency very precisely equal to v_{12} (because of (a)) will emerge. Because of (b), this beam will be made up of nearly parallel rays, the only divergence being due to diffraction. The angular deviation from parallelism will thus be close to λ/a, where a is the thickness of the beam. For a beam of 10 mm thickness with $\lambda = 1000$ nm, this angle is only 10^{-4} radian, about 0·36 minutes of arc.

The laser (light amplifier through stimulated emission of radiation) principle and possibilities were first discussed by Schawlow and Townes in 1958 and the first laser beam was actually produced by Maiman in 1960.

The process of producing the laser medium with an inverted population of excited atoms is known as *optical pumping*. In Maiman's first experiments, the medium was a solid ruby and the pumping process, and hence the light output, was pulsed. A year later Javan, Bennett and Herriott, working with a gaseous mixture of helium and neon as laser medium, produced the first continuously operating laser. The pumping process in this laser makes use of collision processes in the gas. Helium atoms are excited by electron impact to metastable states. These excited atoms transfer their excitation to neon atoms on collision (see Chapter 8, p. 151) and so build up an inverted population of certain neon states which are used as the upper states for laser operation.

It would take us too far afield to discuss further the design of different types of laser. Several which are in successful operation depend on atomic or molecular collision processes for pumping. At the time of writing, the laser techniques are developing at a rapid rate, extending both the available wavelength range, which

Atomic and Molecular Collisions

for some time has been restricted to the infrared and visible part of the spectrum, and the power in the laser beam, which can now be very great.

A description of the applications of lasers already realized and in course of development would be both extensive and beyond the scope of this book. We shall, however, in this chapter be discussing some applications (see p. 248) which depend on the very well defined frequency of a laser beam and also others (see p. 256) which are made possible because of the high intensity available.

13.4. *Excitation of Autoionizing and Autodetaching States*

In Chapter 6, p. 89 it was pointed out that fine structure effects will arise in collision cross-sections for impact of electrons with atoms, considered as functions of electron velocity, due to the possibility of formation of relatively long-lived but nevertheless transitory negative ions through capture of the incident electron. When the unstable complex breaks up, the electron is re-emitted, so the overall process appears to be one of electron scattering.

The transitory negative ions concerned can be regarded as formed by attachment of an electron to an atom in an excited state so that the energy of the ion is less than that of the excited atom by an amount ΔE. The energy of the ion relative to the ground state of the atom is given by

$$(E_n - E_0) - \Delta E, \tag{13.19}$$

where E_n is the energy of the excited state of the atom, E_0 of its ground state. Because (13.19) is always positive, the ion will be unstable towards break-up into the atom in its ground state and an electron with kinetic energy $(E_n - E_0) - \Delta E$, i.e., towards autodetachment. However, because ΔE is positive, the ion will be relatively long-lived before break-up occurs. In elastic scattering, the existence of an autodetaching state of lifetime $1/\Gamma$ will show up in the existence of fluctuations in the cross-sections over an electron energy range of order $h\Gamma$. Experimental confirmation of these expectations was described in Chapter 6 (see particularly figures 6.7 and 6.9).

We can expect that the existence of autodetaching states of negative ions will be observable in the behaviour of the cross-sections for photodetachment from a particular negative ion—say as a function of frequency. Autodetaching states can be excited from the ground state by light absorption, just as are the 'stable' excited states. Thus radiation of frequency given by

$$hv = (E_n - E_0) - \Delta E \tag{13.20}$$

will excite the autodetaching state referred to above. We would therefore expect line absorption due to such excitation to be superposed on the continuous absorption, the frequency width of the lines being of order Γ. However, just as in electron scattering, we must allow for the fact that the scattered amplitude for the direct process must be combined with that for line absorption with due allowance for any phase difference. Interference effects will thereby occur so the net result of the line absorption may be either to increase or decrease the total absorption over the width of the line. Paradoxical as it may seem, the existence of line absorption due to excitation of autodetaching states may show up in a very marked reduction of the total absorption over the line width. Such effects have been observed in photodetachment (see p. 254).

It is not difficult to realize that line absorption will also be superposed on the continuous absorption arising from photoionization of a neutral atom. Consider, as a simple example, the helium atom which contains two electrons. If one of these electrons remains in its lowest level while the other is excited, we obtain a series of excited states of the helium atom, the energies of which are all less than that, 24·58 eV, required to ionize the atom, i.e., to remove an electron from it. We may also consider the excited states of the atom which arise if *both* the electrons are in excited levels. The lowest such state (see figure 13.2) lies above the first ionization threshold by nearly 36 eV. This means that it is unstable towards autoionization in which one electron drops to its lowest level and the other is ejected from the atom with a kinetic energy of 36 eV. The situation is very much the same as for the autodetaching states of negative ions, with similar consequences for the form of the continuous absorption spectrum. Just as in photodetachment, excitation of autoionizing states will produce dark or bright lines on the continuous absoprtion background depending on the phase relations involved. Examples of both are illustrated in figure 13.5.

The autoionizing states we have been considering are states in which two or more electrons are excited. Excitation of an electron from an inner shell of an atom to an outer allowed state will also lead to an autoionizing state for the atom as a whole. Thus, suppose the outer occupied level of an atom is that with total quantum number n and energy E_n. An electron in an inner level of total

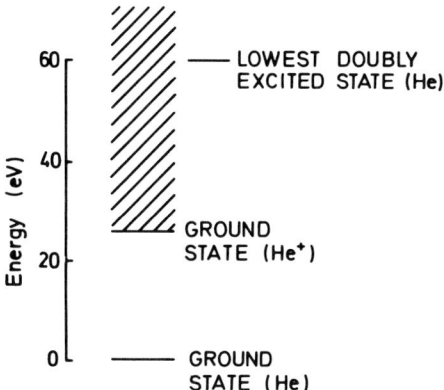

Figure 13.2. Doubly excited levels of an He atom.

Atomic and Molecular Collisions

quantum number m ($< n$) and energy E_m may be excited to an outer unoccupied level of total quantum number s ($> n$) and energy E_s by absorption of radiation of frequency $(E_s - E_m)/h$. This leaves the atom in a state in which a vacancy exists in the level m. If an electron in the level n drops into this vacancy an amount of energy $(E_n - E_m)$ is released and may be taken up by the initially excited electron. If $E_m - E_n - E_s > 0$, the electron will leave the atom with this kinetic energy, i.e., autoionization will occur. Both kinds of autoionizing states have been identified in the absorption (see p. 243) spectrum of the rare gases other than helium, for which it does not arise.

13.5. Relation between Absorption and Photoionization Cross-Sections

Just as for electrons, we may define a total absorption cross-section $Q_a(v)$ for a beam of frequency v passing through a particular gas. This will be such that the initial intensity of the beam I_0 will be reduced, after passing a distance l through the gas, to

$$I = I_0 \exp(-nQ_a l), \qquad (13.21)$$

where n is the number of gas atoms or molecules per unit volume. nQ_a is often referred to as the *absorption coefficient*.

If the gas is monatomic and $v > E_i/h$, where E_i is the first ionization energy of the atom, absorption can only take place through photoionization, so that the photoionization cross-section Q_{pi} is equal to Q_a.

This is not necessarily true if the gas is composed of diatomic or polyatomic molecules, because absorption may occur, even for frequencies $> E_i/h$, through dissociation into neutral fragments. Hence, for molecules $Q_a \geq Q_{pi}$ and it is often important to know both these cross-sections.

13.6. The Experimental Study of Ionization by Single Photons—Introduction

Measurements may be made not only of the total cross-section for the ionization or detachment process as a function of frequency but also of the energy and angular distribution of the ejected electrons. The precision measurement of the energies of the photoelectrons, known as *photoelectron spectroscopy*, is now a subject of great importance. This is because it makes possible the accurate determination of threshold energies not only for the primary ionization or detachment but also for processes in which the ion or atom is left in an excited state. Thus, consider the photoionization of an atom by radiation of frequency v which is greater than $E_i + E_{ex}$ where E_i is the ionization energy of the atom and E_{ex} the excitation energy of the first excited state of the corresponding positive ion. Under these conditions the ion may be left either in the ground state or in the first excited state. If the former, the energy of the photoelectron emitted will be

$$hv - E_i, \qquad (13.22)$$

and if the latter

$$hv - E_i - E_{ex}. \qquad (13.23)$$

Thus the photoelectron spectrum will have two sharp peaks at these energies from observation of which both E_i and E_{ex} may be obtained. This may be extended to cases in which the residual ion or atom may be left in any one of several states and is particularly useful for molecules.

It is possible to obtain quite good information about the primary threshold, and hence E_i, from the variation of the total cross section with frequency but this is much less accurate for the higher thresholds.

The first measurements of photoionization cross sections were carried out with caesium and rubidium vapour as the absorbing media. Although this would seem at first sight to be a rather strange choice it was dictated by experimental convenience. Alkali metal atoms, having low ionization potentials, may be ionized by relatively long-wavelength ultraviolet radiation (see table 13.1), which is not absorbed appreciably in a metre column of air. It is therefore unnecessary to work under vacuum conditions, though the optical components must be of quartz to avoid absorption of the radiation in passage through them.

Mohler and Boechner made the first quantitative measurements of Q_{pi}, the photoionization cross-section, as a function of wavelength, as early as 1929. Four to five years later, Ditchburn and Braddick developed the technique for absolute measurement of the cross-section for metal vapours.

It is only in post-war years that measurements have been extended into the vacuum ultraviolet so that observations could be made for the rare gas atoms. This has been accompanied by the development of photoelectron spectroscopy with high energy resolution. Experimental studies have also been carried out for molecular gases and vapours and even for unstable radicals such as atomic hydrogen, oxygen and nitrogen and the positive ions H_2^+ and He^+.

13.7. Principles Involved in the Measurement of Absorption and Photoionization Cross-Sections

To measure Q_a and Q_{pi}, the usual procedure is the obvious one of transmitting a beam of the radiation of the required frequency through a cell into which the absorbing gas or vapour may be introduced at any desired pressure. If the intensity I of the radiation after passage through the cell may be monitored by some detector, Q_a is obtained from (13.21) in the form

$$Q_a = -\frac{1}{nl}\ln(I/I_0). \qquad (13.24)$$

This does not require absolute measurement of I and I_0 but merely of the ratio, so the detector need not be calibrated.

For measurements with monatomic gases, this is all that is required, since $Q_{pi} = Q_a$. However, if the absorber is molecular, it is necessary to measure the current i_g of electrons produced by the passage of the beam through the absorber in the cell and to calibrate the detector.

If the absorber is monatomic, so each time a photon is absorbed an electron is emitted, the current of photoelectrons produced in passage through the cell is given by

$$i_g = e(I_0 - I)$$
$$= eI_0\{1 - \exp(-nQ_a l)\}. \tag{13.25}$$

I_0 is thus obtained in terms of measurable quantities by

$$I_0 = (i_g/e)/\{1 - \exp(-nQ_a l)\}. \tag{13.26}$$

Once the absolute initial radiation intensity has been obtained in this way, a convenient detector may be calibrated and used to measure Q_{pi} as well as Q_a for molecular absorbers. One such detector is a platinum plate from which photoelectrons are released by photon impact. For calibration, it is necessary to measure the photoelectric yield γ, which is the probability that an electron will be emitted from the plate for each incident photon. If i_p^0 is the total electron current emitted from the plate by the radiation of intensity I_0 which has passed through the cell with no absorber present

$$\gamma = i_p^0/I_0 e, \tag{13.27}$$

with I_0 given from (13.26) with a rare gas absorber.

For molecular gases

$$\frac{Q_{pi}}{Q_a} = \frac{i_g}{e(I_0 - I)} = \frac{i_g}{\gamma(i_p^0 - i_p)}, \tag{13.28}$$

where i_p^0, i_p are the photoelectric currents emitted by the detector with and without the absorbing molecular gas present, and i_g is the current of electrons produced from photoionization of the gas in the cell.

i_g may be measured by including within the cell two plane electrodes parallel to the light beam which passes midway between them. A potential difference is maintained between the plates sufficient to extract all the electrons and positive ions produced by the beam in passing through the cell, but not so great that secondary ionization is produced by the photoelectrons on their way to the negative plate.

13.8. Radiation Sources

In typical experiments of this kind, it is of course necessary to have available a beam of radiation of sufficiently high intensity and well-defined frequency. In general, the primary source will produce radiation covering a wide range of frequency and it is necessary to select from this a narrow band of frequencies by some optical device usually referred to as a *monochromator*. This will usually depend on the use of a diffraction grating to disperse the primary radiation. The smaller the frequency band ultimately used, that is to say the higher the frequency resolution in the experiment, the smaller the fraction of the total intensity of the primary, undispersed radiation from the source which will be used. As it is highly desirable to work with good frequency resolution, it is important that the primary source should be as intense as possible. Also, for measuring rapid variations of the cross-sections with frequency, it is an advantage if the primary radiation is continuous in frequency rather than consisting of a number of strong lines. For the study of photoelectron spectra, on the other hand, observations need only be made at a fixed frequency and a source giving a single strong line is very suitable, provided of course the frequency is high enough to produce photoionization.

No continuous source of radiation in the vacuum ultraviolet was available until in 1956 Tomboulian and his associates drew attention to the possibility of using the radiation emitted by fast electrons when revolving in circular paths in a synchrotron accelerator. Because these electrons are continuously suffering centrifugal acceleration they radiate electromagnetic waves. The theory of this emission was first given by Schwinger in 1949 and shows that the number of photons emitted per second in the wavelength range between λ and $\lambda + d\lambda$ by electrons revolving with energy E with velocity very close to that c of light in a circle of radius R, is given by

$$N(\lambda)\,d\lambda = 3^{3/2}(e^2/12\pi h)(\gamma^4/R^2)F(\lambda/\lambda_c)\,d\lambda. \tag{13.29}$$

Here e is the electron charge, h Planck's constant while

$$\gamma = E/mc^2,$$
$$\lambda_c = 4\pi R/3\gamma^3,$$

m being the electron mass.

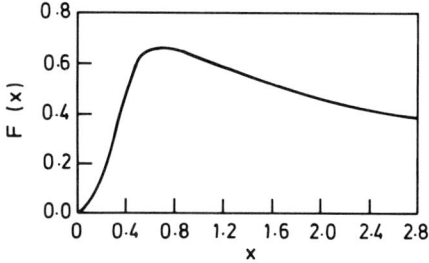

Figure 13.3. The function $F(x)$.

Atomic and Molecular Collisions

Electron energy (MeV)	50	100	150	200	250	300
λ_m (nm)	1867	234·6	69·5	29·3	14·9	8·69

Table 13.2. Variation of the wavelength λ_m for peak emission of power per unit wavelength range, with electron energy, for electrons revolving in a circular orbit of radius 1 m.

The function $F(\lambda/\lambda_c)$ is illustrated in figure 13.3. The peak power is radiated at a wavelength $\lambda_m = 0.42\lambda_c$ which, following (13.29), varies as E^{-3}. On the other hand the peak power radiated varies as E^7. Table 13.2 gives some numerical values showing how λ_m varies with electron energy for an orbit of radius 1 m.

It will be seen that the most intense emission moves from the infrared at 50 MeV into the ultraviolet before 100 MeV and is well down into the X-ray region at 300 MeV. At the same time, the peak power radiated increases by a factor of $7^7 = 8.2 \times 10^5$.

These effects, while very useful for those concerned with the study of radiation phenomena at the short ultraviolet and X-ray (XUV) wavelengths, are very serious for the design of high energy electron accelerators because they represent a power loss which grows rapidly as the electron energy increases. On the other hand, circular machines are now being constructed specifically to provide powerful and experimentally convenient sources of XUV radiation for a wide variety of research applications. Thus at the Science Research Council Laboratory at Daresbury in Cheshire a machine is being constructed in which a current of electrons of 2000 MeV energy of up to 1 A circulates in a ring of radius 5·5 m for several hours, during which XUV radiation of total power 240 kW is emitted. This will provide a useful intensity down to 0·08 nm.

13.9. Photoelectron Spectroscopy

As pointed out earlier, for the study of the energy distribution of the photoelectrons, it is not necessary to have available a continuous source of radiation. Measurements are usually made using a powerful line source covering a narrow band of wavelengths only. In the ultraviolet, the strong helium line at 58·433 nm and the line of ionized helium at 30·379 nm are most frequently used. The line widths, which define the frequency range spanned in each case, are such that the energy spread in the photoelectrons produced is about 35 meV for the longer radiation and 50 meV for the shorter.

By working with X-ray lines, it is possible to investigate the inner shell structure of atoms and molecules, because the photon energy can be high enough to eject electrons from these shells. A great deal of experimental work is being carried out on these lines and it is possible to measure in this way the shift in the energy of an inner shell of an atom when that atom is combined chemically with other atoms. From information of this kind, further insight into the nature of chemical binding is being gained.

Photon Collisions with Atoms and Molecules

It would take us too far afield to pursue this important subject further here, but on p. 248 we shall describe the application of the technique to photodetachment, using laser beams as light sources. At the time of writing, lasers are not yet available at sufficiently high frequency to be able to produce single photoionization, but it should not be long before they do become available, in which case the energy resolution achievable in photoelectron spectra will be substantially increased.

13.10. Some Results of Photoionization Measurements

Figure 13.4 reproduces photographs of vacuum ultraviolet absorption spectra of neon and argon obtained by Madden and Codling in 1963, using radiation from the synchrotron at the National Bureau of Standards in Washington which accelerates electrons to an energy up to 180 MeV in a circular path of radius 0·834 m.

For neon, dark absorption lines due to excitation of autoionizing levels are clearly visible. On the other hand, for argon the autoionizing lines appear brighter than the background, corresponding to absorption windows, a possibility referred to on p. 237. Since these initial observations, very detailed systematic studies have been made for all the rare gases and hundreds of autoionizing levels observed. The nature of many of these levels has been identified.

The photoionization cross sections Q_{pi} of the alkali metals sodium, potassium,

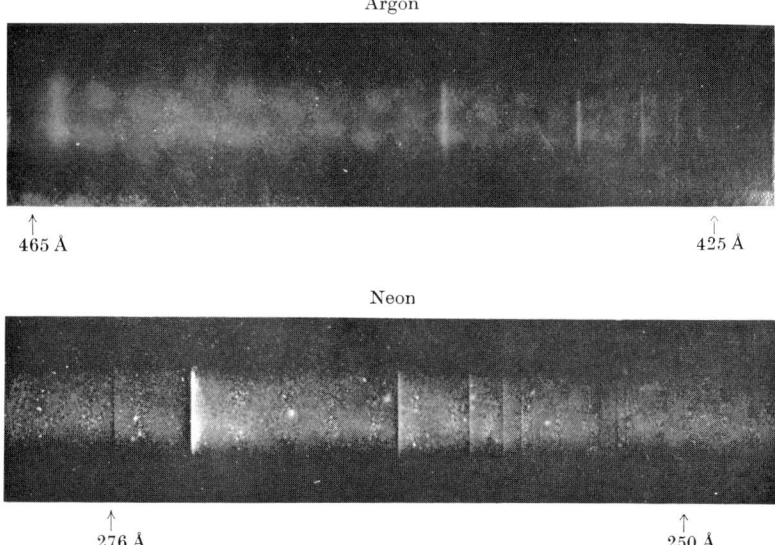

Figure 13.4. Absorption spectra of neon and argon, in the far ultraviolet, observed by Madden and Codling. Note that, while for neon the absorption lines appear dark on a light background, for argon they show up as white on a dark background.

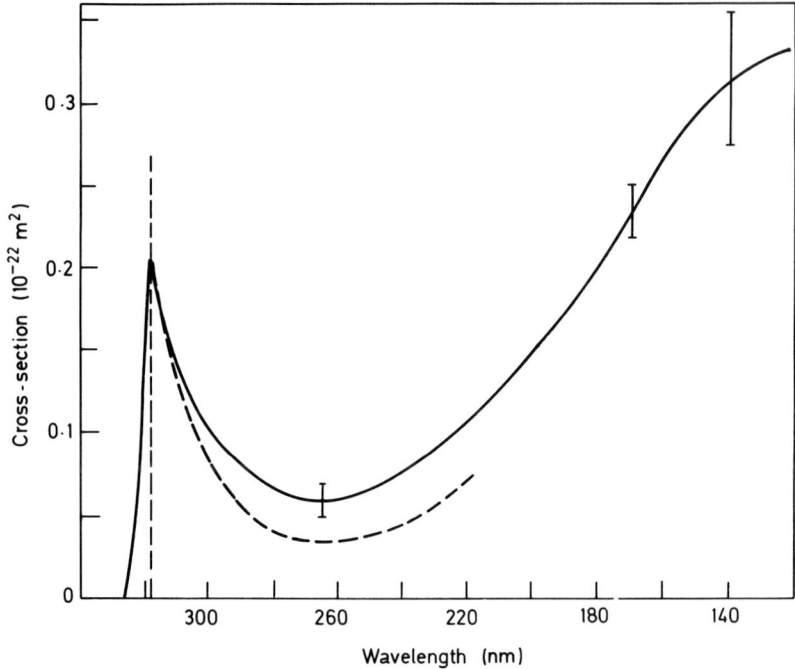

Figure 13.5. Observed photoionization cross-section for caesium. ——, recent observations of Marr and Creek; ----, much earlier observations of Mohler and Boechner with corrected values of the vapour pressure.

rubidium and caesium all exhibit very low minima at frequencies not far beyond the threshold. Figure 13.5 shows the recent measurements for caesium due to Marr and Creek. For comparison, the measurements made nearly 40 years earlier by Mohler and Boechner are also shown. These have been corrected to allow for the improved data on the vapour pressures of the metals as functions of temperature which are available today. When this is done, the early observations are seen to agree remarkably well.

Absorption of light by calcium atoms is of considerable interest in connection with the analysis of observed data on the absorption by matter in interstellar space. Because of the low vapour pressure of calcium, it was difficult to measure the absorption cross-section and at first reliance had to be placed on calculated values. Unfortunately, these were very unsatisfactory because they did not take into account the possible existence of strongly excited autoionized states. Figure 13.6 shows the results of measurements made much later (1960) by Ditchburn and Hudson. The importance of autoionizing states is obvious.

13.11. *Photodetachment by Single Photons*

Interest in photodetachment was first stimulated by the suggestion of Wildt

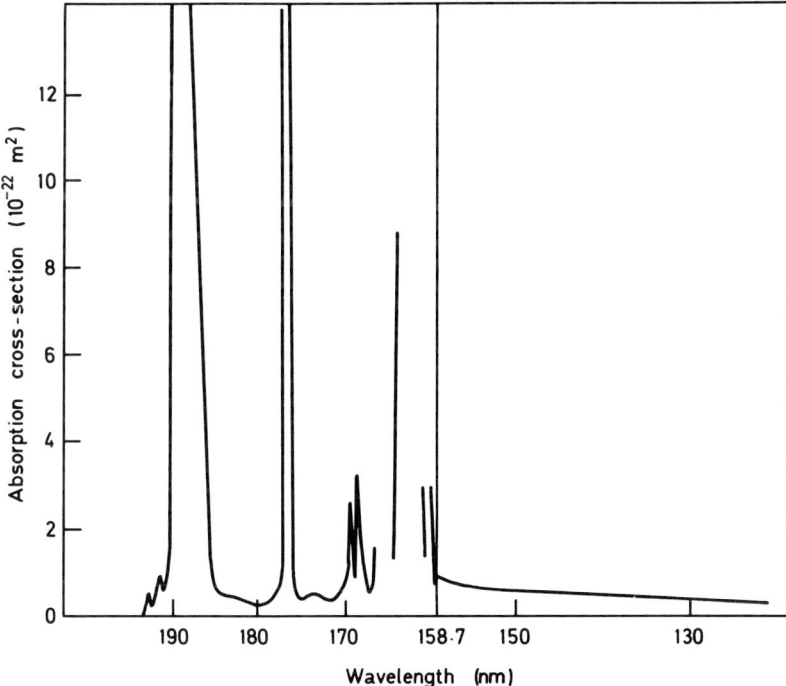

Figure 13.6. Absorption cross-section for calcium, observed by Ditchburn and Hudson. Note the change of the wavelength scale at 158·7 nm.

in 1939 that the main source of opacity in the atmosphere of the Sun at red and infrared wavelengths was absorption by negative hydrogen ions H^-, the process being one of photodetachment

$$H^- + h\nu \rightarrow H + e. \qquad (13.30)$$

It was known at the time that the electron affinity E_a of atomic hydrogen is 0·75 eV, so that the threshold wavelength is 1648 nm. Rough calculations of the relevant absorption cross-section had already been made some years earlier and they provided support for Wildt's suggestion, remarkable as it seemed at the time. That the important emission from the Sun at wavelengths greater than 500 nm should be determined mainly by the properties of such an exotic substance as H^- seemed indeed astonishing.

Improved calculations by Bates and Massey in 1940, and by Chandrasekhar and his colleagues from 1942 onwards, confirmed the importance of H^-. Nevertheless, it was felt to be most desirable to provide further support from direct laboratory measurement of the photodetachment cross-section. This presented formidable difficulty at the time because, unlike the neutral atoms and molecules involved in photoionization experiments, it was not possible to concentrate negative ions in bulk for absorption measurements. The problem was solved in 1954 by Branscomb and Fite following a suggestion by Oldenburg. They were the first to develop the technique of the crossed-beam scattering experiment in which detachment was produced from a beam of H^- ions by crossing with a photon beam. The obvious difficulty here is one of intensity, because the current of detached electrons will be small and difficult to distinguish from background currents arising in various ways within the apparatus.

Thus, suppose the ions are moving with a velocity v and have a photodetachment cross-section Q_d. If the photon beam is of flux density N per unit area per second and illuminates a length l of the path of an ion, the chance of photodetachment is $NQ_d l/v$. For H^- ions of kinetic energy 300 eV, v is $2 \cdot 5 \times 10^5$ m s^{-1}. The power in the photon beam in an early experiment was about $2 \cdot 5$ kW m^{-2} which, for photons of wavelength 500 nm, gives $N = 6 \times 10^{21}$ m^{-2} s^{-1}. Taking l as 0·02 m, and Q_d of the order 10^{-22} m^2 the probability of detachment per ion comes out to be about 6×10^{-8}. At the background pressure of around 10^{-6} torr attainable at the time, the chance of detachment by collision with molecules of the background gas is 100 or more times larger.

This difficulty was overcome by using the phase-sensitive modulation technique. The photon beam was interrupted regularly, or chopped, at a frequency of 450 Hz. The current of detached electrons was amplified by a selective amplifier tuned to this frequency and then measured by a phase-sensitive detector. This means that the detector was so arranged as to respond only to a signal arriving at the correct interval of time after the photon beam had been cut off.

In this way, discrimination between the wanted signal and the background was successfully achieved and provided a valuable technique for use in other crossed-beam collision experiments (see for example Chapter 10, p. 180 and Chapter 12, p. 221).

Many other difficulties had to be overcome, particularly in studying the variation of the cross-section with frequency. Nevertheless, absolute measurements of the H^- cross-section were made by Branscomb and Smith in 1955 and four years later the variation with frequency was measured by Smith and Burch over the wavelength range from 400 to 1300 nm, relative to the cross-section at 528 nm, with a probable error no greater than 2%.

Besides establishing the validity of the calculated cross-sections and of Wildt's suggestion, this pioneer work stimulated a whole new programme of experimental study of photodetachment from atomic and molecular negative ions, which has led not only to remarkably accurate values of electron affinities, but also to much deeper knowledge and insight into the structure and properties of negative ions generally.

The development of lasers made available an important new tool for these studies. Before then, the wavelength resolution of the photon beams was about 10 nm. With a laser beam, however, this could be improved to 0·1 nm. A fixed-frequency laser at present available, such as an argon ion laser (458 nm), has a frequency which is high enough to produce photodetachment from a large number of negative ions—the highest electron affinity of any atom or diatomic molecule is less than 4·0 eV (309 nm) and most are less than 3·0 eV (412 nm). Such lasers provide an excellent source for photoelectron spectroscopy of negative ions while dye lasers, which are tunable in wavelength between 520 and 700 nm, are very suitable for measurement of photodetachment cross-sections as functions of frequency. We now give a brief description of the technique used in carrying out such measurements.

13.12. Measurement of Photodetachment Cross-Sections—Typical Experimental Arrangements

Figure 13.7 shows the general arrangement used in an experiment to measure the photodetachment cross-section as a function of frequency, in which the light source is a tunable dye laser giving rise to a pulsed photon beam.

In these experiments, the cross-section is measured not from the yield of detached electrons but from that of the neutral atoms or molecules. Because it is possible to measure the intensity of a beam of neutral particles only when they have a kinetic energy of some keV, the negative ion beam from a hot cathode discharge source is accelerated to 2000 eV before entering the interaction chamber. Although this chamber is highly evacuated with

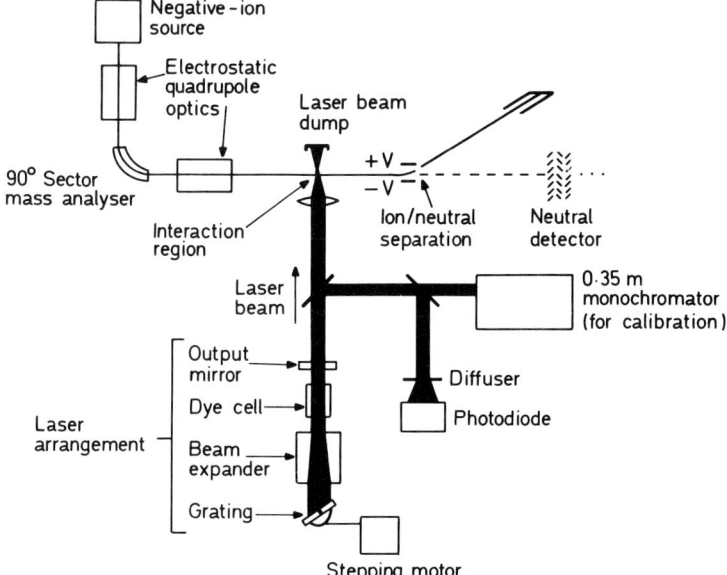

Figure 13.7. Illustrating the arrangement used to measure the photodetachment cross-section of negative ions by crossing a negative ion beam with a laser beam from a tunable dye laser.

powerful pumps (pressure 3×10^{-8} torr), the beam passes through a region between it and the source where the pressure is about 100 times higher. As a result, quite a number of ions lose an electron in collision with neutral gas atoms or molecules in this region, thereby contaminating the beam with neutral particles which would give a spurious signal in the detector. To eliminate this, the negative ion beam is deflected through 5° by an electric field just before it enters the interaction chamber. The neutral contaminant is not deflected and so only the purified beam interacts with the laser beam.

Neutral particles produced by photodetachment pass on undeviated into the secondary emission multiplier detector (Appendix 6), while the negative ions are deflected again by an electrostatic field to enter a Faraday cup so that their current can be measured.

To correct for background effects, the signal in the detector due to a laser pulse is measured at a lapse of time (about 6 μs) equal to that required for the neutral particles formed to travel from the interaction region. From this is subtracted the signal received about 1 μs earlier which must be due to background. At each frequency, the mean is taken over a large number (300) of observations of this kind.

13.13. Measurement of Photoelectron Spectra from Photodetachment

In typical experiments, the energy distribution of electrons ejected from a beam of negative ions by a beam of laser radiation crossing it at right angles, in a direction perpendicular to both beams, is measured.

If M is the mass and E^- the kinetic energy of a negative ion, m the mass of an electron and v the frequency of the radiation, the energy of the ejected electron measured in this system is given by

$$E = hv - E_a - (m/M)E^-, \qquad (13.31)$$

where E_a is the electron affinity of the neutral atom. The last term on the right-hand side arises because of the velocity possessed by the electron due to the motion of the negative ion from which it was ejected.

An argon ion laser, which operates at a wavelength of 485 nm, is a convenient fixed frequency source and it has been used in all the experiments carried out at the time of writing. The first application was made by Brehm, Gusimow and Hall in 1967 to measure the detachment energy of metastable He^- ions. Figure 13.8 illustrates a typical arrangement used in later work.

A beam of negative ions, contaminated by neutral atoms and molecules, photons and electrons is extracted directly from a glow discharge source. After passage through an electron-optical lens L_0, it is deflected by an electrostatic field through 10° to remove the neutral and photon contaminants. Electrons are removed by a weak magnetic field. The deflected beam is then focused through a lens system and collimated before entering a Wien velocity

Photon Collisions with Atoms and Molecules

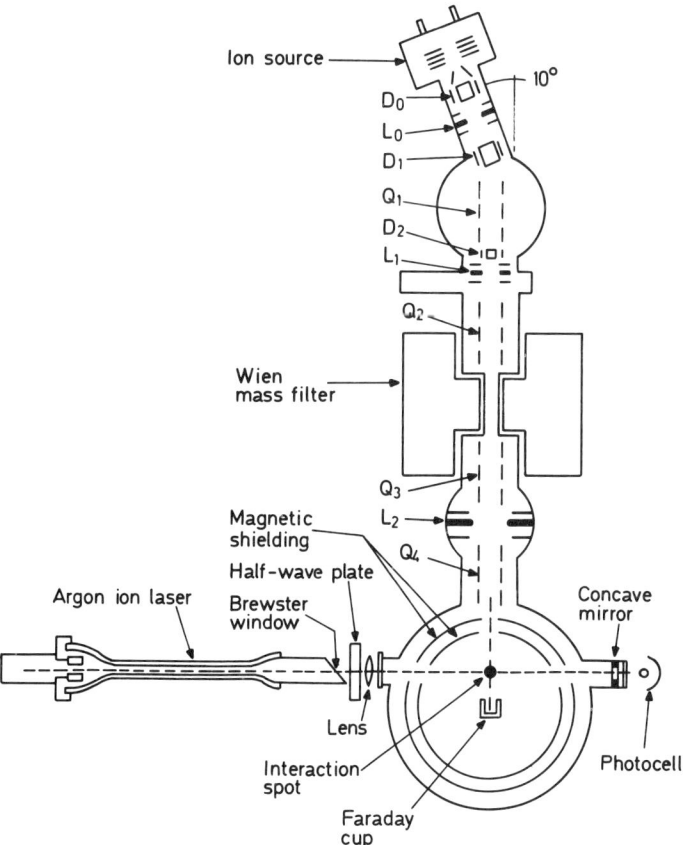

Figure 13.8. Illustrating the arrangement used to measure the energy of the electrons produced by photodetachment from a negative ion beam by light from an argon ion laser. D_0, D_1 and D_2 are electrostatic deflectors. L_0 and L_2 are einzel lens systems and Q_1, Q_2 and Q_3 quadrupole lens systems.

filter (Chapter 7, p. 127) for mass selection. After leaving the filter, the beam is focused into the interaction region to an elliptically shaped spot of area about 1 mm² and collected finally in a Faraday cup. In this region the laser beam is focused to a spot of 0·1 mm diameter and produces a photon flux of 2×10^6 to 5×10^6 kW m^{-2}.

Electrons produced by photodetachment to move in a direction within a cone of semi-angle less than 2' of arc about the direction normal to both intersecting beams, are accelerated into a lens system and injected into a hemispherical analyser (see Chapter 6, p. 94). With the geometry used, the theoretical width of the energy peak at half height is about 40 meV. Electrons transmitted through the analyser are accelerated to 112 eV before entering a high-gain electron multiplier so their flux can be measured.

Atomic and Molecular Collisions

The interaction region, the analyser and the particle multiplier are enclosed in two permalloy magnetic shields to reduce the magnetic field to 1 mG. Some results obtained with this equipment are described below.

13.14. Photodetachment from H^-—Absorption in the Sun's Atmosphere

The first crossed-beam experiments were directed towards the measurement of the cross-section for photodetachment from H^- as a function of frequency from the threshold, in order to check the theoretical calculations which were so significant for the interpretation of the opacity of the solar atmosphere. Figure 13.9 shows the observed cross-sections as functions of wavelength in comparison with theoretical values calculated by elaborate and presumably reliable approximations. The observed results are due to Branscomb and Smith, who measured the absolute value at 528·0 nm, and to Smith and Burch who made accurate measurements of cross-sections at other wavelengths relative to this value. In both cases, conventional light sources were used.

The agreement between theory and experiment is very good and confirms the reliability of the data for use in solar applications.

It may be shown from statistical arguments that the number $n(H^-)$ of H^- present per unit volume in the solar atmosphere is given to a very good approximation by

$$n(H^-) = \tfrac{1}{4} n_e n(H) \exp\{(E_a/kT)\} h^3 (2\pi m k T)^{-3/2}, \tag{13.32}$$

where n_e, $n(H)$ are the respective concentrations of free electrons and neutral H atoms, E_a the electron affinity of H, k the Boltzmann constant, T the temperature and m the electron mass. Substituting numerical values and introducing the

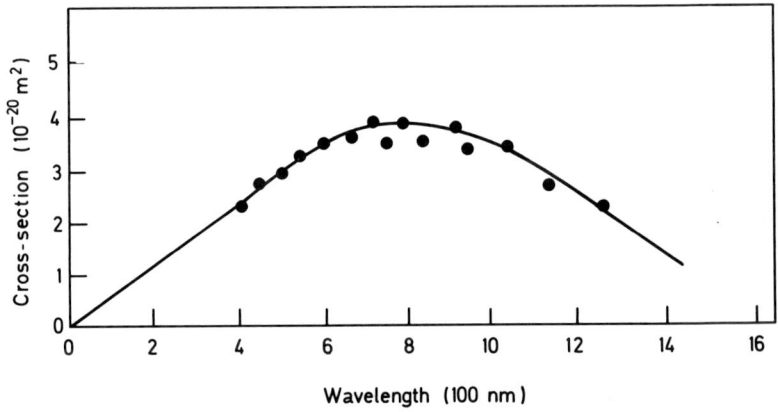

Figure 13.9. Photodetachment cross-section for H^-.
———, calculated; ●, observed.

Photon Collisions with Atoms and Molecules

electron pressure $p_e (= n_e kT)$, it is found that

$$n(H^-) = p_e n(H) \phi(T) \tag{13.33}$$

where

$$\log_{10} \phi(T) = -0\cdot 12 + \frac{5040}{T} E_a - 2\cdot 5 \log_{10} T, \tag{13.34}$$

E_a being measured in eV. The absorption coefficient of the solar atmosphere for radiation of frequency v is given by

$$\alpha = n(H^-) Q_d(v),$$

where Q_d is the photodetachment cross-section for radiation of frequency. Following (13.33), this may be written

$$\alpha = n(H) p_e \chi, \tag{13.35}$$

where $\chi = \phi(T) Q_d(v)$ is the absorption coefficient per hydrogen atom for unit electron pressure. Figure 13.10 shows the calculated value of χ assuming the absorption is due to photodetachment from H^- and that $T = 6300$ K. In the same figure the value of χ derived from solar observations is also shown.

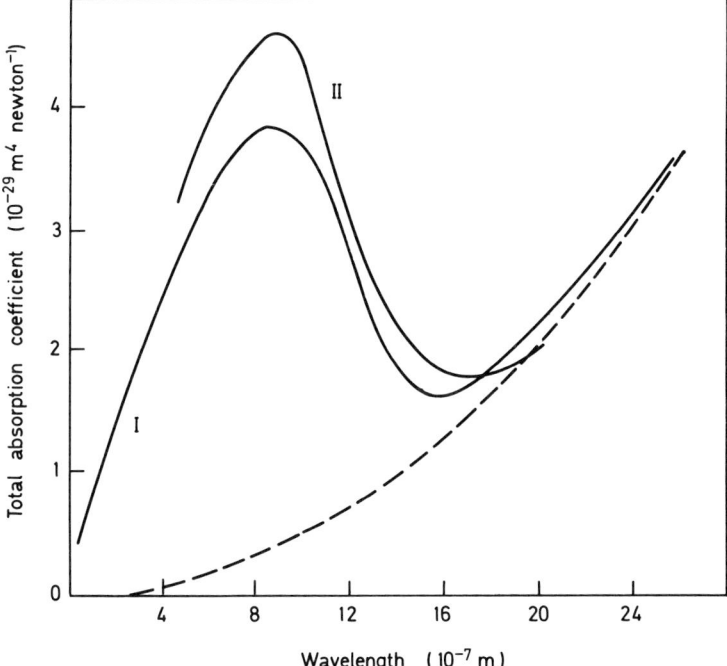

Figure 13.10. Total absorption coefficient per hydrogen atom per unit electron pressure in the atmosphere of the sun.

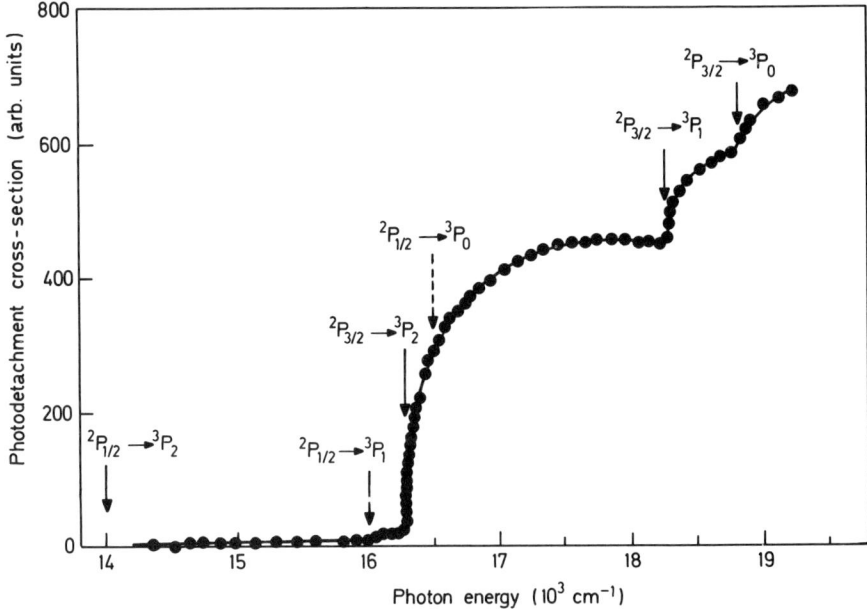

Figure 13.11. Photodetachment cross-section of Se⁻ observed by Hotop, Patterson and Lineberger. The threshold energies for transitions from the respective fine structure levels $2P_{1/2}$ and $2P_{3/2}$ of Se⁻ to the respective levels $^3P_0$, $^3P_1$ and $^3P_2$ of Se are indicated.

There is quite good agreement over the wavelength region from 400 to 1400 nm but at longer wavelengths the calculated values become increasingly too small. This is because no allowance has been made for the contribution from processes in which a free electron, moving in the field of a hydrogen atom, absorbs a photon so that its kinetic energy is increased. Such a process is referred to as a free–free transition and is possible only in the presence of a field which is accelerating (decelerating) the otherwise free electron. When allowance is made for this, good agreement is obtained with the observed data as shown in figure 13.11.

13.15. Photodetachment from O⁻

The experimental study of photodetachment from O⁻, carried out by Branscomb and Smith in 1955 and extended by Branscomb, Burch, Smith and Geltman in 1958, is a classic example of what could be done before the advent of laser sources. Prior to the photodetachment experiments, the electron affinity of atomic oxygen, a quantity of importance in ionospheric and other applications, was taken to be 2·2 eV on the basis of experiments by other techniques which were presumed to be reliable. The experimental determination of the threshold wavelength for photodetachment showed that the true value is

1·465 ± 0·005 eV. This has been confirmed in later experiments with higher frequency resolution obtainable with laser sources and is now used as a reference standard.

13.16. *Photodetachment from Se⁻ and Si⁻*

The most precisely determined electron affinity E_a at the time of writing is that of selenium. From the threshold, determined from the measured photodetachment cross-section as a function of wavelength, using a tunable dye laser as light source, it was found by Hotop, Patterson and Lineberger that

$$E_a(\text{Se}) = 2{\cdot}0206 \pm 0{\cdot}0003 \text{ eV}. \tag{13.36}$$

The accuracy attained depended very much on the high frequency resolution of the laser source because the interpretation of the data was complicated by the fact that the ground states of both Se⁻ and Se exhibit fine structure. Transitions can therefore take place between the different fine structure levels for which the thresholds are all different, giving rise to the photodetachment cross-section shown in figure 13.11. With the high resolution source the separate thresholds may be resolved.

Figure 13.12 shows the observed energy distribution (photoelectron spectrum) of the electrons detached from negative ions of silicon, by light from an argon ion laser. This was obtained by Kasdan, Herbst and Lineberger using the equipment described on p. 249. In this case detachment takes place not only from Si⁻ ions in the ⁴S ground state but also from ions in the ²D and ²P metastable excited states. The different peaks seen in figure 13.12 occur at the threshold energies for transition from one or other of these states to either the

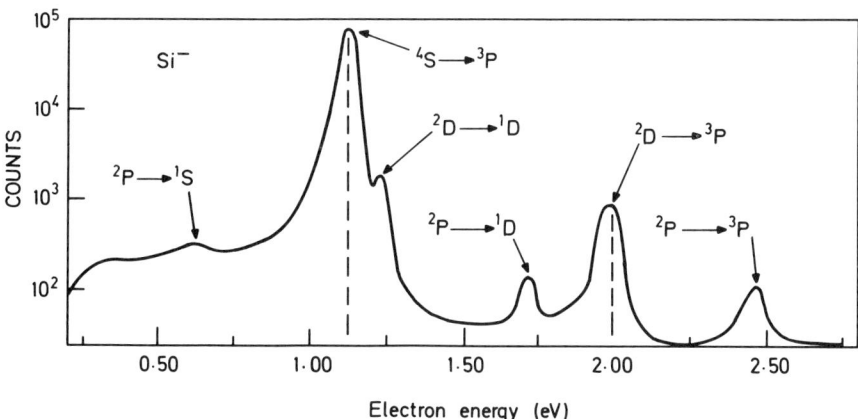

Figure 13.12. Energy distribution of the photoelectrons produced by photodetachment from Si⁻ by radiation from an argon ion laser. The peaks arise from transitions between different levels ²P, ²D of the ground term of Si⁻ to those ⁴S, ²D and ²P of the ground term of Si, as indicated.

^{1}S ground state or the ^{1}D and ^{3}P excited states of the neutral atom Si. These identifications may be made from knowledge of the energies of the latter states from spectroscopic measurement. It is then possible to obtain

$$E_a(\text{Si}) = 1{\cdot}385 \pm 0{\cdot}005 \text{ eV}. \tag{13.37}$$

13.17. Window Resonances in Photodetachment from Cs$^-$ and Rb$^-$

Figure 13.13 shows photodetachment cross-sections for the negative ions of caesium and rubidium measured by Patterson, Hotop, Kasdan, Norcross and Lineberger. These exhibit very clearly the same kind of window resonances as those observed in the absorption of ultraviolet light by argon atoms (see figure 13.4). Thus, in both cases the photodetachment cross-section exhibits a deep minimum at a certain frequency (and for Cs$^-$, two deep minima) which corresponds to a very weak absorption by the negative ion concerned. The explanation is essentially the same as for absorption by neutral atoms. At frequencies

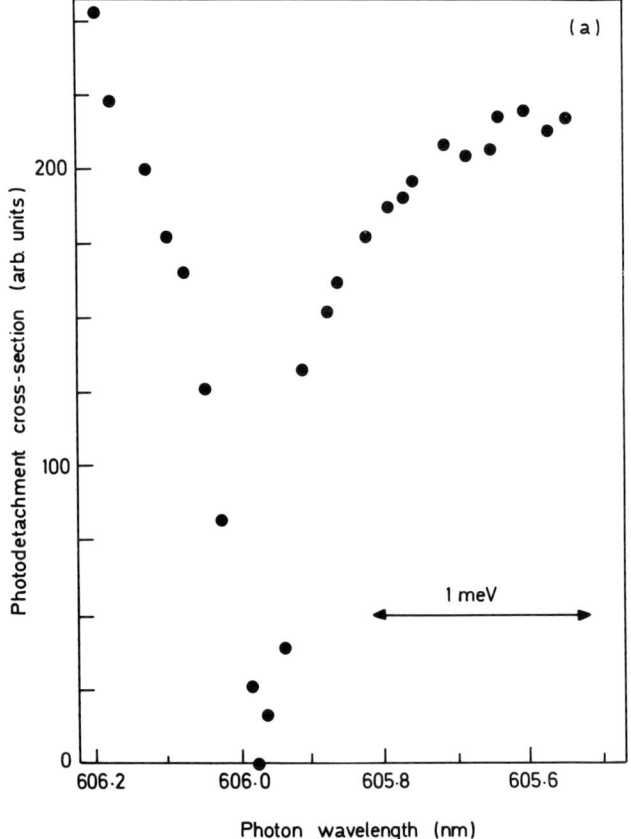

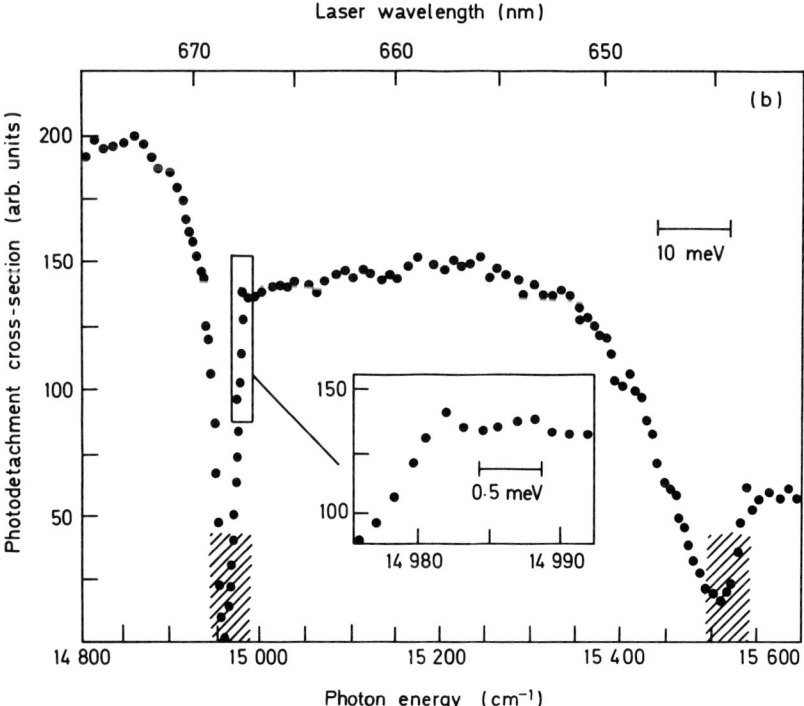

Figure 13.13. Photodetachment cross-sections observed by Patterson, Hotop, Kasdan, Norcross and Lineberger for (a) rubidium and (b) caesium, showing the 'window' resonances. The inset in (b) shows the variation with wavelength on a gently expanded scale around the long wave side of the window resonance.

near the minima, absorption occurs both through direct photodetachment and through excitation of an autodetaching state of the negative ion. Interference between the amplitudes for these two processes reduces the net probability to a very low value.

13.18. *Photoelectron Spectrum of* O_2^-

Figure 13.14 illustrates the photoelectron spectrum observed by Celotta, Bennett, Hall, Siegel and Levine for O_2^-, using an argon ion laser source at 485·0 nm. In this case, the spectrum arises mainly from transitions between the lowest vibrational level associated with the ground state of O_2^- and different vibrational levels of the ground and first excited electronic states of the neutral molecule. It is possible to identify the different transitions and hence to determine the electron affinity of O_2, the energy required to make a transition between the lowest vibrational levels of O_2 and O_2^-. This is found to be

$$E_a(O_2) = 0·438 \pm 0·001 \text{ eV} \quad (13.38)$$

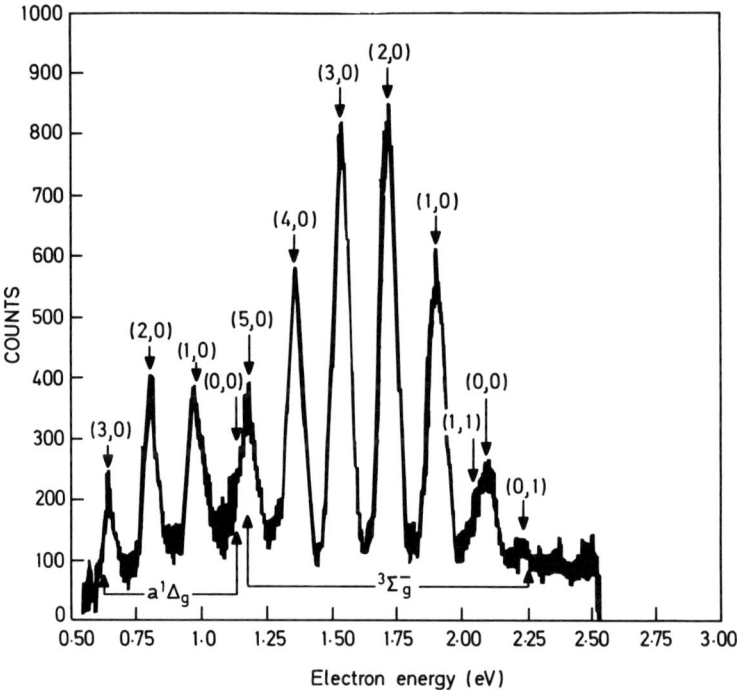

Figure 13.14. Energy distributions of the photoelectrons produced by photodetachment from O_2^- by radiation from an argon ion laser. The peaks arise mainly from transitions between ground vibrational states of O_2^- and different vibrational levels of the ground ($3\Sigma^-$g) and excited ($a^1\Delta_g$) states of the neutral molecules. The final and initial vibrational quantum numbers n, m respectively are indicated in each case as (n, m). There are indications (see (1, 1) and (0, 1)) of transitions from the first excited vibrational state of O_2^-.

a value of some importance in the theory of the lower ionized regions in the Earth's atmosphere (Chapter 14, p. 273).

13.19. Multiphoton Processes

We have already referred earlier (Chapter 11, p. 206) to three-body collisions. Until the invention and development of lasers, no one thought seriously about the possibility of studying collisions in which an atom or molecule and two photons were involved—the chance that a second photon would be in the neighbourhood of the first would be far too small. Thus in black body radiation at 10 000 K, the number of photons per m³ with wavelength within 1 nm of 1000 nm is only about 10^{16}, so the mean distance between photons is of order 2×10^{-3} mm, very much larger than atomic dimensions. With lasers, very much greater photon densities may be achieved. Near the focus of a laser beam, the number of photons per m² per second may be as high as 10^{35}. Since the photons are all travelling at the velocity of light their concentration per m³ will be as high as 3×10^{26} and the mean distance between them around 10^{-9} m, which is comparable with atomic dimensions. With these intensities, it is possible to study not only atomic processes involving simultaneous interaction with two photons but also with several photons.

As a further indication of the intensity of the focused laser beam, the amplitude of the alternating electric field in such a beam may be calculated. The energy density in an alternating electromagnetic field of frequency v and amplitude E is $E^2/8\pi$ and the photon density $E^2/8\pi h v$, so it is a simple calculation to show that the amplitude of the electric field is close to $4 \cdot 5 \times 10^9$ V m^{-1} for photons of energy 2 eV. This is to be compared with the electric field due to a proton at the most probable distance $(0 \cdot 53 \times 10^{-10}$ m) of the electron in a hydrogen atom, 5×10^{11} V m^{-1}.

A number of processess involving collisions with more than one photon have been observed using focused laser radiation, the first being due to Hall, Seman

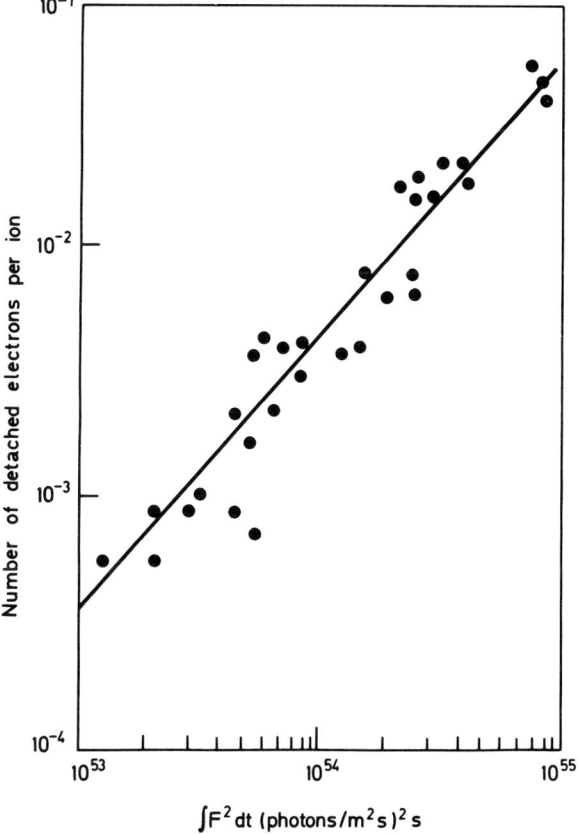

Figure 13.15. Variation of the number of electrons detached per ion with the integral $\int F^2 \, dt$ of the square of the laser flux over the duration of the pulse, in two-photon detachment from I$^-$.

and Branscomb, who detached electrons from I^- ions with light from a ruby laser. The photon energy of this light is only 1·79 eV, while the electron affinity of iodine is 3·06 eV, so that at least two photons must be absorbed simultaneously. In such a case, the rate will be proportional to the square of the photon flux F. The experiments were carried out using laser pulses in which case the fraction of electrons detached from the I^- ions per pulse should vary as $\int F^2 \, dt$, the integral of the square of the flux over the duration of the pulse. Figure 13.15 shows that the observed results are consistent with this relation.

The probability of a photodetaching transitions per ion per second may be written

$$W = \delta F^2, \qquad (12.39)$$

where δ has the dimensions $L^4 T$. From their observations, Hall, Robinson and Branscomb found

$$\delta = 1\cdot80 \times 10^{-57} \, m^4 \, s,$$

which is quite close to the value estimated theoretically.

At about the same time, Voronov and Delone reported measurements of the rate of ionization of neutral xenon atoms, also by ruby laser light. In this case, the ionization energy is 12·13 eV, so that at least seven ruby laser photons must be absorbed to provide sufficient energy. Many other experiments have since been carried out in which multiphoton ionization has been observed. Attention is now being paid to the importance of well controlled experimental conditions to make detailed interpretation possible.

CHAPTER 14
atomic collisions in the earth atmosphere, the solar corona and interplanetary space

We have already referred to some applications of the physics of atomic collisions to problems in astrophysics. Thus in Chapter 13, Section 13.14, we described how the form of the solar emission spectrum in the visible, a matter of vital importance for life on the earth, is determined by the rate of photodetachment of electrons from negative hydrogen ions in the bright emitting surface of the Sun.

Again in Chapter 3, Section 3.6, attention was drawn to the radiative transition between the hyperfine structure levels of atomic hydrogen which gives rise to the emission line at a wavelength of 0·21 m. Observations of this line in radio emission received in different directions from space have made it possible to map out the distribution of atomic hydrogen in many regions of astronomical importance. This includes our own galaxy which has been shown from such studies to have a spiral structure.

There are many aspects of the behaviour of the Sun's outer atmosphere which depend on the rates of atomic collision processes. This applies also to many other astronomical objects as well as to the matter in the space between the stars which, while extremely tenuous, is nevertheless important for understanding stellar evolution.

The Earth's ionosphere which, until the advent of satellite communication, provided the only means by which radio signals could be transmitted around the Earth, is a partly ionized region of the Earth's atmosphere. Its properties depend very much on the rates of collision processes involving both charged and neutral species. This applies also to the ionospheres of planets such as Venus, Mars and Jupiter which have retained atmospheres of appreciable density.

In this chapter, we shall select a few examples from these wide-ranging applications which illustrate the value of the data now available from laboratory and theoretical research on the physics of electronic, ionic and atomic collisions. Although the examples are concerned with very different environments, they have one thing in common. In all cases, the rate of recombination between ions and electrons plays a key role.

We deal first with the ionosphere of the Earth. Next we consider the extreme outer atmosphere of the Sun, the solar corona, and finally journey into the depths of interstellar space to consider how the molecular fragments observed there in recent years have been synthesized.

14.1. The Earth's Ionosphere

In 1902, Marconi showed that, contrary to expectation at the time, it was possible to communicate by radio over long distances. Before his experiments, in which he received the Morse letter S in Newfoundland, 1800 miles away from his radio transmitter at Poldhu, near Lands End in Cornwall, it was believed that waves radiated from an aerial on the Earth's surface would pass straight out into space. Thus at a distance of 1800 miles on the Earth's surface, they would pass 100 miles or more overhead because of the curvature of the Earth. It is true that there will be some spreading by diffraction, but the ratio of the wavelength of the radio waves to the radius of the Earth is so small that this must be unimportant. Why then did Marconi succeed?

The explanation is the presence of a layer in the Earth's atmosphere which reflects back to the ground the radio waves which would otherwise escape into space. As long ago as 1883, Balfour Stewart had suggested that electric currents flowing in the high atmosphere might be responsible for the regular ('quiet') variations of the Earth's magnetic field at ground level. Kennelly and Heaviside independently suggested that it is this electrically conducting region which reflects back the radio waves. This explanation was far from universally accepted and for some time many still clung to the belief that diffraction was responsible. However, by the 1920s, this explanation had worn very thin. Thus, radio amateurs had been assigned a band of short waves to operate with, in the belief that worldwide communication would not be possible in this band because of the ineffectiveness of diffraction at such short wavelengths. On the contrary, they were soon in worldwide communication with each other.

Early in the 1920s, direct scientific evidence of the reality of the reflecting layer or layers came from the experiments of Appleton and his collaborators in Britain, followed shortly after by those of Breit and Tuve in the United States. In these experiments, the reflection of radio waves back from the high atmosphere was directly observed and the height of reflection determined as a function of the wavelength.

To interpret these radio-probing observations in terms of the properties of the reflecting medium use must be made of the fact that electromagnetic waves entering, from free space, a region in which there is a concentration n of free electrons of charge e and mass m, per unit volume, will suffer refraction, with refractive index

$$\mu = 1 - ne^2/4\pi m v^2, \qquad (14.1)$$

where v is the frequency of the wave. Since $\mu < 1$, a ray will be bent away from the normal on entering the region and total reflection will occur when the sine of the angle of incidence exceeds μ. In practice, the electron concentration will vary with height. As n increases upwards, the incident ray is turned more and more towards the horizontal until again total reflection occurs.

It may be shown that, to a good approximation, a ray transmitted vertically is reflected back when it reaches the height at which the refractive index given by (14.1) vanishes, i.e., when

$$n = 4\pi m v^2/e^2.$$

Proceeding upwards into the atmosphere, we would expect that at first n will increase to reach a maximum n_m at some height h_m after which it will decline gradually to zero. In such a case, the maximum frequency which can be reflected is given by

$$v_m = e^2 n_m / 4\pi m.$$

From these considerations, it is clear that the variation of n with height can be determined, in principle, from the height at which reflection occurs for radio waves of different frequencies. A great number of observations of this kind have been made since the early 1920s. The simple theoretical conclusions we have outlined need to be rendered more precise, to take account of the Earth's magnetic field as well as other features we have ignored.

It was soon found that the broad picture given above is generally correct and the region of the Earth's atmosphere in which the free electron concentration is high enough to affect the propagation of radio waves was designated the *ionosphere*. In addition to the free electrons, the ionosphere also includes positive and negative ions so that, if n_e, n^+, n^- are their respective concentrations

$$n_e + n^- = n^+.$$

However, the ions are so massive compared with the electrons that they have little effect on the propagation of radio waves in the frequency range with which we are concerned.

Ground-based radio-probing studies soon showed that the Earth's ionosphere consists not only of a single ionized layer but, in the daytime, of three, referred to as the E, F_1 and F_2 layers. At night the F_1 and F_2 merge leaving the E and a single F layer. Figure 14.1 illustrates the typical variation of n_e with height in the atmosphere during daytime. While the electron concentration and its variation with height varies with time of day, season, geographical and geomagnetic coordinates etc, the mean altitudes of the maxima in the E, F_1 and F_2 layers are found to be, in temperate latitudes, at 120, 220 and 300 km respectively. The corresponding mean values of the maximum electron concentrations are 1.5×10^{11}, 2.5×10^{11} and 10^{12} per m^3. The F_2 layer is a very much thicker layer than the lower layers, the electron concentration falling at heights above the maximum only as $\exp(-h/70)$ where h is in km. For the E and F_1 layers the corresponding rates of fall are as $\exp(-h/10)$ and $\exp(-h/30)$ respectively.

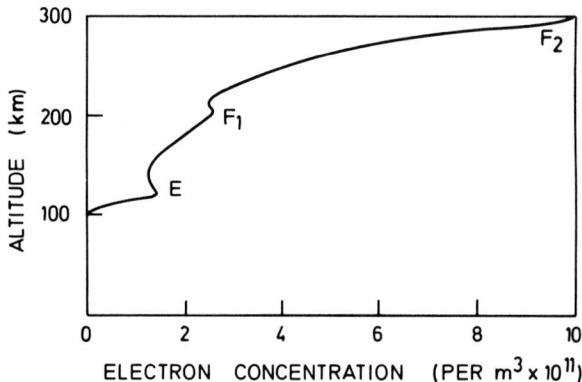

Figure 14.1. Typical variation of n_e with height in the atmosphere during daytime.

It was soon realized that the ionosphere is subject to solar control, suggesting that it arises from photoionization of atmospheric atoms and molecules by ultraviolet and shorter-wavelength radiation from the Sun. These radiations are absorbed in the high atmosphere, so very little information was available about their intensity before World War 2. At any time t, electrons will be produced by photoionization and lost either by recombination to positive ions to produce neutral atoms or molecules (see Chapter 7, p. 100), or by attachment to neutral atoms or molecules to form negative ions (Chapter 7, p. 112). In the latter case, while the negative charge remains unneutralized, the large mass of the negative ion will prevent it from affecting the propagation of the radio waves. Without prejudice regarding the loss processes for electrons, we may write, at any height and time t,

$$\frac{dn_e}{dt} = q - \alpha n_e^2, \tag{14.2}$$

where q is the rate of production of photoelectrons by sunlight at the height concerned. If the electrons are lost by direct recombination to positive ions and no complications due to negative ions arise, α would be the recombination coefficient as defined in Chapter 7, p. 100 and would be expected to be independent of the gas pressure and the electron concentration. On the other hand, if electrons are lost primarily by attachment, the loss rate would be $\beta n_e n$ where β is the attachment coefficient to the neutral atoms and molecules and n their concentration. In this case, (14.2) would only be valid if $\alpha = \beta n/n$, proportional to the gas pressure and to $1/n_e$. Other possibilities include allowance for detachment of electrons from negative ions and for the fact that several species of neutral and ionized atoms and molecules are present. We shall return to these a little later, but at this stage consider the implications of (14.2).

Equilibrium will be reached between production and loss of free electrons

Collisions in Atmosphere, Corona and Space

when dn_e/dt vanishes i.e. when

$$q = \alpha n_e^2, \tag{14.3}$$

$$n_e = (q/\alpha)^{1/2}. \tag{14.4}$$

Now q, considered as a function of height, will be proportional to the intensity I of ionizing radiation and the concentration n of ionizable atoms and molecules. At great heights in the atmosphere I will be nearly equal to the incident intensity, but n will be very small. Low in the atmosphere n will be large but the intensity of the radiation will be very small due to absorption above. q must therefore have a maximum at some intermediate altitude h_m. If α is independent of altitude it follows from (14.4) that n_e will also have a maximum at h_m. In this way, the formation of an ionized layer can be understood, provided the effective recombination coefficient is independent of altitude.

So far so good, but it is also necessary to understand why there are three layers, not just one, and why the layer thickness increases so markedly in going from E to F_2. The assumption of the constancy of α also needs to be justified.

14.2. Theory of the Ionosphere before 1950

For a complete understanding of the properties of the ionosphere detailed information needs to be available about the following:

(a) The intensity and composition of the solar radiation in the ultraviolet and X-ray wavelength bands.
(b) The density, temperature and composition of the neutral atmosphere as a function of height.
(c) The absorption and photoionization cross-sections of atmospheric atoms and molecules as a function of radiation frequency.
(d) The rates of recombination and other reactions involving ionized and neutral atmospheric constituents.

Given (a), (b) and (c) the rate of production of free electrons and of different species of positive ions may be calculated while, from (b) and (d), the rates of loss of the various charged constituents may be derived. Even so, the picture would still be incomplete, because no account has been taken of the redistribution of charged particles due to diffusion and drift motions. However, these effects are not dominant in the F_1 and even less so in the E region although they exert a major influence on F_2.

Before World War 2, there was no information available about (a) and only very sketchy knowledge about (b). Equally, there was almost no experimental information about any of the reaction rates involved in (d) and very little about (c). Despite these limitations, some progress was made from theoretical considerations. In particular, it was shown by Bates and Massey that, in the E region and above, negative ions play only a negligible role. While it is true that at an altitude of 120 km the rate at which electrons attach radiatively (see Chapter 7, p. 110) to O atoms to form O^- ions is much faster than their rate of recombination with positive ions, detachment from the O^- due to solar radiation (photodetachment, see Chapter 13, p. 252) occurs so rapidly that at any time only a very small fraction of electrons is attached.

The identification of the recombination processes effective in the different ionospheric regions presented a considerable problem at that time. It was known from the form of the diurnal variation of the electron concentration in the E layer that α was of the order 10^{-14} m^3 s^{-1}. Reference to Chapter 7, p. 111 shows that this is 10^3 or more times larger than for radiative recombination. After thorough examination of other possibilities, in 1947 Bates and Massey proposed that, in the E region, the process is one of dissociative recombination (see Chapter 7, p. 111), it being assumed that the main positive ions are molecular. Although no estimates of the rate of this process were then available, it seemed clear that no other possibilities remained.

Assuming dissociative recombination to be the dominant loss process for electrons in the E region, Bates and Massey examined how the effective recombination coefficient could vary at greater altitudes. The crucial factor is that the recombination process considered can *only* occur to molecular ions. As the altitude increases, the relative concentration of molecular ions will decrease. This is because the rates of processes leading to destruction of molecules, such as photodissociation, will fall off less rapidly than those which lead to reassociation of the atomic fragments into molecules. At great altitudes, such as those at which the F_2 layer is found, suppose that recombination occurs through dissociative recombination to some rare molecular ion XY^+. This ion will be formed from the neutral molecule by charge transfer from an atomic ion A^+ say, which may for example be an O^+ ion, i.e., by

$$A^+ + XY \to A + XY^+. \tag{14.5}$$

The rate of this process will be $\gamma n(A^+) n(XY)$ where γ is the charge transfer rate coefficient, $n(A^+), n(XY)$ the respective concentrations of A^+ and XY. Recombination then follows through

$$XY^+ + e \to X' + Y'', \tag{14.6}$$

at a rate $\alpha n_e n(XY^+)$.

At sufficiently high altitudes the effective rate of recombination will be limited by the slowness of the reaction (14.5)—once an XY^+ ion is formed it will be neutralized through the process (14.6). Under these circumstances the effective loss rate of electrons will be given by the rate $\gamma n(A^+) n(XY)$ of (14.5). Most of the ions will be A^+ so $n(A^+) \simeq n_e$ and we may write

$$\frac{dn_e}{dt} = q - \gamma n_e n(XY). \tag{14.7}$$

Collisions in Atmosphere, Corona and Space

In equilibrium

$$n_e = q/\gamma n(XY) \qquad (14.8)$$

which may be contrasted with (14.4). γ will be independent of height but $n(XY)$ will fall off with height at great heights at least as fast as q. No maximum in n_e would be expected under these conditions. However, at sufficiently great heights loss due to radiative recombination will begin to exceed that due to dissociative recombination to the increasingly rare ion XY^+. Under these conditions (14.4) would apply and n_e would fall thereafter as the height increased.

These considerations provide a means for understanding the great thickness of the F_2 layer and later work has confirmed the picture in its essential aspects. Quantitative interpretation of ionospheric behaviour, which is now possible to a degree, has had to await the major developments of laboratory and theoretical research in atomic and molecular physics, and of space research, which makes possible *in-situ* observations of ionospheric properties and of the solar ultraviolet and X-radiation in the high atmosphere.

14.3. *Post-war Research and the Ionosphere*

Since World War 2, very great progress has been made in detailed knowledge and understanding of the ionosphere.

We have already outlined some of the experimental methods which have provided information on recombination, attachment and ionic reaction rates as well as of absorption, photoionization and photodetachment cross-sections, applicable to the ionosphere. This has already gone far to provide the information required in (c) and (d) above, but of equal importance has been the use of rocket propulsion to carry equipment to great heights in the atmosphere for making observations of ionospheric and other atmospheric properties and of the solar radiation.

This work became possible as a result of the development of powerful rocket motors by the Germans during World War 2, culminating in the bombardment of London by the V2 missiles. These were conveyed to their targets from launching sites some hundreds of kilometres (km) distant across the Channel by rocket propulsion. The trajectories of these missiles, which weighed 1000 kg, attained heights of as much as 100 km but, if fired vertically upwards, they were capable of reaching a height of 160 km, well into the E region of the ionosphere. It was apparent that the V2 rocket motors were powerful enough to transport a useful payload of scientific instruments to make *in-situ* observations of properties of the high atmosphere. Shortly after the end of the war, American scientists used captured V2 rockets for just this purpose. The twelfth flight, which took place on 10 October 1946 was a historic one. For the first time, spectra were taken of solar ultraviolet radiation, which is absorbed in the atmosphere above 30 km, the pioneering work being due to Johnson, Purcell and Tousey. During the same flight, measurements were made of atmospheric pressure up to 72 km.

The possibilities of the new technique of rocket exploration were thus demonstrated at an early stage and steps were taken to develop new rockets for research purposes, to be available before the supply of captured V2s ran out. The decade from 1947 was a period in which the application as well as further development of the technique expanded rapidly.

Although the availability of high-power rocket motors was essential, it was by no means sufficient. It was necessary to have suitable control mechanisms for the rocket flight, systems for tracking the rocket so that its position at any time during flight could be determined and means for recovering the data which were automatically observed by the instruments carried. Fortunately, the rapid development of electronics made all these requirements feasible. As it was normally difficult to recover a payload in any useful way after the rocket had returned to Earth, electronic systems were devised which translated data observed onboard into coded radio signals which during the rocket flight were transmitted back to a receiver system on the ground for decoding—a procedure known as telemetry.

While breaking entirely new ground, the use of rockets launched on more or less vertical trajectories suffered from many disadvantages. One of the most serious of these is the short time the rocket is in flight, thereby limiting the time during which observations can be made to a few minutes at most. Since many of the physical properties of the atmosphere are highly variable, with time of day, location on the Earth's surface and state of the Sun, a great number of systematic observations are required, extending over long times and a wide range of geographical locations. Short-time rocket flights are singularly unsuited for such a programme.

Nevertheless, a great deal of vital pioneering work was carried out with sounding rockets. Figure 14.2 shows the British rocket Skylark which just came into operation for scientific research in 1957 and, at the time of writing is still in use after many successful flights.

The next major step forward came in 1957 with the launching of the first artificial satellite, Sputnik I, by the Soviet Union. This opened up the possibility of observing atmospheric and surface properties on a global basis as the satellite revolves in its orbit with a period around 90 minutes. Data obtained from the instruments on board could be transmitted back to ground by telemetry as already developed for sounding rockets. Since 1957 a great deal has been done to exploit these possibilities.

At first the application of satellites as space laboratories for studying atmospheric properties suffered from one serious limitation. Because of friction with the atmosphere, a satellite in a circular orbit located at the altitude of the E region of the ionosphere will have a very short lifetime before plunging into the dense lower atmosphere where it burns up. Until very recently, most satellite studies of the ionosphere and the ambient neutral atmosphere were

Figure 14.2. A full scale model of the Skylark rocket at the IGY exhibition held in the rooms of the Royal Society, London in 1957. The launching platform for Skylark rockets, a model of which is shown in the background, was composed of Bailey Bridge parts and proved very satisfactory.

confined to much greater altitudes, usually above 250 km. Thus, again choosing from a very great number of satellites, one of the first to measure electron concentrations and mean energies as well as positive ion composition was Ariel I, the first joint American and British satellite shown in figure 14.3. It was launched in April 1962 into an orbit in a plane inclined at 54° to the equator. The orbit was eccentric, the maximum and minimum distances from the Earth

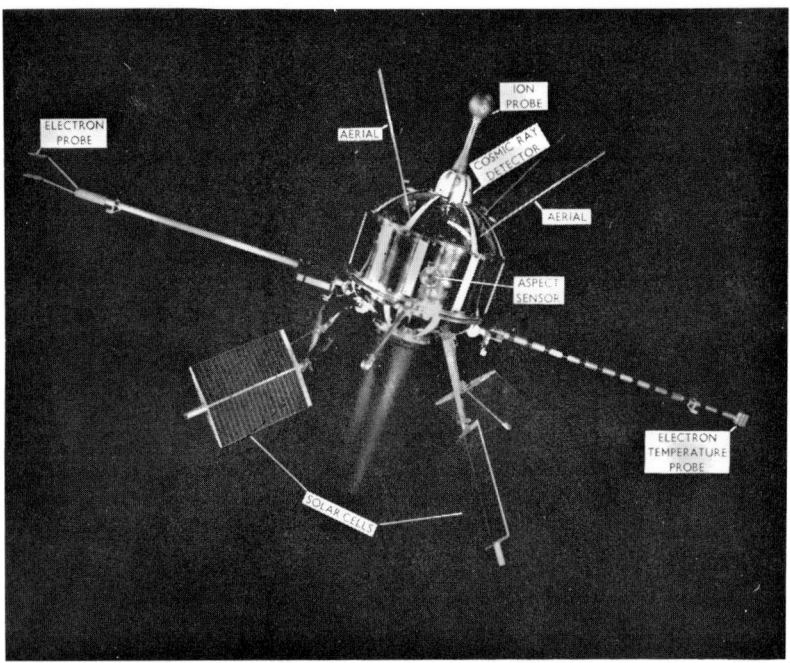

Figure 14.3. The first Anglo-American satellite Ariel 1.

being 1200 and 400 km respectively. With such an orbit only the outer F_2 region of the ionosphere could be studied.

During this period, rocket-borne equipment of increasing sophistication was used to study the E and F_1 regions. In particular, the importance of making simultaneous observations of related atmospheric and solar quantities was realized. However, in the early 1970s a series of Atmospheric Explorer satellites were planned specifically to make systematic and comprehensive studies of the atmosphere down to altitudes as low as 100 km. For this purpose, atmospheric decay effects, which are severe at these altitudes, are compensated by the thrust from small booster rockets carried aboard the satellite which are operated near perigee (the closest point of approach to the Earth). In addition, the payload includes a wide variety of instruments which make it possible to observe simultaneously atmospheric and ionospheric properties as well as the solar radiation which controls them. Thus the first of the Explorer series, Explorer C (launched in December 1973) carries instruments to measure the total atmospheric density, the neutral and ion composition, the neutral gas temperature, the electron and ion concentrations and mean energies (or temperatures), the solar radiation in the wavelength range from 4 to 185 nm over the orbital range of the satellite. The normal perigee is at 150 km but it is possible to make excursions to even lower altitudes for limited periods. The rate of accumulation

Collisions in Atmosphere, Corona and Space

of data from operation of one of these satellites is enormous and the provision of adequate facilities to analyse it, let alone interpret it, is a major task—so great in fact, that the impact of the results from the first Atmospheric Explorer is just beginning to be felt. Before long it should go far to establishing in detail a comprehensive picture of the main ionosphere and its relation to the sun.

14.4. The Present Position about Ionospheric Photochemistry

It has been pointed out earlier that, for a full interpretation of the behaviour of the ionosphere, we need to take account of the importance of atmospheric motions, both of the neutral and ionized constituents, in modifying the distribution of the ionization in space and time. From studies of these motions, carried out partly from ground-based observations and partly using rocket sounding, it appears that, in the inner ionosphere, they are of comparatively little importance in the E region at say 120 km altitude. As the altitude increases, so does the relative importance of atmospheric motions, until they are of primary significance in determining the behaviour of the F_2 region.

For these reasons, it is desirable to have available for the initial analysis of the ionospheric photochemistry as much systematic information as possible about the E region. Unfortunately, as explained above, this region has, until very recently, been largely inaccessible to satellite observations. With the Atmospheric Explorer Satellite in orbit, this situation is rapidly changing and at the time of writing, detailed analysis and interpretation of the immense amount of data flooding out from this project has already begun. However, a great deal of information has been obtained about the E region from rocket flights, so that it has been possible for some time to discuss in some detail the photochemistry of this region in relation to the observed data. The first results from the Atmospheric Explorer do not modify the conclusions derived in this way except in detail. We shall accordingly use the rocket data in our further discussion. At much higher altitudes, a great deal of information has been available from satellite observations for some time.

In table 14.1 we give some typical values for the concentrations of different neutral constituents at four altitudes, obtained from analysis of rocket sounding data. Values are also given for the temperature of the neutral gas at these

Altitude/km	Temperature T/K	Concentration of neutral particles/m^{-3}			
		O	N_2	O_2	NO
130	335	3.2×10^{16}	1.1×10^{17}	1.1×10^{16}	1.6×10^{13}
160	620	4.2×10^{15}	9.3×10^{15}	7×10^{14}	—
220	937	7.5×10^{14}	6.1×10^{14}	3.3×10^{13}	—

Table 14.1. Concentrations of neutral atoms and molecules and gas temperatures at three altitudes in the upper atmosphere, derived from rocket observations.

altitudes. We must expect considerable variability of all atmospheric parameters at these altitudes, so the figures given are to be regarded rather as typical, or characteristic, mean values.

Up to about 100 km, the atmospheric composition remains much the same at ground, but, just above this level, the oxygen changes quite rapidly from being mainly molecular (O_2) to mainly atomic (O). This is because, at these altitudes, the rate of dissociation of O_2 by the solar radiation exceeds the rate of reassociation of the O atoms to form O_2. The latter process can only occur in the presence of a third atom or molecule, M say,

$$O + O + M \rightarrow O_2 + M, \qquad (14.9)$$

so its rate varies as the square of the atmospheric pressure. At low pressures it therefore falls below that of photodissociation, which is simply proportional to the pressure. In addition to the change in the form of the oxygen, diffusive separation begins to set in above about 120 km, so that the lighter O atoms begin to float above the heavier N_2 molecules. At 160 km the O concentration is roughly equal to that of N_2 and it becomes the major constituent at higher altitudes.

Unlike oxygen, nitrogen remains mainly in the molecular form to quite high altitudes. However, through observations made with mass spectrometers in the Atmospheric Explorer satellite, it has been found that even nitrogen becomes mainly atomic above 380 km.

Figure 14.4 shows some early results obtained from the Atmospheric Explorer which show very clearly the way the concentrations of the different constituents

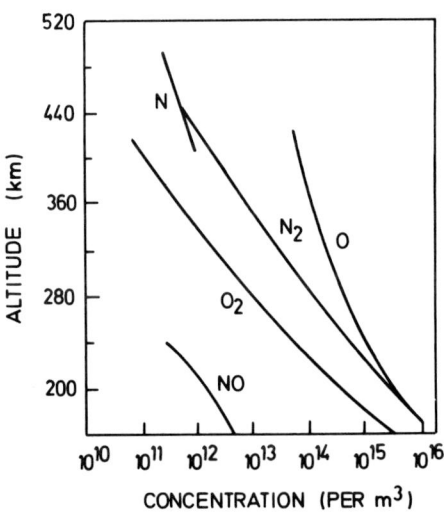

Figure 14.4. Variation with altitude in the Earth's atmosphere of the concentrations of N_2, O_2, N, O and NO observed with instruments on the Atmospheric Explorer C satellite.

vary with height out to 450 km altitude. The curve shown for N has been obtained by downward extrapolation from 400 km assuming diffusive equilibrium. The observed concentrations of the important minor constituent NO are also shown. Again, these results, while typical, must be taken as illustrative rather than applying quantitatively at any particular time or geographical position.

Figure 14.5 shows a typical set of data on the height variations of the electron concentration n_e and the concentrations of different positive ions. These data, obtained to a large extent from analysis of results from rocket-borne instruments, refer to conditions near the minimum of the solar cycle when solar conditions are least disturbed by sunspots etc. The E region appears as a small 'ledge' in the electron concentration curve.

To interpret these observations, use may be made of the accompanying data on the short wave spectrum of the Sun, obtained from instruments carried both on rockets and on satellites. Taken together with photoionization cross-sections obtained from laboratory experiment and from theory, it is possible to calculate the rate q at which electrons are produced per unit volume as a function of altitude. Typical results obtained are given in table 14.2. The equilibrium concentration n_e of electrons at any particular altitude follows from (14.4) but with allowance for the fact that more than one positive ion is present. Thus, if we distinguish these ions by suffices $1, 2, \ldots$, the recombination coefficient α in (14.2) is given by

$$\alpha = \alpha_1 f_1 + \alpha_2 f_2 + \ldots \tag{14.10}$$

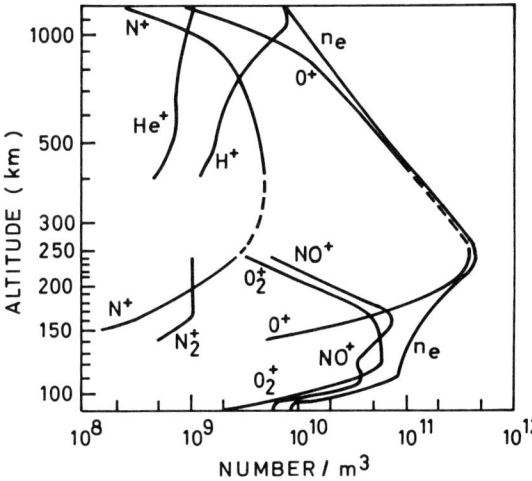

Figure 14.5. Variation with altitude in the Earth's atmosphere of the concentrations n_e of electrons and of the main positive ions derived from measurements made with instruments carried in rockets and (at the higher levels) in satellites.

Atomic and Molecular Collisions

	Ionization rate/$m^{-3} s^{-1}$		
Altitude/km	O^+	N_2^+	O_2^+
130	$7\cdot3 \times 10^8$	$2\cdot1 \times 10^9$	$8\cdot5 \times 10^8$
160	$7\cdot7 \times 10^8$	$1\cdot5 \times 10^9$	$3\cdot0 \times 10^8$
220	$2\cdot5 \times 10^8$	$2\cdot3 \times 10^8$	$2\cdot6 \times 10^7$

Table 14.2. Primary rates of production of O^+, N_2^+ and O_2^+ ions, in the upper atmosphere, by solar photo-ionization, derived from rocket observations.

where f_n is the fraction of ions in the form distinguished by the suffix n. From figure 14.5 the fractions f_n may be obtained at any altitude.

At 130 km we thus have

$$\alpha = 0\cdot6\alpha_1 + 0\cdot4\alpha_2 \qquad (14.11)$$

where 1 and 2 distinguish O^+ and NO^+ ions respectively. The laboratory values for α_1 and α_2 at 300 K, which is not far from the neutral gas temperature at 130 km, are 3 and 4×10^{-13} $m^3 s^{-1}$ respectively (see Chapter 7, p. 123), so we expect that α in (14.10) should be close to $3\cdot4 \times 10^{-13}$ $m^3 s^{-1}$. From table 14.2, the total rate of electron production q is $3\cdot7 \times 10^9$ $m^3 s^{-1}$, so that the equilibrium electron concentration $(q/\alpha)^{1/2}$ should be close to 10^{11} m^{-3}. Referring to figure 14.5, we see that this is indeed very close to the mean value derived from direct observations with rocket-borne instruments.

Analysis of the situation at higher altitudes is less decisive because, on the one hand, atmospheric motions soon become important and, on the other hand, the temperature (or mean energy) of the neutral gas and of the charged particles rises rapidly to values well above room temperature (300 K) (see table 14.1). Laboratory information about the dependence of reaction rates on temperature is still not adequate for proper allowance to be made for this. However, there is no reason to believe that a consistent picture of ionospheric behaviour at these higher altitudes will not be obtained when more basic information is available. In particular, the general validity of the interpretation of the effective loss rate for free electrons as a function of height out to great altitudes, sketched out on pp. 264, is now well established.

Turning now to the interpretation of the main features of the ion composition as a function of height, as shown in figure 4.5, we note first the striking feature that the most abundant positive ion is NO^+ even though NO is only a very minor neutral constituent (see table 14.1 and figure 14.4)

Again it will be seen by reference to the rates of formation of different ions by photoionization given in table 14.2 that only N_2^+, O^+ and O_2^+ are formed in this way. We must therefore ascribe the redistribution of the positive ionization which gives the equilibrium concentrations shown in figure 14.5 to the occurrence of ionic reactions. Because NO has a much lower ionization energy

(9·4 eV) than the mean atmospheric atoms and molecules, NO^+, once formed, cannot lose its charge by charge-transfer reactions such as

$$NO^+ + O \rightarrow NO + O^+,$$
$$NO^+ + N_2 \rightarrow NO + N_2^+.$$

All of these reactions can only proceed if the kinetic energy of relative motion exceeds a threshold value equal to the excess of the ionization energy of O (13·6 eV) or N_2 (15·6 eV) above that of NO. In the ionosphere, even at the highest altitudes, the mean kinetic energy of the ion is much below threshold. The NO^+ concentration therefore builds up until it becomes the dominant ion over a large altitude range.

From measured values of ionic reaction rates (see Chapter 11, p. 212), it appears that the main source of NO^+, particularly at the higher altitudes is through the reaction

$$N_2^+ + O \rightarrow NO^+ + N. \tag{14.12}$$

At lower altitudes

$$O^+ + N_2 \rightarrow NO^+ + N, \tag{14.13}$$

and

$$O_2^+ + N_2 \rightarrow NO^+ + NO, \tag{14.14}$$

make significant contributions.

N_2^+ ions are lost mainly through the reaction (14.12) and O^+ through (14.13). At low altitudes, an appreciable contribution to the loss rates comes from charge transfer with O_2

$$N_2^+ + O_2 \rightarrow N_2 + O_2^+, \tag{14.15}$$
$$O^+ + O_2 \rightarrow O + O_2^+. \tag{14.16}$$

14.5. *The Lower Ionosphere*

Below 100 km, primary photoionization of NO becomes important because Lyman α radiation (see Chapter 3, p. 46) from the Sun can penetrate down to 80 km or so. This radiation has a quantum energy of 10·17 eV sufficient to ionize NO but not O_2. NO^+ becomes progressively more dominant as the altitude falls. At 80 km, for example, it is about 50 times more abundant than O_2^+.

Observations taken with rocket-borne mass spectrometers have shown that, at altitudes below 80 km or so, water molecules form clusters with the positive ions, the most abundant ions being of the form $H^+(H_2O)_n$ with $n = 2$ or more, i.e., hydrated protons. The determination of the sequence of reactions leading to the production of these clusters, from primary NO^+ ions, has required a very considerable amount of laboratory research. It appears likely that the sequence is

$$NO^+ + N_2 + M \rightarrow NO^+ \cdot N_2 + M,$$

$$NO^+ \cdot N_2 + CO_2 \rightarrow NO^+ \cdot CO_2 + N_2,$$

M being a third body, or

$$NO^+ + CO_2 + M \rightarrow NO^+ \cdot CO_2 + M,$$

followed by

$$NO^+ \cdot CO_2 + H_2O \rightarrow NO^+ \cdot H_2O + CO.$$

Further hydration of $NO^+ \cdot H_2O$ then occurs to produce $NO^+(H_2O)_3$ which then reacts as follows

$$NO^+(H_2O)_3 + H_2O \rightarrow H^+(H_2O)_3 + HNO_2.$$

Below about 75 km, the concentration n^- of negative ions becomes comparable with that n_e of the electrons. At all higher altitudes, the rate of destruction of negative ions through reactions of the kind (see Chapter 11, p. 211)

$$O + O^- \rightarrow O_2 + e,$$

$$O + O_2^- \rightarrow O_3 + e,$$

is so fast compared to the rate of formation that the ratio n^-/n_e is very small. However below 75 km the formation rate by the three body reaction

$$O_2 + O_2 + e \rightarrow O_2^- + O_2,$$

which increases as the square of the partial pressure of O_2, becomes fast enough to make up for the rapid rates of loss. Just as with the positive ions, the negative ions are clustered, $NO_3^-(H_2O)_n$ with n greater than 2, being among the abundant species.

Study of these low regions of the ionosphere must depend on the use of rocket techniques. Considerable effort is being devoted to this work because a detailed knowledge of the chemistry, including the ionic chemistry, of the atmosphere below 100 km is important for the estimation of the importance of various man-made pollutants.

We must, at this stage, conclude our brief description of the application of atomic physics to the interpretation of ionospheric phenomena. The subject is now so large that we have been obliged to select only a limited number of illustrations of the progress which has been made through the availability of new laboratory and theoretical data obtained from *in-situ* observation made possible through rocket propulsion.

Collisions in Atmosphere, Corona and Space

14.6. *Ionization and Recombination in the Solar Corona*

Our Sun is a comparatively small and dull star—astronomically it is classified as a red dwarf star of the fifth magnitude. Its energy is derived from the fusion of hydrogen nuclei to form helium nuclei. The temperature at the centre of the Sun is about 26 million degrees and the pressure about 10^{10} times that of the Earth's atmosphere at ground.

The bright emitting surface of the Sun is known as the *photosphere*. If this is regarded as the boundary surface, the diameter of the Sun is $1\cdot 6 \times 10^6$ km. The spectral distribution of the emission is determined largely by the absorption of H^- ions as explained in Chapter 13, p. 250. It has a maximum in the green at a wavelength of 500 nm and the peak intensity corresponds to that emitted by a black body at a temperature near 6000 K.

While the broad spectral distribution of the radiation emitted by the Sun in the visible, as well as the near infrared and ultraviolet, is not very far from that expected from such a black body, this description cannot be extrapolated to much shorter wavelengths. This is primarily because of the effects of the solar atmosphere, which extends far beyond the photosphere. Immediately above the photosphere, there is a layer of gas about 650 km thick. Line absorption by the atoms of this gas gives rise to the Fraunhofer spectrum of the Sun as observed on the Earth and ground levels. This layer, often referred to as the *reversing layer* because it superposes an absorption spectrum on the photosphere emission, merges upwards with the *chromosphere* which is several thousands of kilometres thick. Its temperature is considerably higher than the photosphere and its presence leads to some increased emission of ultraviolet light.

The solar atmosphere extends even further, as was first realized from observations made during total solar eclipses. Thus the photograph reproduced in figure 14.6 which was taken during the totality phase of such an eclipse, shows a bright glow extending well beyond the dark observed disc. This bright glow is part of the solar corona which extends out to some solar radii to merge gradually into interplanetary space.

The spectrum of the corona in the visible includes bright emission lines superposed on a continuum. One of the most intense of these lines, at a wavelength of 530·3 nm, known as the green coronal line, was not identified until 1945. During that year Edlèn was able, for the first time, to reproduce the line in the laboratory and found that it arises from transitions between certain energy levels of iron atoms which have lost no less than 12 electrons! As a great deal of energy is necessary to remove so many electrons and as the likely source of this energy is that of temperature motion, it follows that the coronal temperature must be very high, of the order of 10^6 K. Under these conditions it will consist almost exclusively of ionized matter.

Suppose that $n(s)$ is the concentration of Fe atoms from which s electrons

Atomic and Molecular Collisions

Figure 14.6. Photograph taken during the solar eclipse of 30 August 1905 showing the solar corona surrounding the screened solar disc.

have been removed, Fe^{s+} ions as they are usually denoted. These ions will be produced by further ionization of $Fe^{(s-1)+}$ ions by electron impact and lost by recombination to electrons. Thus we have

$$\frac{dn(s)}{dt} = \beta n(s-1)n_e - \alpha n(s)n_e. \tag{14.17}$$

α is the recombination coefficient and β an ionization coefficient given by

$$\beta = \int Q_i(v) v f(v) \, dv. \tag{14.18}$$

Q_i is the cross-section for removal of a further electron from an $Fe^{(s-1)+}$ ion by impact of an electron of velocity v and $f(v) \, dv$ is the fraction of electrons with velocities between v and $v + dv$. Since $f(v)$ depends on the coronal temperature T, so also will β. α, which differs from β in that Q_i is replaced by the recombination cross-section $Q_r(v)$, will also depend on T.

In equilibrium $\dfrac{dn(s)}{dt} = 0$ and we have

$$\frac{n(s-1)}{n(s)} = \frac{\alpha}{\beta}. \tag{14.19}$$

It is possible from observations of the relative intensities of the coronal emission

Collisions in Atmosphere, Corona and Space

lines to derive $n(s-1)/n(s)$ for suitable values of s. If α and β can be determined as functions of T the relation (14.19) enables T to be determined.

This procedure was first suggested by Woolley and Allen in 1947. At that time they assumed that recombination must be radiative (see Chapter 7, p. 109), in which case, provided the electron velocity distribution took the Maxwellian form (Appendix 1), α could be calculated as a function of T. While it was not possible to calculate β directly, it could also be derived to a good approximation by an empirical procedure based on observed cross-sections for single electron ionization of various atoms (Chapter 6, p. 104). From this analysis a coronal temperature close to 10^6 K was indeed obtained.

It is possible to estimate the temperature in other ways. Because of the random temperature motion, the emitting ions will possess a spread of velocities along the line of sight determined by the temperature. The radiation from an ion moving with such a velocity v will show a frequency shift due to the Doppler effect which will depend on v. The result is that each emission line will be broadened by an amount depending on the temperature. Measurement of the widths of coronal spectrum lines therefore provides a fairly direct means of determining this temperature.

The temperature obtained by this alternative procedure was found to be smaller than that derived by the procedure of Woolley and Allen by an amount which could not be accounted for by uncertainty in the determination of the rate coefficients α and β. The matter was finally resolved by Burgess, following a suggestion due to Unsöld, that a further mechanism of recombination, known as *dielectronic recombination*, might be important.

We have discussed earlier the nature of autoionizing states of atoms (Chapter 3, p. 55) which have energies in excess of the ground state of the corresponding positive ion and so can decay via the process

$$A'' \to A^+ + e. \tag{14.20}$$

If the ion A^+ is in the ground state, the ejected electron carries away the excess energy ΔE as kinetic energy. Provided $\Delta E > E_{ex}$ the ion may alternatively be left in an ionic state with excitation energy E_{ex}, in which case the electron will be ejected with kinetic energy $E - E_{ex}$. In Chapter 13, p. 243, the effect of the existence of these states on the cross-section for photoionization is described.

Because the process (14.20) is reversible electrons of energy ΔE can be captured into the autoionizing state. When the state decays, the electron is ejected once more. This will contribute to the elastic scattering, if the electron leaves with its initial energy ΔE, or to inelastic scattering if it leaves with a smaller energy $\Delta E - E_{ex}$. The nature of these contributions has been discussed in Chapter 6, p. 90.

The possible importance of the autoionizing states for recombination arises because, during the finite mean lifetime τ_a of the autoionizing state, a radiative transition may take place to a lower state, so stabilizing the neutral system. That is to say recombination takes place due to the sequence

$$e + A^+ \to A'',$$
$$A'' \to A' + h\nu, \qquad (14.22)$$

where A' is an excited state, stable towards autoionization. If τ_r is the mean time before a radiative transition can occur, the probability that (14.22) will result from the initial electron capture will be $\tau_r/(\tau_r + \tau_a)$.

For recombination in a plasma at temperature T, the contribution, in this way, from a particular autoionizing state will be proportional to the chance that an electron in the plasma has the energy ΔE, i.e., to $\exp(-\Delta E/kT)$ where k is the Boltzmann constant. In almost all cases the lowest autoionizing state lies only a little below the first excited state of the *ion*. ΔE is then of the order of a few eV and the exponential factor is very small at room temperature.

Conditions are very different in the solar corona because the temperature T is of the order 10^6 K corresponding to a mean temperature energy of 100 eV. For recombination of an electron to an ion such as Fe^{15+} under these conditions, the contribution by processes such as (14.21) and (14.22), summed over the accessible initial autoionizing states in which the electron may be captured, considerably exceeds that due to direct radiative recombination. Thus at 10^6 K the respective contributions are $1 \cdot 1 \times 10^{-16}$ and $4 \cdot 8 \times 10^{-18}$ m^3 s^{-1}.

When dielectronic recombination is taken properly into account, the discrepancy between the temperatures deduced by the two methods disappears. This is an example where apparent contradictions in the interpretation have stimulated further research in atomic physics which has revealed factors of importance hitherto disregarded.

14.7. *Molecule Formation in Interstellar Space*

The space between the stars is by no means empty as may be seen from a photograph such as that shown in figure 14.7. This shows a portion of the Milky Way in the Southern Hemisphere. Besides the great numbers of individual stars there are also light and dark clouds of considerable extent. The dark areas arise from absorption of starlight by gas atoms and molecules and by dust grains in the clouds. From observation of the degree of reddening associated with the absorption, the relative contributions from the grains may be estimated.

The study of the interstellar medium is of importance not only for its own sake but also because of the light it might throw on the nature of the evolution of stars. As long ago as 1904, the existence of calcium ions, Ca^+, in interstellar space was revealed from observations made by Hartmann of the spectrum of a double star. In the star system the plane of motion was nearly in the line of sight. As well as absorption lines in the spectra of the stars, which showed

Collisions in Atmosphere, Corona and Space

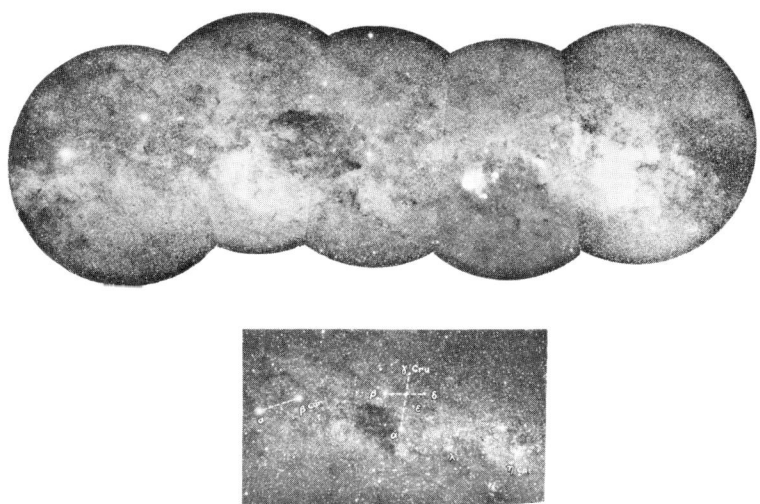

Figure 14.7. A composite photograph of the Milky Way in the southern sky taken with the Schmidt camera at Mt. Stromlo, Canberra, Australia by A. W. Rodgers. The lower photograph taken by A. R. Hogg with a small camera and a red-suppressing film provides a key to the identification of the stars alpha and beta centauri, the Southern Cross, lambda centauri and the eta carina (η car) nebula. Immediately below, to the left of the Southern Cross, is the dark region known as the Coalsack.

a Doppler shift due to their motion in the line of sight, other lines were observed showing no Doppler shift. These were therefore identified as due to absorption by interstellar atoms, neutral or ionized. A number of other atoms have since been identified.

So long as observations were confined to the visible wavelength region, it was not possible to observe the most abundant of all elements, hydrogen. This is because H atoms do not absorb visible radiation. With the introduction of radioastronomy, it became possible to observe the line emitted in the transition between the hyperfine structure levels of the ground state (see Chapter 3, p. 45) at a wavelength of 0·21 m. This was first achieved by Ewen and Purcell in 1951 and has since become a most important technique for plotting the atomic hydrogen distribution in different directions in space. In this way, the spiral structure of our own galaxy has been established.

The first molecular absorption was identified in 1937 by Swings and Rosenfeld in the wavelength region around 400 nm as due to CH^+ and CH. The presence of CN was also established though it is much less abundant. About a quarter of a century later radio techniques established the existence of OH through observation, in both emission and absorption, of the characteristic line at 0·18 m. With the development of millimetre wave astronomy, the first polyatomic molecules were observed, H_2O in 1968 and NH_3 a year later. In 1970, observations were made of the 2·6 mm line of CO, which showed that this molecule is remarkably abundant in interstellar space. Detailed observations of this radiation, which may be made using large optical telescopes in daytime, are proving of much value for deriving the properties of the interstellar medium in different directions in space.

Many other molecules have been identified, including formaldehyde (H_2CO), methyl cyanide (CH_3CN), HCN, CS_2, COS and SiO. CH has also been observed from radio observations at a wavelength near 9 cm. Special interest attaches to the observation of molecular hydrogen H_2. The only absorption bands which could be observed for this purpose are in the far ultraviolet between 100 and 111·5 nm. Since radiation of this wavelength is strongly absorbed in the atmosphere, no observations could be made until spacecraft became available to transport the necessary equipment above the absorbing region. Observations in the late 1960s using spectrographs in rockets showed that H_2 is not present in appreciable concentration in the general interstellar medium. In 1970, however, Carruthers observed the H_2 absorption bands in the direction of ξ Persei. Along this direction the absorption of visible radiation by dust is quite appreciable. The number of H_2 molecules in the column along the line of sight of cross-section $1 m^{-2}$ (the column density) was found to be $1·3 \times 10^{24}$ comparable with that of the interstellar H atoms (4×10^{24} m^{-2}) determined from the ultraviolet Lyα absorption by these atoms. The evidence now available supports the idea that H_2 is relatively abundant in dense absorbing regions.

To calculate the fraction of hydrogen in the molecular form, we must consider the rates of the processes which lead to their production and destruction.

It seems likely that association of H atoms to form H_2 takes place largely on the surface of dust grains. If there are n_1 H atoms and n_d dust grains per unit volume, the rate at which molecules will be formed can be written

$$\pi n_1 n_d \rho^2 v S, \qquad (14.23)$$

where ρ is the mean radius of a grain, v the mean velocity of a gas atom and S the chance of molecule formation when an H atom collides with a grain. A typical dust grain has a mean radius of $0·17\ \mu m$ and mass $m_d = 4 \times 10^{-14}$ g. The mass ratio of dust to gas in a typical low-density interstellar cloud in which the hydrogen is nearly atomic, is about 10^{-2}, so

$$n_1/n_d \simeq 10^{-2} m_H/m_d, \qquad (14.24)$$

where m_H is the mass of an H atom. This gives $n_d/n_1 \simeq 4 \times 10^{-13}$ so, since $n_1 \simeq 10^7\ m^{-3}$,

$$n_d \simeq 4 \times 10^{-6}\ m^{-3} \qquad (14.25)$$

The gas temperature is about 100 K, so that $v = 1·6 \times 10^3$ m s^{-1}. Theoretical reasons may be given for assuming that the sticking factor S is not far from unity. We therefore have for (14.23)

$$1.4 \times 10^{-10} n_1 n_d \, \text{m}^3 \, \text{s}^{-1}. \tag{14.26}$$

At first it was difficult to identify an effective process of dissociation. However, in 1967, Stecher and Williams pointed out that radiation with wavelength between 96 and 101 nm could dissociate the molecule. In atmospheric regions of interstellar space, the radiant intensity is about 10^{-14} times that which would be emitted by a black body at a temperature of 10^4 K. Following Stecher and Williams, the rate of dissociation of H_2 by this radiation then comes out to be

$$10^{-10} n_2 \, \text{s}^{-1}, \tag{14.27}$$

where n_2 is the concentration of H_2. Equating this to (14.26), we find

$$n_2/n_1 \simeq 1.4 n_d,$$
$$= 5.6 \times 10^{-6}. \tag{14.28}$$

This shows conclusively that very little hydrogen will be in molecular form in a low-density (lightly absorbing) cloud.

On the other hand, the situation will be very different in a dense cloud. The strong absorption of radiation in such a cloud will reduce the rate of dissociation below $10^{-10} n_2 \, \text{s}^{-1}$ by an amount which increases rapidly towards the centre of the cloud. Thus, consider a cloud of mass equal to 500 times that of the Sun with a gas atom concentration of $5 \times 10^7 \, \text{m}^{-3}$, in which the chance of absorption of radiation by the dust grains in passing from the periphery to the centre is 1/3. The fractional concentration of H_2 rises from about 10^{-3} at a depth below the surface of $0.1R$, where R is the cloud radius, to nearly 0.1 at the centre. A rough estimate indicates that about a fraction 0.4 of interstellar hydrogen is molecular.

In a dense absorbing cloud in which the core contains mainly molecular hydrogen, the chief source of atomic hydrogen is from cosmic radiation. This consists of fast charged particles, mainly protons, which can break up molecules on impact. Estimates may be made of the rate of production of atoms in this way, assuming the intensity and composition of the cosmic rays to be the same as in the neighbourhood of the Earth. This assumes that there is little absorption of the rays by the dust grains.

14.8. *The Problem of the CH^+ and CH Production*

The first paper which studied critically, and as comprehensively as possible at the time, the processes which give rise to molecules in interstellar space was published by Bates and Spitzer in 1951. At that time, the only molecules which had been identified were CH^+ and CH, and to a lesser extent CN. Bates and Spitzer concerned themselves with the production of CH^+ and CH. Despite the increase in knowledge of basic reaction rates many of their conclusions still remain valid. As we shall see, we are still not clear about the reaction sequences which are responsible. Until these are clear, we cannot be sure about the mechanism of production of the many other molecules, including such abundant ones as CO which have been observed mainly by radio techniques.

Analysis of the observational evidence shows that the average value of the ratio $N(CH)/N(H)$ of the column density of CH^+ to that of H is of order 10^{-9}, or even greater, while $N(CH)/N(CH^+)$ varies between 1 and 10.

We begin by considering the situation in a region in which the hydrogen is atomic. Following Bates and Spitzer, the only significant process of formation of CH^+ is one of radiative capture

$$C^+ + H \to CH^+ + h\nu. \tag{14.29}$$

Although this is a very slow process, it is much faster, at the very low pressure in interstellar space, than any three-body reaction. Again, it is unlikely that any CH^+ formation will occur on dust grains—because of photoelectric emission from the grains due to the background radiation field they are probably positively charged, so C^+ ions would be repelled. C atoms could produce CH on the grain surfaces but it is difficult to see how they could form CH^+.

The rate coefficient β for (14.29) is too small for it to be measured in the laboratory by techniques available at the time of writing. The best theoretical estimates, which do not differ very much from the original calculations of Bates and Spitzer, give 3×10^{-23} m^3 s^{-1}.

Once formed, CH^+ will be destroyed by radiative electron capture

$$CH^+ + e \to CH + h\nu, \tag{14.30}$$

and by dissociative recombination

$$CH^+ + e \to C + H, \tag{14.31}$$

as well as by chemical reactions with O and possibly other atoms.

The rate of production of CH^+ per unit volume per second from (14.29) is $\beta n(C^+)n(H)$, where $n(X)$ is the number of atoms or molecules X per unit volume. The rates of destruction by (14.30) and (14.31) are proportional to $n(CH^+)n(e)$, the constants of proportionality being the corresponding rate coefficients γ_1 and γ_2. Hence, in equilibrium,

$$\beta n(C^+)n(H) \geq (\gamma_1 + \gamma_2)n(CH^+)n(e), \tag{14.32}$$

so

$$\frac{n(C^+)}{n(H)} \leq \frac{\beta n(C^+)}{(\gamma_1 + \gamma_2)} n(e). \tag{14.33}$$

Collisions in Atmosphere, Corona and Space

In typical conditions $n(e) \simeq 2n(C^+)$ so

$$\frac{n(CH^+)}{n(H)} \lesssim \frac{1}{2}\frac{\beta_1}{\gamma_1 + \gamma_2}. \tag{14.34}$$

To reproduce the observed ratio of CH^+ to H, taking $\beta = 3 \times 10^{-23}$ m^3 s^{-1}, we must have

$$\gamma_1 + \gamma_2 < 3 \times 10^{-15} \text{ m}^3 \text{ s}^{-1}. \tag{14.35}$$

On the other hand, for most processes of dissociative recombination which have been studied experimentally, the rate coefficient corresponding to γ_2, at ordinary temperatures, is about 100 times larger, the sole exception being He_2^+ (see Chapter 7, pp. 121). Although γ_2 will fall with the temperature, it is still likely to exceed 3×10^{-15} m^3 s^{-1} by a considerable factor at 100 K. To clear up this difficulty, laboratory information, very difficult to come by, is required.

We turn next to the neutral radical CH under the same conditions. Because of its low ionization energy (11·26 eV), carbon is present mainly as C^+, so that radiative association between C and H atoms is not important. Instead CH will be produced mainly by (14.30) and destroyed, in a weakly absorbing region, by photoionization

$$CH + h\nu \to CH^+ + e. \tag{14.36}$$

The rate of (14.35), assuming the radiation field as on p. 281, comes out to be

$$4 \times 10^{-10} n(CH)$$

Hence, in equilibrium,

$$\gamma_1 n(CH^+) n(e) = 4 \times 10^{-10} n(CH), \tag{14.37}$$

so

$$\frac{n(CH^+)}{n(CH)} = \frac{4 \times 10^{-10}}{\gamma_1 n(e)}. \tag{14.38}$$

Once again, there is uncertainty about the critical rate coefficient γ_1. Bates and Spitzer estimated it to be 7×10^{-17} m^3 s^{-1}, in which case the observed data require

$$0 \cdot 1 < \frac{n(CH^+)}{n(CH)} = \frac{6 \times 10^6}{n(e)} < 1. \tag{14.39}$$

i.e., $n(e)$ must lie between 6×10^6 and 60×10^6 m^{-3} s^{-1}. The free electrons arise by ionization by starlight, of elements with ionization energy less than that of hydrogen. Estimated concentrations are very much smaller than those required to satisfy (14.39). Even with later, increased, estimates of γ_1, up to 5×10^{-16} m^3 s^{-1}, the value of $n(e)$ required is still excessive.

We must, however, take account of the fact that the molecule formation must also occur in absorbing clouds in which much of the hydrogen is in the molecular form. In that case, the problem of producing sufficient CH is a little eased, while that for CH^+ is rendered more difficult. Thus Black and Dalgarno suggest that the basic reaction in place of (14.30) is now

$$C^+ + H_2 \rightarrow CH_2^+ + h\nu, \tag{14.40}$$

followed by dissociative recombination directly to form CH

$$CH_2^+ + e \rightarrow CH + H, \tag{14.41}$$

or via CH_3^+ formation

$$CH_2^+ + H_2 \rightarrow CH_3^+ + H, \tag{14.42}$$

$$CH_3^+ + e \rightarrow CH + H + H. \tag{14.43}$$

The observed CH abundance in a particular dark cloud, that in front of ζ Ophiuchi (figure 14.7), can be reproduced if the rate coefficient for (14.40) is about 5×10^{-22} m^3 s^{-1} at 22 K. The best estimates, based on laboratory measurements of the rate of the reaction

$$C^+ + H_2 + He \rightarrow CH_2^+ + He,$$

are considerably smaller, 3×10^{-23} m^3 s^{-1}. Although the discrepancy here is somewhat less than for the case in which H_2 is absent, we must still await more reliable estimates of the rate of (14.40) before being sure that the proposed reaction sequence is the correct one. It could be that CH molecules are formed mainly on grains but, even assuming optimum conditions, the rate of such formation is barely enough to yield the observed abundances.

In a region in which the hydrogen is mainly molecular, the production of CH^+ is inhibited because the reaction (14.29) is no longer significant. One may speculate that the dissociative recombination coefficient for CH^+ is small enough so that CH^+ is formed with the observed abundance in weakly absorbing regions, and that the rate of (14.40) is fast enough for the CH to be formed, again in the observed abundance, in the cores of dense absorbing clouds. Dissociative recombination would be of key importance in both cases.

These examples serve to show how important is an accurate knowledge of the rates of atomic and electronic collision processes for the interpretation of phenomena on the very large scale, both in time and in space, characteristic of astrophysics. Under these conditions, some very slow reactions become of key importance. It is a severe challenge to laboratory research in atomic physics to measure the rates of these reactions. Great though the progress has been in experimental technique, some of the key reactions we have been discussing still remain outside the scope of experiment (at least at the time of writing!).

APPENDIX 1
the kinetic theory of gases—effusive flow

We recall here some of the concepts and formulae of the kinetic theory of gases which have been used in the main text.

Consider a perfect gas containing N molecules in a volume V at an absolute temperature T. By perfect we imply that the behaviour of the gas is not influenced appreciably by any interaction between the molecules except in so far as this leads to continual redistribution of momentum and energy among them. According to the kinetic theory, the gas pressure p is given by

$$pV = \tfrac{1}{3} N m \overline{c^2}, \tag{A1.1}$$

where m is the mass of a gas molecule and $\overline{c^2}$ is the mean value of the square of the velocity c of a molecule.

The temperature T of the assembly of molecules is proportional to the mean kinetic energy of a gas molecule, so we may write

$$\tfrac{3}{2} kT = \tfrac{1}{2} m \overline{c^2}, \tag{A1.2}$$

where k is a universal constant known as the *Boltzmann constant*.

From (A1.1) and (A1.2) we have

$$pV = NkT. \tag{A1.3}$$

According to Avogadro's law, equal volumes of different gases at the same temperature and pressure contain equal numbers of molecules. Let N_0 be the number of molecules in a mass of gas equal in grams to the molecular weight of the gas—Avogadro's number $(6 \cdot 02 \times 10^{23})$—and V_0 the volume occupied by these molecules. According to Avogadro's law, V_0 will be the same for all gases and is known as the gram molecular volume. We then have from (A1.3)

$$pV_0 = N_0 kT$$
$$= RT, \tag{A1.4}$$

where $R = N_0 k$ is the universal gas constant. (A1.4) is in agreement with observation. The observed value of R is $8 \cdot 317 \, \text{J} \, (\text{g mol})^{-1} \, \text{K}^{-1}$ so $k = 1 \cdot 380 \times 10^{-23}$ J K^{-1}.

The velocity distribution of the gas molecules is of the form derived by

Maxwell. If $f(c)\,dc$ is the fraction of the molecules possessing a velocity of magnitude between c and $c + dc$, then

$$f(c) = 4\pi(m/2\pi kT)^{3/2} c^2 \exp(-\tfrac{1}{2}mc^2/kT), \tag{A1.5}$$

which of course satisfies the condition

$$\int f(c)\,dc = 1.$$

Thus the lower the temperature, the narrower the velocity distribution.

The mean value of the velocity c is given by

$$\bar{c} = \int c f(c)\,dc,$$
$$= (8/\pi)^{1/2}(kT/m)^{1/2}, \tag{A1.6}$$

while the root mean square velocity, $(\overline{c^2})^{1/2}$, is given by

$$c_{rms} = 3^{1/2}(kT/m)^{1/2} \tag{A1.7}$$

and the most probable velocity c_m by

$$c_m = 2^{1/2}(kT/m)^{1/2}. \tag{A1.8}$$

The *mean free path* l of a molecule in the gas is the mean distance it travels between collisions. If the effective cross-section for a collision between two molecules is Q, independent of the relative velocity of the molecules, then it may be shown that

$$l = 2^{-1/2}/nQ, \tag{A1.9}$$

where n is the number of molecules per unit volume. When the collision cross section depends on relative velocity Q in (A1.9) must be replaced by $\bar{Q}$ which is a certain mean value of Q calculated using the velocity distribution function. In general $\bar{Q}$ will differ from $Q(c_m)$ only by a factor of order unity.

Effusive Flow and Molecular Beams

In the flow of a gas at normal pressures, the mean free path of the gas molecules is small compared with the linear dimensions of the flow system. When the gas pressure is reduced, the mean free path increases until conditions are reached in which the situation is reversed. Flow in which these conditions apply, the free path being greater than the dimensions of the flow system, is known as *effusive flow*. It is of considerable importance in the experimental investigation of atomic collisions because of the low working gas pressures.

In particular, an atomic or molecular beam is one in which collisions are not occurring within the beam, a situation which differs very much from a gas jet at ordinary or higher pressures (cf. Appendix 3).

An effusion source for a molecular beam (see Chapter 9, p. 181) consists of a small chamber containing the gas at a pressure of a few torr. Gas issues from the chamber from a small slit of width around 0·02 mm and height up to 1 cm. The gas flow is effusive because the slit width is smaller than the mean free path of the gas molecules.

The number δJ of molecules emerging per second from the slit travelling at an angle between θ and $\theta + \delta\theta$ relative to a normal to the plane containing the slit is given by

$$\delta J = n\bar{c} \cos\theta A_s \sin\theta \delta\theta \qquad (A1.10)$$

where A_s is the area of the slit.

Integrating (A1.10) over all angles θ from 0 to $\pi/2$ we obtain the total number of molecules emerging from the slit in all directions per second

$$J = \tfrac{1}{4} n\bar{c} A_s. \qquad (A1.11)$$

The velocity distribution of molecules in a molecular beam is not given simply by (A1.5) because the probability of a molecule emerging from the source slit is proportional to its velocity. (A1.5) must therefore be replaced by

$$f_B(c) = \frac{1}{2}\left(\frac{m}{kT}\right)^2 c^3 \exp(-\tfrac{1}{2}mc^2/kT). \qquad (A1.12)$$

so that, in units $(kT/m)^{1/2}$ the mean velocity c_B, the root mean square velocity $(c_{B2})^{1/2}$ and the most probable velocity c_m become $\tfrac{3}{4}(2\pi)^{1/2}, 2$ and $3^{1/2}$ respectively.

APPENDIX 2
high vacuum technique

The attainment of very low pressures in experimental chambers is of vital importance for the success of atomic collision experiments. Thus in a typical beam experiment in which the particles in the beam are fired through a selected gas, the gas pressure must be such that the particles make at most a single collision with a gas molecule on their way through the collision chamber. This requires that, for a typical path length of say 0·5 m, the gas pressure should not exceed 10^{-3} torr. Thus if Q is the cross-section for the type of collision concerned, the chance of a beam making a collision when passing through a distance l in a gas containing N molecules per m³ will be NQl. Taking Q of the order 10^{-19} m² and l 1 m requires that N be less than 10^{19} molecules per m³ in order that the chance of making a collision be less than unity. For N to satisfy these conditions, the gas pressure at 300 K must be less than 10^{-6} of an atmosphere, i.e., less than 10^{-3} torr. Furthermore, it is essential in many experiments that the purity of the gas under study should be very high. This means that the background pressure produced in the chamber before introducing the gas must be much lower still, 10^{-6} torr or less. Finally, in experiments in which the collisions studied are those occurring between particles in two intersecting beams, the background pressure must be reduced even further in order that the signals arising from collisions of either beam particles with residual gas should not obscure completely the wanted signals.

The standard procedure for obtaining a high vacuum is to employ a dual system consisting of a high vacuum pump and a backing pump. The former will only operate when the pressure has been reduced sufficiently by the backing pump.

Backing pumps are mechanical in operation and will reduce the pressure in an experimental system to around 10^{-2} torr. Against such a pressure a condensation pump may be used. The first such pump used mercury as the working liquid. Essentially, a suitably directed stream of mercury vapour drives the gas issuing from the chamber to be evacuated towards the entrance to the backing pump. Back diffusion of mercury vapour into the experimental chamber is prevented by including a trap cooled in liquid air or nitrogen between the main mercury stream and the chamber. This trap also condenses out any condensable vapours from the chamber. This arrangement will only work if the back pressure, which tends to produce back diffusion of the gas into the experimental chamber, is low enough. With suitable design, a pressure as low as 10^{-6} torr may be reached with this system.

As alternative working liquids, certain high-boiling oils are now extensively used. They have higher molecular weight and volume than mercury and so are more effective in driving out the residual gas. Thus the pumping speed, defined as the volume of gas extracted per second at the prevailing pressure, is considerably higher for oil condensation pumps. Care must be taken, however, to avoid contamination of the experimental equipment by thin oil films, so that it is often desirable to operate with a liquid-air cooled trap, as with mercury, between the vapour pump and the experimental chamber.

In many cases, the background pressure may be reduced by first baking the glass envelope to a temperature just below that which produces softening while at the same time heating metal parts to a dull red heat with an eddy current heater. This gets rid of much of the absorbed gas which otherwise continuously evolves during the experiment.

The pressure may be further reduced well below 10^{-6} torr by using a mercury pump with two cooled vapour traps in series, so that one can be baked while the other is filled with liquid air. However this procedure is very slow and to reach pressures of 10^{-9} torr a 'getter' pump operating at much higher pumping speeds should be used. This depends on the ability of a thin film of a getter material, such as titanium, to absorb gases, including even the rare gases. The absorption process is much more efficient if the gas molecules are bombarded with electrons—ionized atoms or molecules are accelerated to the film by a suitable voltage while the dissociation products, even if neutral, will be chemically more reactive.

The absolute measurement of very low pressure presents considerable difficulty. Until 1961, reliance was placed entirely on McLeod gauges, the operation of which is described in many text books. However, it had not been realized that the gauge is subject to an error which is especially important for gases of high atomic number and is difficult to correct for. In recent years capacitance manometers have been introduced as an alternative when the conditions are such that McLeod gauges are unsuitable. These manometers depend on the measurement of the capacitance of a condenser system in which one electrode is a thin membrane separating the high vacuum system from another system at a higher pressure.

APPENDIX 3
supersonic flow through a nozzle

The velocity V of sound in a gas is given by

$$V = (dp/d\rho)^{1/2}, \tag{A3.1}$$

where p, ρ are the gas pressure and density respectively. If the gas is flowing at subsonic speed, its velocity will increase as it passes through a converging nozzle as in figure A3.1. However, the reverse will be the case if the flow is supersonic. When these conditions prevail, the velocity will decrease on passage through the converging nozzle; to produce an increase in velocity, the nozzle must be diverging (see figure A3.1).

To see how this arises, we note first that, if v is the velocity of the gas flow where the sectional area of the nozzle is A then, to satisfy the conservation of mass of the gas,

$$A\rho v = \text{constant.} \tag{A3.2}$$

Consider next the gas within a small circular cylinder PP′ of length δx in the flow direction, with cross-section s as in figure A3.2. Let p, v be the pressure and flow velocity at P, $p + \delta p, v + \delta v$ the corresponding quantities at P′. The force on the mass δm of gas within the cylinder in the sense of the flow is

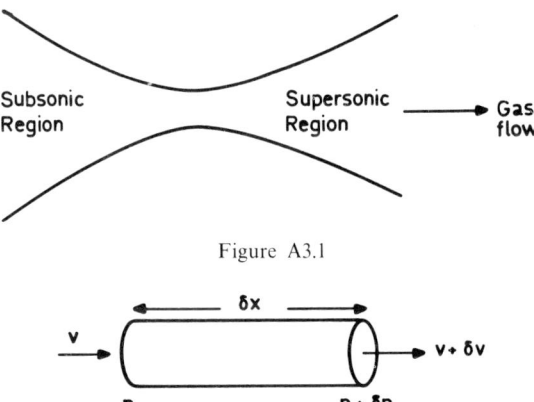

Figure A3.1

Figure A3.2

Supersonic Flow through a Nozzle

$-s\delta p$ so that

$$\delta m \frac{\delta v}{\delta t} = -s\delta p,$$

or, since $\delta m = \rho s \delta x$

$$\rho \frac{\delta v}{\delta t} = -\frac{\delta p}{\delta x}.$$

But $\delta x = v\delta t$ so that

$$\rho \delta(\tfrac{1}{2}v^2) = -\delta p,$$

or, using (A3.1), to eliminate p

$$\frac{1}{\rho}\delta\rho = -\frac{1}{V^2}\delta(\tfrac{1}{2}v^2). \tag{A3.3}$$

Also by logarithmic differentiation of (A3.2)

$$\frac{\delta A}{A} + \frac{\delta\rho}{\rho} + \frac{\delta v}{v} = 0.$$

Substituting for $\delta\rho/\rho$ from (A3.3)

$$\frac{\delta A}{A} = -\frac{\delta v}{v}\left(1 - \frac{v^2}{V^2}\right).$$

It follows that, when $v < V$, v increases when A decreases but when $v > V$ it increases when A increases.

For this reason, a supersonic nozzle has the shape shown in figure A3.1.

APPENDIX 4
the motion of charged particles in magnetic fields

A particle of charge e moving with velocity v in a uniform magnetic field of strength B and direction perpendicular to that of v will experience a force evB in the direction perpendicular to both v and B and in the sense shown in figure A4.1. It follows that the charge will describe a circle of radius r given by

$$\frac{mv^2}{r} = Bev,$$

m being the mass of the particle (figure A4.2). Thus $r = mv/Be$ and the time t of revolution in the circular path will be

$$t = \frac{2\pi r}{v} = 2\pi m/Be.$$

The force due to the magnetic field may be balanced by application of an uniform electric field of strength E in a direction normal to v and B, in the

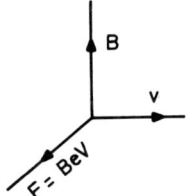

Figure A4.1

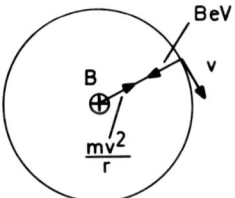

Figure A4.2

appropriate sense and of magnitude

$$E = Bv.$$

This means that the path of a particle entering a region bathed in uniform electric and magnetic fields of strengths E and B in directions normal to each other, in a direction perpendicular to both E and B, will be a straight line along this direction if

$$u = E/B.$$

For other velocities the path will be cycloidal. This is the basis of the Wien mass filter (see Chapter 7, p. 127).

APPENDIX 5
electron and ion optics

A very important part of most experiments on the scattering of charged particles is the preparation of a beam which is either well collimated, that is to say is one in which the particles are pursuing paths which are nearly parallel, or which is focused at some point where interaction takes place. These are the same requirements for light beams in optical instruments, in which they are achieved by the use of lens systems. Using suitable electric and/or magnetic fields, it is possible to construct electron (or ion) lenses which affect an electron (or ion) beam in much the same way that a conventional lens affects a light beam. This is provided the electron (or ion) beam is not too intense. Unlike photons, the charged particles in a beam repel each other. This tends to spread an otherwise collimated beam to a significant extent unless the particle concentration in the beam is small.

To follow the analogy to optics a little further, consider a charged particle travelling with velocity u through a region in which the electrostatic potential is V until it impinges at an angle of incidence α on an infinite plane interface between this region and a second region in which the electrostatic potential is V' (figure A5.1). After passage across the interface it will enter this region with velocity u' at an angle of 'refraction' α'. Since the component of velocity of the particle parallel to the interface will not be changed

$$u \sin \alpha = u' \sin \alpha'.$$

Also if the electrostatic potential is chosen to be zero when the particle's velocity is zero we have

$$u/u' = (V/V')^{1/2},$$

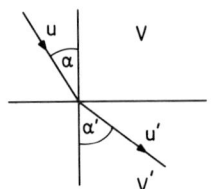

Figure A5.1

Electron and Ion Optics

so

$$V^{1/2} \sin \alpha = V'^{1/2} \sin \alpha'.$$

This is exactly the same as Snell's law of refraction for light if the refractive index is taken as being proportional to the square root of the electrostatic potential.

In practice, the electrostatic potential will not change discontinuously as does the refractive index in most optical systems but will vary gradually along the path of the beam. Largely because of this, determination of the cardinal points of an electron (or ion) lens is much more complicated than for an optical lens.

As an example of a simple electrostatic lens, we choose two parallel plane electrodes with apertures through which the particles enter in a direction normal to the planes. The planes are maintained at different potentials V_1, V_2 respectively (see figure A5.2). This is not very different from a simple experimental slit system.

We consider a parallel beam of electrons (or ions) entering the aperture from the left at its midpoint in a direction normal to the plane electrodes. Provided the width of the beam is small compared with the width of the apertures, it will be focused at a point distant f_1 from the midpoint where

$$\frac{1}{f_1} = \frac{1}{4V_2^{1/2}} \int V^{-1/2} \frac{d^2V}{dz^2} dz,$$

z being measured along the direction of the beam. V is the electrostatic potential at the point z.

If $\frac{d^2V}{dz^2} > 0$, f_1 is positive and the focus is real. The system then acts as a converging lens. On the other hand, if $\frac{d^2V}{dz^2} < 0$, f_1 is negative and after passage through the aperture the beam diverges as if it came from a focus at a distance f_1 to the left of the midpoint.

Equally well, if the beam enter from the right, the focus will be at a point

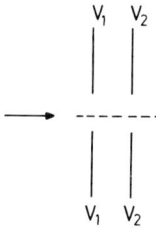

Figure A5.2

f_2 where

$$\frac{1}{f_2} = \frac{1}{4V_1^{1/2}} \int V^{-1/2} \frac{d^2 V}{dz^2} dz.$$

The cardinal points of the system considered as a lens depend on the detailed variation of the electrostatic across the aperture. This is characteristic of electron (or ion) optics, so that accurate measurement of electric field distributions is required before the optical behaviour of the system can be determined.

Elaborate electrostatic lens systems are used in modern experiments, as described for example in Chapter 6, p. 97, and in Chapter 12, p. 225.

Focusing by magnetic fields or by combined electric and magnetic fields may also be used. The magnetic velocity selectors described in Chapter 6, p. 92, provide examples of the former.

APPENDIX 6
the detection and measurement of small fluxes of electrons, ions and neutral atoms

Charged particles are usually collected in a Faraday cage, which is a hollow metal cylindrical container. It is essential that all particles which enter are collected and that no secondary particles produced can leave. The entrance hole to the cage must therefore be small, while secondary electrons may be suppressed by an electrode, with suitable aperture, in front of the cup, insulated from it and maintained at a negative potential with respect to it.

A single electron per second corresponds to a current flow of $1 \cdot 6 \times 10^{-19}$ A. In many scattering experiments, only a few thousand scattered particles may be collected per second in the Faraday cage. For the detection of such small currents thermionic amplification is used. However, this is most effective for amplification of alternating voltages, so the measurement is converted to one of an a.c. voltage in a so-called 'vibrating reed electrometer.' The current from the cage is used to load one plate of a small condenser which is connected to the grid of the thermionic tube and earthed through a very large resistor (of order 10^{12} Ω). The other plate of the condenser is earthed directly and vibrated at a frequency of about 10^3 Hz, so that the capacity of the condenser and hence the voltage between the plates is oscillating with the same frequency.

Since World War 2, increasing use has been made of secondary emission multipliers which are capable of current amplification by factors of the order 10^6. Such a device consists of a number of metal electrodes, usually referred to as dynodes. These electrodes are arranged so that secondary electrons, produced by impact of a primary beam on the first electrode, are accelerated by a potential of the order 100–300 eV on to the second electrode, thereby producing further 'secondary' emission from that electrode. These electrons in turn are accelerated to a third electrode and so on. Provided that each electron incident on the next dynode produces f secondary electrons from that dynode, where $f > 1$, the incident current will be amplified by a factor of order f^n where n is the number of dynodes. The electrodes must be shaped so that the electrons emitted from one are focused on the next. By connecting the final dynode to a pulse amplifier, it is possible to record the pulse due to arrival of single particles at the first electrode.

The secondary emission coefficient f depends very much on the nature of the surface so that special care must be directed towards ensuring that surface conditions do not change. To avoid this difficulty, use is made of a photomultiplier in which the first dynode is replaced by a photocathode from which electrons are emitted when photons of suitable wavelength are incident. The whole electrode system is then enclosed in a glass envelope under ultra-high vacuum, so surface conditions should not change. Charged particles are counted by arranging for them to impinge on a suitable phosphor from which they cause photon emission. The photon pulse is then recorded via the photomultiplier.

The use of multipliers is not restricted to detection of charged particles. Fast beams of neutral atoms will produce secondary electron emission and hence may be detected.

Calorimetric methods are conveniently used for the detection of fast beams of ions or neutral atoms provided the intensity of the beam is not too small. The beam is allowed to fall on a metal disc attached to a thermoelectric device such as a thermistor. Powers of order as low as 10^{-6} W may be measured in this way. This corresponds to a current of particles with energy 10 keV, of 10^{-10} A. The detectable current varies inversely as the kinetic energy of the particles.

A very sensitive thermal detector for very low energy atomic beams is the bolometer developed by Cantini, Dondi, Scoles and Torello (see Chapter 10, p. 194). The chopped atom beam to be monitored impinges on a doped germanium crystal which is electrically insulated but in thermal contact with a high purity copper substrate maintained at the temperature of liquid helium. The chopped beam produces temperature oscillations in the crystal which are converted to oscillatory resistance changes in the copper and thence to electrical signals which may be amplified by a low frequency amplifier. At a chopping frequency of 30 Hz, a beam of 3×10^8 particles per second impinging on the germanium crystal with a mean energy corresponding to a temperature as low as 4 K may be detected.

APPENDIX 7
mass analysers and selectors

We have often referred in the main text to the use of mass spectrographs either to determine the masses of charged particles in a beam or to select particles of a chosen mass from such a beam. It is not difficult to see how this may be achieved, at least in principle.

Suppose that, by application of an electric field, the particles in a beam, all possessing the same charge e, are all accelerated to the same kinetic energy E. The velocity of a particle of mass M in the accelerated beam will be $(2E/M)^{1/2}$, so that mass analysis or selection can be reduced to that of velocity, techniques for which have been described in Chapter 6, Section 6.3.

For example, the accelerated beam may be passed through a region bathed in a uniform magnetic field of strength B perpendicular to the direction of the beam. A particle of mass M with kinetic energy E will describe a circular path in this region, of radius r_M given by Mv/Be, where v is its velocity $(2E/M)^{1/2}$. If V is the voltage applied to accelerate the beam, then $E = eV$ so

$$r_M = \left(\frac{M}{e}\right)^{1/2}\left(\frac{2V}{B^2}\right)^{1/2}. \tag{A7.1}$$

Hence, if the beam after acceleration enters the magnetic field at P in a direction normal to the line PQ in figure A7.1, the particles within it of the desired mass M will leave the field again at a point P' where $PP' = 2r_M$. It is then only necessary to make PQ a solid boundary, except for slits at P and P', to ensure that only particles of the required mass may pass out through P' into some experimental chamber. Alternatively, for mass analysis, the magnetic field may be varied and the current passing through P' observed as a function of B. If a transmitted peak is obtained when B has the value B_1 say, then the height of

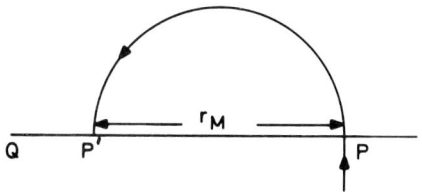

Figure A7.1

this peak is, ideally, proportional to the fraction of the beam current carried by particles of mass M_1, where

$$r_{M_1} = \left(\frac{M_1}{e}\right)^{1/2} \left(\frac{2V}{H_1^2}\right)^{1/2}. \tag{A7.2}$$

A common design of mass spectrograph is based on these principles. Of course, care must be taken to maintain the collimation of the beam by a suitable lens system, so that acceleration by the electric field does not cause serious angular dispersion. Questions of the resolution of the velocity analysis need to be carefully considered. In particular, the initial velocity of the particles before acceleration may need to the taken into account. This is especially so when it is required to measure the relative intensities of the currents of particles of different mass in a beam. Unless special precautions are taken the peak heights in the mass analysis cannot be assumed to be proportional to these intensities if the particles of different mass possess initial velocities which are significantly different.

While semi-circular magnetic field selection is often used, it is also possible to use other types of magnetic analysis.

It must be remembered that we have assumed that all the particles have the same charge e. In fact, the selection is one of particles with a chosen value of the ratio Q/M of charge Q to mass M. The possible existence of doubly, or multiply, charged ions in a beam sometimes leads to ambiguity of interpretation unless the resolution is high. Thus for $Ar^{2+} Q = 2e$, $M \simeq 40 M_H$ where M_H is the mass of a proton so $Q/M \simeq \frac{1}{20}\frac{e}{M_H}$. For $Ne^+ Q = e$, $M \simeq 20 M_H$ so again

$$Q/M \simeq \frac{1}{20}\frac{e}{M_H}.$$

Velocity analysis depending on a quite different principle is used in radio-frequency mass analysers which have the advantage of being very compact. This makes it possible to use them to sample ions in a plasma without producing major changes due to insertion of the instrument (see for example Chapter 7, p. 120).

Suppose that an ion beam accelerated to an energy E is allowed to pass through a number of equally spaced apertures to which an alternating electric potential $V_0 \cos \omega t$ is applied, ω being the angular frequency. An ion arriving at the first aperture at an instant when the potential is in the correct phase to accelerate it towards the next aperture will also arrive at that aperture at a time to receive further acceleration if

$$2s/v = 2\pi/\omega, \tag{A7.3}$$

$2s$ being the spacing between successive apertures and v the velocity of the particle.

Mass Analysers and Selectors

It follows that particles of the beam with velocity satisfying (A7.3) will gain the maximum additional energy ΔE by passage through the apertures. By applying a suitable retarding potential after passage of the ions through the last aperture, all ions except those which have gained energy close to the maximum may be rejected. The accepted current measured as a function of ω will then exhibit peaks when (A7.3) is satisfied.

We have assumed in this analysis that V_0 is such that the maximum value of ΔE is small compared with the incident energy of the ions, so that their velocity remains close to its initial value throughout.

Instruments of this kind, containing up to 24 apertures, may be made with total path length for the ions no more than 2 cm and of diameter around 1 cm. In a typical 24-aperture instrument, the maximum value of ΔE may be around 30 times eV_0.

A further type of mass analyser employs an alternating quadrupole electric field. This field is generated between four parallel circular cylindrical electrodes of radius a arranged as in figure A7.2, the shortest distance between opposite cylinders being $a/1\cdot 2$. All four cylinders possess electric charges of the same magnitude, but for the upper and lower pair, the sign of the charge is opposite to that for the transverse pair (see figure A7.2).

If we choose rectangular co-ordinates with the x-axis midway between the pairs of cylinders parallel to their axes, the y-axis horizontal and the z-axis vertical (see figure A7.2), then for the quadrupole system the electrostatic potential close to the x-axis can be written

$$W(y^2 - z^2)/a.$$

W is independent of position but may vary with time. In a quadrupole mass analyser it has the form

$$W = U + V \cos \omega t,$$

where U and V are constants.

It may then be shown that particles of charge e and mass M entering the region between the cylinders will be transmitted without collision with the

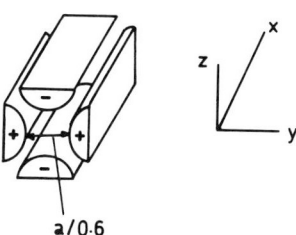

Figure A7.2

electrodes only when

$$V = 0.18\, M\omega^2 a^2/e, \qquad U = 0.166\, V. \qquad (A7.4)$$

These conditions are not sensitive to the velocity of the charge, its initial direction of motion or the point of entry into the systems. This offers substantial practical advantages in using the quadrupole arrangement for mass selection. For such purposes, the length L of the cylinders must be large enough to ensure that particles with mass and charge which do not satisfy (A7.4) will strike one of the electrodes before they can emerge at the exit end. This requires that

$$L \gg a\pi(V_0/V),$$

V_0 being the potential through which the particles are accelerated before entering the analyser.

INDEX

Aberth 225
Absorption, in solar atmosphere 245, 251
   line 230
   of ultrasonic waves 145
Adlard 137
Afterglow, flowing 207
   of electric discharge 114
   in Ar and Ne 120
   in He 120, 122
Allen 277
Allowed transitions 43
Amplitude, of wave motion 13
Andrick 97, 99, 102
Angle of projection 8
Angle of scattering, for rare gas atom
   collisions 163–5
   in CM system 7
   in laboratory system 8
   relation to deflection function 162
Angular distribution, in He–He collisions 197–8
   in Na–Hg collisions 191
   in resonance scattering 90, 98, 101
   of scattered ions 225, 226, 228
   of scattered electrons, in various gases 83, 85
      measurement of 80
      theory of 83–6
Angular frequency 14
Angular momentum, allowed values of 26
   and electron spin 41
   combination of 42
   definition 12
Appleton 260
Argon atoms (Ar), ionization energy of 53
   scattering of slow electrons in 75–80
   structure of 53
   photoionization of 243
Argon molecular ions (Ar_2^+) 64
   in discharge afterglows 116, 120
Ariel 1 267, 268
Arnot 81
Associative detachment 205, 211
Atmospheric Explorer satellite 268
Atmosphere, upper, electron reactions in 262, 271–2
   ionic reactions in 273–4
   neutral composition of 269, 270
Atomic beams, sources of 181, 182, 189, 196, 286
   velocity selection in 183, 189, 196
Atomic number 39
Atomic polarizability 50

Attachment 108, 109
   dissociative 112, 116, 124
   in ionosphere 262
   measurement of probability of 125
   radiative 110
   three-body, in O_2 129
Auerbach 153, 154
Autodetachment 58
   and resonant scattering 89
Autodetaching states, excitation of by photons 236
Autoionization 56, 104
Autoionizing states, and dielectronic recombination 277
   excitation of by photons 236
   of calcium 244
Avogadro's law 285

Baede 153, 154
Bailey 73, 75, 76
Balmer series 44
Bandel 98
Bates 245, 263, 264, 281, 282
Beck 158
Bennett 235, 255
Bennewitz 195, 196
Biondi 121, 125
Black 284
Boechner 239, 244
Bolometer, detector of He atoms 194, 295
Boltzmann constant 119, 233, 285
Braddick 239
Branscomb 246, 252, 257
Brehm 248
Breit 260
Briglia 103, 128, 131
de Broglie 1, 21
Bromine atoms (Br), electron affinity of 57
   ionization energy of 53
Bromine molecules (Br_2), dissociation energy of 62
   electron affinity of 155
   vibrational relaxation time in 147
Buck 184, 188, 192
Buffer gas 117
Bullard 50, 81, 82, 83
Burch 246, 252
Burgess 277
von Busch 187
Busse 195, 196

Cantini 298
Caesium atoms (Cs), electron affinity of 57
 ionization energy of 53, 153
 photoionization of 244
Caesium negative ions (Cs^-), photodetachment from 255
Calcium atoms (Ca), absorption of light by 244
Carbon atoms (C), electron affinity of 57
 ionization energy of 53
Carbon negative ions (C^-), formation of 132
CH in interstellar space 279, 283
CH^+ in interstellar space 279, 282
CH_2^+ in interstellar space 284
CN in interstellar space 279
Carbon monoxide (CO), dissociative attachment in 131
 dissociation energy of 62
 in interstellar space 279
 vibrational relaxation time in 147
CO^+, dissociation energy of 64
Carruthers 280
Celotta 255
Centre of mass coordinates 5
Chadwick 176
Chandrasekhar 285
Chanin 125
Chantry 127, 128, 131, 133
Charge transfer, and ion mobilities 203
 and production of fast atom beams 141
 between slow ions and atoms 204, 211
 in afterglow experiments 117
 in ionosphere 264
 probability of 224
 symmetrical 213, 215, 221
 asymmetrical 221, 222, 223
Chlorine atoms (Cl) electron affinity of 57, 153
 ionization energy of 53
Chlorine molecules (Cl_2), dissociation energy of 62
 electron affinity of 155
 vibrational and rotational constants of 65
 vibrational relaxation time in 147
Chromosphere, of Sun 275
Cluster ions 117, 206, 212, 273
Codling 243
Coffinet 148
Collision complex 90
Collision frequency 38, 215
Compton effect 20
Condensation pump 288
Continuous spectrum 27, 45
Corona of Sun 108, 109, 275
 temperature of 277
Coronal line 275
Cosmic rays 281
Creek 244
Cross-section
 concept of 9
 differential, definition 10
 classical 167
 in terms of angular momentum 13
 in terms of impact parameter 11
 wave mechanics 29, 35
 for excitation of bound states 105, 106
 impact of two rigid spheres 11
 inelastic collision of electrons 88
 ionization by electrons 103, 104
 momentum loss (diffusion) 70, 140
 photodetachment 247–8
 photoionization 238, 240
 viscosity 140
 total, classical 141
 elastic for atom–atom collisions 140
 for charge transfer 219, 222
 for atom–atom collisions 187, 195, 196
 for attachment 126, 127
 slow electrons in gases 73–5

Dalgarno 284
Daly 221
Davison 21
Deflection function 158, 161
Delone 258
Detector, attachment 135
 bolometer 194, 298
 calorimetric 298
 electron multiplier 98, 245, 249
 for atom beams 141
 phase-sensitive 186
 photomultiplier 224
 secondary emission multiplier 248, 297
 surface ionization 154, 186
Deuterium atom (D), binding energy of 51
Deuterium molecule (D_2), vibrational and rotational constants of 65
 vibrational and rotational relaxation in 148
Diffraction, grating 241
 of electrons 21
 of radio waves 260
Diffusion, of electrons in discharge afterglows 116
 in ionosphere 263
 of gases 140
Diffusion (condensation) pump 96, 288
Direct scattering, of electrons 89
Dispersion, of ultrasonic waves 144
Dissociation, of molecules by electron impact 89
Dissociation energy, definition 61
 of diatomic molecular ions 64
 of diatomic molecules 61–2
Ditchburn 239, 244
Dohmann 193, 195, 196
Dondi 176, 193, 194, 195, 298
Drift velocity of electrons 69, 71, 72
 of negative ions 125
Ducoing 149
Dunoyer 180
Dust grains, in interstellar space 278, 280

Index

Edlèn 275
Effusive flow 181, 286
Ehrhardt 97, 99, 102
Eigenfunctions 24
Eigenvalues 24, 25
Einstein 20
Electron affinity, of atoms 57
 and dissociative attachment 112
Electron attachment detector 135
Electron configuration in atoms 53
 doubly-excited 56
 terms of 54
Electron exchange 86
Electron multiplier 98, 225, 249
Electron spin 41, 54
Electrostatic lens 295
Energy, allowed, for molecular rotation 66
 molecular vibration 65
 particles in Coulomb fields 28
 simple harmonic oscillator 25
Energy analysers, electrostatic cylindrical 93, 97, 224, 225
 hemispherical 94, 95, 249
 magnetic 92
 Wien filter 127
Energy distribution, of ions from dissociative attachment 129, 132
Energy width, in scattering resonances 91, 98
Engländer–Golden 103
Everhart 224
Ewen 279

Faraday cylinder (cup or cage) 81, 220, 221, 224, 297
Farrar 196, 197, 198
Felseth 151
Fite 246
Fluorescence, resonance 150
 sensitized 151
Fluorine atoms (F), electron affinity of 57
 ionization energy of 53
Fluorine molecules (F_2), dissociation energy of 62
 vibrational and rotational constants of 65
 vibrational relaxation time in 147
Forbidden transitions 43
Ford 156
Fractional energy loss in collision 7, 69
Franck 47
Franck–Condon principle 66
Free–free transitions 252
Frequency, of electric oscillation in a cavity 118
 of vibration in a molecule 65
 of wave motion 13
 threshold for photodetachment 231
 for photoionization 231
Frommhold 121

Galaxy, spiral structure of 45
Gas chromatography 136
Geltman 252
Germer 21
Getter pump 289
Glory singularity 167
 undulations in atom–atom collisions 174, 187–190
 in symmetrical charge transfer 223
Gold atoms (Au), electron affinity of 57
Golden 98
Goodman 136, 138
Goulden 136, 139
Gray 180, 182
Gusimow 248

Half-ball angle 4
Hall 248, 255, 257, 258
Hartmann 278
Heaviside 260
Heddle 107
Helbing 153
Helium atoms (He), beam of, velocity selection in 193
 collision of, with He^+ 218
 drift velocity of electrons in 72
 excited states of 55
 excitation of, by electrons 106–7
 interaction of, with He 198
 ionization energy of 52
 ionization of, by electrons 104
 probability electron density in 51
 total cross-section of, for slow electrons 75–80
 resonance effects in 98
Helium molecular ions (He_2^+), dissociation energy of 64
Helium negative ion (He^-), autodetachment from ^4P state of 58
Herbst 253
Herriott 235
Hertz 47
Hotop 252–5
Hudson 244
Hundhausen 191
Hydrogen atoms (H), absorption of energy by 46
 allowed energy levels of 42
 atomic polarizability of 50
 collisions of, with H^+ 215, 216, 227
  with H^- 223
 continuous spectrum of 45
 electric field of 50
 electron affinity of 57
 excited, radiation from 43
 hyperfine structure in spectra of 45
 in interstellar space 280
 interaction of, with H 61
  with H^+ 214
 series in spectrum of 44

Atomic and Molecular Collisions

Hydrogen molecules (H_2), dissociation energy of 61
  excited electronic states of 62
  in interstellar space 280
  radiative dissociation of 283
  vibrational and rotational constants of 65
  vibrational and rotational persistence in 148
Hydrogen molecular ion (H_2^+), dissociation energy of 63
Hydrogen negative ion (H^-), and solar absorption 245
Hyperfine structure, in hydrogen spectrum 45

Impact parameter, definition 10
  classical scattering theory 11, 158
Inner shell ionization 89
Interaction energy, between atoms 60, 61, 157, 192, 197–8
  determination of 196–7
  between ions and atoms 62, 63, 201, 215
  definition of 6
  effective in electron–atom scattering 86
  of electrons with H atoms 50
  van der Waals 60
Interference, in scattering due to symmetry 175, 195
  in symmetrical charge transfer 215–7
  of waves 16
Interstellar matter, absorption by 244
  calcium ions in 278
  dust grains in 278
  diatomic molecules in 239
  polyatomic molecules in 279–80
Inverted population 234
Iodine atoms (I), electron affinity of 57
  ionization energy of 53
Iodine molecule (I_2), electron affinity of 155
  vibrational relaxation time in 147
Iodine negative ion (I^-), double photon detachment from 258
Ion pair production 153
Ion sources 218
Ionic reactions, in discharge afterglow 118
  flowing afterglow 207–210
  ionosphere 273
  lower ionosphere 274
Ionization, by electrons 102–4
  by metastable atoms 152
  by photons 231
  in solar corona 276
Ionization energy 46
  of various atoms 46, 52, 53
Ionosphere, discovery of 260
  E, F_1 and F_2 layers of 261
  electron concentration in 271
  ion composition of 271
  lower 271
  recombination in 271

Isotopes 39
  of H 41
  of He 176, 196, 199, 227

Javan 235
Joffrin 149
Johnson 265

Kantrowitz 180, 182
Kasdan 253–5
Keesing 107
Kennelly 260
Kinetic energies, of ions from dissociative attachment 113
Kollath 80, 82, 83
Krypton atoms (Kr), ionization energy of 53
  structure of 53
  total cross-section of, for slow electrons 75–80
Krypton molecular ion (Kr_2^+) 64
Kurepa 107
Kuyatt 95

Laboratory coordinates 5
Lagergren 132
Langevin 202
Laser 152, 232, 234–5
Lee 196–8
Levine 255
Lineberger 252–5
Lipsky 136
Lithium atoms (Li), electron affinity of 57
  ionization energy of 52
  probability charge density in 52
Loesche 188
Lorents 225
Los 153–4
Lovelock 136–7
Lyman series 44

Mach number 182, 189, 193, 196
Madden 243
Manometers, capacitance 289
McLeod gauge 289
Maggs 137
Maiman 235
Marconi 260
Marr 244
Mass analyser, in afterglow experiments 117
  electric quadrupole 301
  magnetic 299
  radio-frequency 120, 300
Mass number 39
Massey 80–83, 193, 245, 263–4
Mean free path 286
Metastable states, excitation of, by electrons 102, 107
  quenching of 151
  ionization by 152
  in afterglows 152, 209

Index

of helium 55, 152
of oxygen 151
reactions between 152
Microwave probing, of a discharge afterglow 118
Mielczarek 95
Mobility, of ions and gases 202–4
Mohler 239, 244
Mohr 193
Monochromator 241
Mott 176
Multiple ionization 88
Multiphoton processes 251

Nearly-adiabatic collisions 142
Negative ions 41, 57
  detachment energy of 57
  drift velocity of 125
  formation of 110, 112
Neon atoms (Ne), ionization energy of 53
  photoionization of 243
  structure of 53
  total cross-section of, for slow electrons 75–80
  resonance effects in 101
Neon gas, afterglow in 119
Neon molecular ions (Ne_2^+) 64
  in discharge afterglows 120
Neutrons 39
Night sky radiation 151
Nitrate ion (NO_3^-), in lower ionosphere 274
Nitric oxide (NO), dissociation energy of 62
  electron affinity of 155
  in flowing afterglow 209
Nitric oxide ions (NO^+), dissociation energy of 64
  dissociative recombination to 123
  in ionosphere 272
  in lower ionosphere 273
Nitrogen atoms (N), in flowing afterglow 209
  in upper atmosphere 270
  ionization energy of 53
Nitrogen molecules (N_2), dissociation energy of 62
  vibrational and rotational constants of 65
  vibrational relaxation times in 147
Nitrogen molecular ions (N_2^+), dissociation energy of 64
  dissociative recombination to 123
  in the ionosphere 272
Nitrogen peroxide (NO_2), electron affinity of 155
Nitrous oxide (N_2O), dissociative attachment in 132
Norcross 254–5

Oldenberg 246
Optical excitation function 105
Optical pumping 235
Orbiting 168, 201, 202, 206

Oxygen atoms (O), electron affinity of 57, 129. 252–3
  in flowing afterglows 209
  in upper atmosphere 270
  ionization energy of 53
Oxygen atomic negative ions (O^-), in ionosphere 274
  photodetachment cross-section of 253
Oxygen atomic positive ions (O^+), in ionosphere 273
  radiative recombination to 110
Oxygen molecules (O_2), dissociative attachment to 128
  dissociation energy of 62
  electron affinity of 155, 255
  vibrational and rotational constants of 65
  vibrational relaxation times of 147
Oxygen molecular negative ions, in ionosphere 274
  formation of 129
Oxygen molecular positive ions, dissociation energy of 64
  dissociative recombination to 123
  in discharge afterglows 116
  ionosphere 273
Oxygen ions, polyatomic, in discharge afterglows 116

Pack 71
Parson 196–8
Paschen series 44
Patterson 252–5
Pauli principle 51, 52, 54
Pauly 176, 184, 188, 191–5
Penning ionization 152, 209
Perel 221, 224
Periodic Table 41
Persistence of vibration 145
  rotation 147
Pesticides 108, 114
  detection of, by attachment 136
Phase, shifts in quantum scattering theory 34
  calculated, for rare gases 79
  variation of, with angular momentum 36
  of wave motion 13
  velocity 79
Phase-sensitive detection 186, 194, 196, 221, 246
Phelps 71, 125
Phosphorus atoms (P), electron affinity of 57
Photocathode 298
Photodetachment 230
  by laser beams 246
  effect of autodetachment on 237
  from H^- 245–6, 250
  from O^- 252
  from Se^- and Si^- 253
  measurement of cross-sections for 247
Photoelectric yield 240

Photoelectron spectroscopy 238, 242, 248
Photoionization 230
  effect of autoionization on 237, 243
  in flowing afterglows 209
  of atmospheric molecules by solar radiation 231, 262
Photomultiplier detector 224
Photosphere of Sun 275
Planck 20
Planck's constant, magnitude of 1
Planck distribution of radiation 20, 233
Plane polar coordinates 158
Polar aurora 151
Polarization, of hydrogen atom by electrons 50
Positive ions 41
Potassium atoms (K), collisions of with Br_2 154
  with Kr 187–8
  electron affinity of 57
  ionization energy of 53
Potential energy, curve 61
  effective, for centrifugal force 26, 168
  function 6
Powell 203
Probability distribution, of charge in states of H atoms 47
  density, definition 24
Protons, in atoms 39
  hydrated 273
Purcell 265, 279

Quenching, of metastable states of O 151
  resonance radiation 149

Radiation sources 241–3
Radiative transitions 43
Radio waves, long-range transmission of 260
Rainbow angle 165, 191
  optical 165
  supernumerary 171, 174
  wave theory of 170
Ramsauer 68, 73, 80, 82–3
Ramsauer–Townsend effect 68, 73
  theory of 76
Rapp 103, 128, 131
Rate coefficient 9
Rearrangement collisions 205, 211–2
Recombination 108
  dielectronic 109, 277
  dissociative in
    ionosphere 262, 264, 271–2
    Ne 121
    N_2, NO and O_2 124
  measurement of 115–120
  nature of 111
  to CH^+ 282
  in solar corona 276
  radiative 109–111
    to CH^+ 282
  three-body in helium 122

Reduced mass 6
Refractive index 18
  for material waves 23
  of ionized atmosphere 260
Relaxation time, for vibrational excitation 145, 147
Resonance radiation, in flowing afterglow 209
  quenching of 149
Resonant scattering, of electrons by atoms 90
Reynolds 136, 138
Robinson 258
Rosenfeld 279
Rotational excitation, in molecule–molecule collisions 141, 144
Rothe 153
Rubidium atoms (Rb), electron affinity of 57
  ionization energy of 53
  photoionization of 244
Rubidium negative ion (Rb^-), photodetachment from 254
Rutherford 32

Scattering, by a centre of force
  classical theory of 10
  quantum theory of 32
  semi-classical theory of 171
  in ion–atom collisions 216
  of electrons by atoms,
    angular distribution of 80, 83–7
    direct 89
    fine structure in 95, 97
    measurement of 80
    resonant 89
    theory of 83–7
  of waves 17
Raman 232
Schafer 196–8
Schawlow 235
Schlier 187
Schröder 195–6
Schrödinger 1, 21
Schulz 98, 127–8, 131–2
Schwinger 241
Scoles 176, 193–5, 298
Secondary emission multiplier detector 248, 297
Selenium atoms (Se), electron affinity of 57, 253
Selenium negative ions (Se^-), photodetachment cross-section of 252–3
Seman 255
Shadow scattering 17
  in wave mechanics 30
Shock waves, heating by 119
  persistence of vibration in 146
Siegel 255
Silicon atoms (Si), electron affinity of 57, 253
Silicon negative ions (Si^-), photoelectron spectrum of 253
Silver atoms (Ag), electron affinity of 57
Simpson 95

Index

Skylark rocket 266–7
Smith 246, 252
Sodium atoms (Na), collisions of with Hg 188
  electron affinity of 57
  ionization energy of 53
  structure of 53
Solar atmosphere, absorption coefficient of 251
Spherical polar coordinates 25
Spitzer 281–2
Spontaneous emission coefficient 46, 105, 232
Sputnik I 266
Sputter source, of fast neutral atoms 154
Stamatovic 131–2
Stecker 281
Stern 180
Stewart, Balfour 266
Stimulated emission of radiation 233–4
Strunck 187
Stuhl 151
Sulphur atoms (S), electron affinity of 57
Sulphur hexafluoride (SF_6), attachment of slow electrons in 134
  in attachment detector 137–8
Sulphur hexafluoride ions (SF_6^-) 132
Supersonic flow, through a nozzle 291
Surface ionization detector 154, 186, 189
Swings 279
Synchrotron 241

Telemetry 266
Three-body reactions, associative 206
  of electrons in O_2 129
Tomboulian 241
Torrello 176, 193–5, 298
Tousey 265
Townes 235
Townsend 68, 73, 75–6
Transfer, of electronic excitation 151
  vibration and rotation 142
Tritium atom (T) 41
Trujillo 185
Tuve 260
Tyndall 203

Uncertainty principle 28
  autoionization 56
  in scattering 31
van der Waals constant 157
  force 60, 179
Velocity distribution, Maxwellian 286
  of molecules in a beam 287
Velocity, of sound in a gas 145, 290
Velocity selection 183, 189, 193
Vernon 211

Vibrational excitation in molecule–molecule collisions 143
  relaxation 145–8
Vibrating reed electrometer 225, 297
Viscosity of gases 140
  helium 199
Voronov 258
V2 rockets 265

Water vapour, effect of on three-body attachment to O_2 130
Wave equation 18
  for material waves 23
Wave front 15
Wave function 18
  for material waves 24
  relation to probability density 24
Wave group 14
  velocity of 15
Wavelength 14
  local 19, 23, 171
Wave mechanics 21
  relation to classical mechanics 23, 28
Wave motion 13
  group velocity of 15
  phase velocity of 14
  plane 15
  scattering of 17
  simple harmonic 13
  spherical 15
Wave number 14
Welge 151
Wheeler 156
Wien filter 127, 291
Wildt 244–5
Williams 281
Window resonances, in photodetachment 254–5
  in photoionization 237, 243
Woolley 277

Xenon atoms (Xe), multiphoton ionization of 258
  structure of 53
  total cross-section of, for slow electrons 76–80
Xenon molecular ions (Xe_2^+) 64

Zero point energy 61
  in H_2 61
Zonal harmonics 26
  in quantum scattering theory 35

LIBRARY OF DAVIDSON COLLEGE

Books on regular loan may be checked out for **two weeks**. Books must be presented at the Circulation Desk in order to be renewed.

A fine is charged after date due.

Special books are subject to special regulations at the discretion of the library staff.